Next Generation Energy and Propulsion Systems for Transportation Electrification

Next Generation Energy and Propulsion Systems for Transportation Electrification

Editors

Jinhao Meng
Dongdong Zhao
Daniel Stroe
Ji Wu
Qian Xiao
Zeyu Chen
Yanan Wang
Ruixin Yang
Stefano Bracco

MDPI

Basel • Beijing • Wuhan • Barcelona • Belgrade • Novi Sad • Cluj • Manchester

Editors

Jinhao Meng
Xi'an Jiaotong University
Xi'an
China

Dongdong Zhao
Northwestern Polytechnical University
Xi'an
China

Daniel Stroe
Aalborg University
Aalborg
Denmark

Ji Wu
Hefei University of Technology
Hefei
China

Qian Xiao
Tianjin University
Tianjin
China

Zeyu Chen
Northeastern University
Shenyang
China

Yanan Wang
Shandong University
Jinan
China

Ruixin Yang
Beijing Institute of Technology
Beijing
China

Stefano Bracco
University of Genoa
Genoa
Italy

Editorial Office
MDPI
St. Alban-Anlage 66
4052 Basel, Switzerland

This is a reprint of articles from the Special Issue published online in the open access journal *Energies* (ISSN 1996-1073) (available at: https://www.mdpi.com/journal/energies/special_issues/energy_transportation_electrification).

For citation purposes, cite each article independently as indicated on the article page online and as indicated below:

Lastname, A.A.; Lastname, B.B. Article Title. *Journal Name* **Year**, *Volume Number*, Page Range.

ISBN 978-3-7258-1227-1 (Hbk)
ISBN 978-3-7258-1228-8 (PDF)
doi.org/10.3390/books978-3-7258-1228-8

Contents

energies

MDPI

Article

An Open Circuit Voltage Model Fusion Method for State of Charge Estimation of Lithium-Ion Batteries

Quanqing Yu [1,2], Changjiang Wan [1], Junfu Li [1], Lixin E [1], Xin Zhang [1], Yonghe Huang [1] and Tao Liu [1,*]

1 School of Automotive Engineering, Harbin Institute of Technology, Weihai 264209, China; qqyu@hit.edu.cn (Q.Y.); 20S030123@stu.hit.edu.cn (C.W.); lijunfu@hit.edu.cn (J.L.); 180120103@stu.hit.edu.cn (L.E.); 181210327@stu.hit.edu.cn (X.Z.); 181320208@hit.edu.cn (Y.H.)

2 National Engineering Laboratory for Electric Vehicles, School of Mechanical Engineering, Beijing Institute of Technology, Beijing 100081, China

* Correspondence: liutao@hitwh.edu.cn

Abstract: The mapping between open circuit voltage (OCV) and state of charge (SOC) is critical to the lithium-ion battery management system (BMS) for electric vehicles. In order to solve the poor accuracy in the local SOC range of most OCV models, an OCV model fusion method for SOC estimation is proposed. According to the characteristics of the experimental OCV–SOC curve, the method divides SOC interval (0, 100%) into several sub-intervals, and respectively fits the OCV curve segments in each sub-interval to obtain a corresponding number of OCV sub-models with local high precision. After that, the OCV sub-models are fused through the continuous weight function to obtain fusional OCV model. Regarding the OCV curve obtained from low-current OCV test as the criterion, the fusional OCV models of $LiNiMnCoO_2$ (NMC) and $LiFePO_4$ (LFP) are compared separately with the conventional OCV models. The comparison shows great fitting accuracy of the fusional OCV model. Furthermore, the adaptive cubature Kalman filter (ACKF) is utilized to estimate SOC and capacity under a dynamic stress test (DST) at different temperatures. The experimental results show that the fusional OCV model can effectively track the performance of the OCV–SOC curve model.

Keywords: electric vehicles; lithium-ion batteries; open circuit voltage; state of charge; model fusion; adaptive cubature Kalman filter

Citation: Yu, Q.; Wan, C.; Li, J.; E, L.; Zhang, X.; Huang, Y.; Liu, T. An Open Circuit Voltage Model Fusion Method for State of Charge Estimation of Lithium-Ion Batteries. *Energies* **2021**, *14*, 1797. https://doi.org/10.3390/en14071797

Academic Editor: Giovanni Lutzemberger

Received: 6 February 2021
Accepted: 18 March 2021
Published: 24 March 2021

1. Introduction

In response to the various policies of sustainable development, the development of electric vehicles (EVs) with batteries as the main power source has become the theme of the automotive industry. Because of numerous appealing characteristics, lithium-ion batteries (LiBs) have been an indispensable part of EVs [1]. As a significant indicator, the state of charge (SOC) dominates the range anxiety of drivers. Precise and reliable SOC estimation allows battery management system (BMS) to protect the battery, prevent overcharge/over-discharge, extend the battery life, and make reasonable control strategies to save energy [2,3]. Therefore, SOC estimation always attracts the attention of academia and the industrial community.

Various approaches of SOC estimation have been developed [4–7]. The classical approach to estimate SOC is the ampere-hour integral method which directly originates from the definition of the SOC. The precision of ampere-hour integral method is seriously limited by the initial SOC error, sensor error and battery aging [8]. Model-based methods have been extensively used due to their self-correcting merits. The essence of model-based methods is the mapping between SOC and other battery parameters like impedance spectroscopy and open circuit voltage (OCV). The Kalman filter is commonly used to estimate SOC in various model-based methods with the bearable amount of calculation cost and the supply of state uncertainty. The Kalman filter algorithm family includes the classical

Kalman filter, extended Kalman filter (EKF) [9,10], unscented Kalman filter (UKF) [11], cubature Kalman filter (CKF) [12] and all the extensive form of Kalman filter. EKF needs to solve the Jacobi matrix, which limited its computation speed and remains just 1 order Taylor expansion precision [13]. UKF needs to set reasonable noise parameter and sigma point configuration parameter, which is dependent on specific battery working conditions. CKF generates the cubature points to approximate the state transfer function. With fewer parameters that need to be adjusted and minimal calculation costs, CKF remains in 3 order Taylor expansion precision. Therefore, CKF is generally more practical than EKF and UKF. In order to solve the possible divergence of the filter, the adaptive extended Kalman filter (AEKF) and adaptive unscented Kalman filter (AUKF) have been studied [14–16]. In this paper, the adaptive cubature Kalman filter (ACKF) is introduced to estimate SOC.

For the model-based method, the precision of the OCV–SOC function model (OCV model) which reflects the mapping between SOC and OCV seriously determines the performance of SOC estimation. The incremental OCV test and the low-current OCV test are usual experiments to obtain experimental OCV–SOC characteristic curve (OCV curve) [17,18]. In the same time-consuming condition, the incremental method is slightly suitable for specific battery material system at a certain temperature [17]. However, the low-current OCV test provides intact OCV–SOC data points (OCV points), and the measured voltage can maintain close-to-equilibrium status if the current rate is extremely small [19]. Therefore, the OCV curves obtained by the low-current OCV test are more suitbale used as reference models (OCV curve models) for various OCV models.

OCV curve presents distinct characteristic with the change of battery material system it belongs and is influenced by ambient temperature, aging status and current rate [20]. OCV points obtained from OCV test can be sampled reasonably as control points to fit OCV curve. Lots of OCV function models have been proposed to express OCV curve [8]. The polynomial function, power function, logarithm and exponent are alternative choices to constitute the OCV model [21–25]. Among these functions, polynomial functions were commonly used to fit the OCV curve. By comparing five OCV models, Hu et al. [26] concluded that the sixth-order polynomial function was considered as the most accurate OCV model among them. In practice, most OCV models were adopted to fit incomplete OCV curve whose SOC range is between 10% and 90%. By setting cut-off voltage, the SOC range is commonly regarded as nominal 0–100% to ensure safety. So, the battery capacity is cut down by the control of the BMS, and not fully utilized. If the accuracy in the marginal region of the OCV model can be improved as much as possible, it will be helpful for BMS to further extend the cut-off voltage and increase the available battery capacity. Furthermore, establishing OCV models which satisfy high fidelity in the full SOC range is worth studying.

In order to solve the aforementioned problems, this paper attempts to make the following contributions:

- An OCV model fusion method is proposed to obtain fusional OCV model which may match the characteristic of OCV curve in complete SOC range. The OCV model fusion method is applied for a $LiNiMnCoO_2$ (NMC) battery and a $LiFePO_4$ (LFP) battery. OCV fitting curves with high precision are obtained at temperature of 10 °C, 25 °C and 40 °C, respectively.
- CKF and ACKF are utilized to estimate SOC and capacity, and the effect of the fusional OCV model on SOC and capacity estimation is evaluated by comparing with the OCV curve model. Besides, the adaptability of the ACKF algorithm for OCV model errors is verified.

Notably, this paper is concerned with how to obtain the OCV model with high fidelity based on limited but persuasive control points obtained from an experimental test.

The rest of this paper is organized as follows. Section 2 presents the battery model and OCV model fusion method. Section 3 introduces the ACKF and capacity estimation algorithm. Section 4 presents the experimental validation. Finally, conclusions are given in Section 5.

2. Battery Modeling and Open Circuit Voltage (OCV) Curve Fusion Method

2.1. Battery Modelling

Battery models, the basis of battery algorithms, mainly include electrochemical mechanism models [27] and equivalent circuit models (ECMs) [28]. The electrochemical mechanism models are highly accurate but limited by unsustainable computation load. The ECMs describe dynamic characteristic and operation mechanism of batteries by using circuit network which comprises traditional resistance, capacitor, and constant voltage source. An resistor-capacitor (RC) network is commonly used to characterize the dynamic features of batteries. Among various ECMs, the first-order RC ECM as shown in Figure 1, also called the Thevenin model, shows the best balance between complexity and accuracy [29,30].

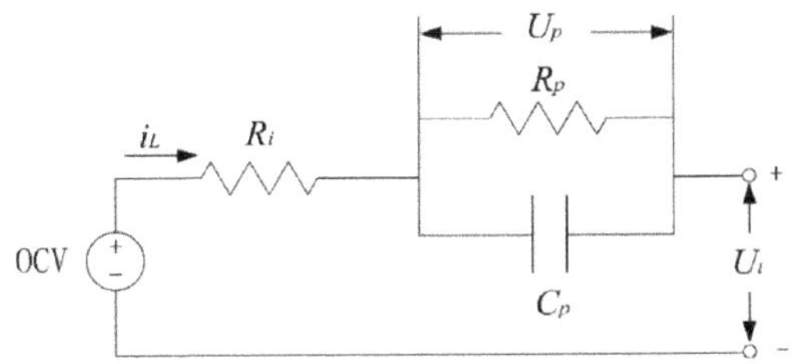

Figure 1. Thevenin model.

OCV denotes the terminal voltage which has a non-linear relationship with SOC; i_L denotes the current of the battery; R_i denotes the internal resistance, and characterizes the contact resistance among battery electrode material, electrolyte, diaphragm resistance and various parts; R_p denotes polarization resistance and C_p denotes polarization capacitance, the parallel connection of C_p and R_p reveals the dynamic characteristics of the battery; U_t denotes the terminal voltage; U_p denotes the potential difference of the RC network which is called polarization voltage.

The mathematical expression between U_t and i_L is:

$$U_t = OCV - U_p - i_L R_i \tag{1}$$

The mathematical expression between U_p and i_L is:

$$\dot{U}_p = -\frac{1}{C_p R_p} U_p + \frac{1}{C_p} i_L \tag{2}$$

After discretizing the above equation, following equation is summarized as:

$$\begin{cases} U_{p,k} = U_{p,k-1}\exp\left(-\frac{\Delta t}{R_p C_p}\right) + i_{L,k-1} R_p \left(1 - exp\left(-\frac{\Delta t}{R_p C_p}\right)\right) \\ U_{t,k} = OCV_k - U_{p,k} - R_0 i_{L,k-1} \end{cases} \tag{3}$$

where the subscript k denotes the sampling step, Δt denotes the step size. OCV_k is a nonlinear function of SOC. By calculating Equation (4), SOC is obtained.

$$SOC_k = SOC_{k-1} - \frac{\eta_i \Delta t}{C_a} i_{L,k} \tag{4}$$

where η_i denotes the coulombic efficiency of cell [31], C_a denotes current maximum available capacity which is directly relevant to battery aging and ambient temperature et al., so it is not fixed like nominal capacity. η_i is generally defaulted as 1.

The terminal voltage U_t and the current i_L in the Thevenin model can be measured by the voltage sensor and the current sensor respectively. In the meanwhile, the model parameters R_i, R_p, C_p and non-linear function OCV_k have to be determined so that subsequent battery algorithm can be sustained. Because parameters are seriously influenced by the variable factors like battery temperature, aging status, charge or discharge current etc., recursive least squares method with forgetting factor (FFRLS) is utilized to identify parameters R_i, R_p, C_p in real time [32].

2.2. OCV Model Fusion Method

The mapping between OCV and SOC is a basic part of battery modeling. During the process of OCV model selection, the non-linearity of OCV curve brings out the toughest part. Based on the control points obtained from the OCV test, the coefficients of the OCV model can be solved after carrying out the curve fitting algorithm. The curves presented by the solved OCV models can be named as OCV fitting curves. Reasonable selection of the OCV model can maximize experiment effect as much as possible. From the perspective of structure, the components of OCV models basically include generalized polynomial, logarithm, exponent and power function. The polynomial functions satisfy the needs of general condition, but perform poorly in the local SOC range as shown in Figure 2a. Besides, polynomial functions may cause under fitting or overfitting if the number of control points do not match the degree of polynomials. The OCV model which consists of reasonable combination of polynomial, logarithm, exponent or power function may perform well in the global SOC range. However, it is still inevitable that the precision of OCV model in the local SOC range declines seriously, as shown in Figure 2b,c.

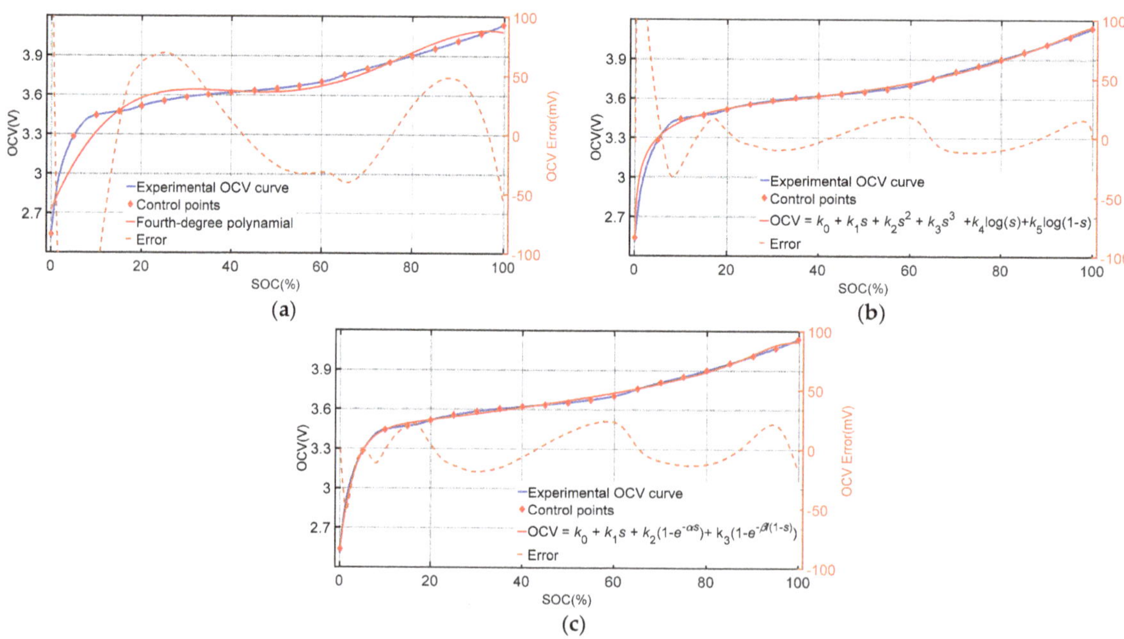

Figure 2. Conventional open circuit voltage (OCV) models for $LiNiMnCoO_2$ (NMC) battery (21 control points, s denotes state of charge (SOC), $k_{0,1,2,3,4}$, α, β denote fitting coefficients). (**a**) Fourth-degree polynomial. (**b**) Polynomial and logarithm. (**c**) Exponent and linear function.

With further thinking about problems above, the reasons why it is difficult to perfectly fit the complete OCV curves can be summarized:

- From the perspective of battery characteristics, the marginal region of some OCV curves may be polarized, and the changing trend of the OCV curves may be transformed within a small SOC range. It is difficult for the OCV model to fully take into account the characteristics of the OCV curve.

- From the perspective of practical application, some algorithms are sensitive to the error of OCV fitting curve. For example, the OCV curve of a LFP battery may have several large flat regions. If SOC is inferred from the OCV based on OCV fitting curve which is stored in a table, the error of OCV will lead to larger error of the SOC due to the deviation of flat regions. Therefore, the requirement for the OCV model's accuracy is strengthened.

The problems caused by former reason is the focus of this paper. Predictably, if a single OCV curve segment with stable changing trend is extracted for fitting, then a high-precision OCV fitting curve segment can be obtained correspondingly. However, it is a question of combining all these OCV fitting curve segments. In order to obtain OCV fitting curve which can maintain high fidelity in the local SOC range as much as possible, an OCV model fusion method is proposed. Figure 3 demonstrates the flowchart of the OCV model fusion method which fits an OCV curve of NMC battery.

For different types of battery, the general steps are as follows:

1. Separate out OCV sub-intervals: according to the characteristics of OCV curve, the global SOC interval (0, 100%) can be divided into several local sub-intervals. In order to ensure the smoothness of fusional curve, each sub-interval exists overlap with neighboring sub-intervals.
2. Assign OCV sub-models: according to the characteristics of the OCV curve in the local SOC sub-interval, each sub-interval corresponds to a specific OCV sub-model.
3. Curve segment fitting: according to practical conditions, collecting the control points in each sub-interval. After fitting, the OCV fitting curve segments of all sub-models are obtained.
4. Assign weight: different global weight functions are assigned to corresponding OCV sub-models. The function should convert weight from high to low continually when the SOC gradually away from sub-interval in the overlapped region. Logistic function is suitable for defining conversion above.
5. Fuse: according to weight functions, all OCV sub-models can fuse into a fusional OCV model. The final OCV fitting curve can be expressed by using equation as follows:

$$OCV(s) = \frac{\sum_{i=1}^{n} W_i(s) OCV_i(s)}{\sum_{i=1}^{n} W_i(s)} \tag{5}$$

where s denotes SOC, $OCV_i(s)$ denotes the OCV value of sub-model i at s, $W_i(s)$ denotes the corresponding weight at s. The final fusional OCV model can be directly used for subsequent algorithms.

Theoretically, the proposed OCV model fusion method is universal for any type of battery as long as the OCV model can be used to fit the OCV curve. As representative commercial lithium-ion batteries, the LFP battery and NMC battery are used to verify the effectiveness of the proposed fusion method. Detailed application of the OCV fusion method in the two batteries is presented.

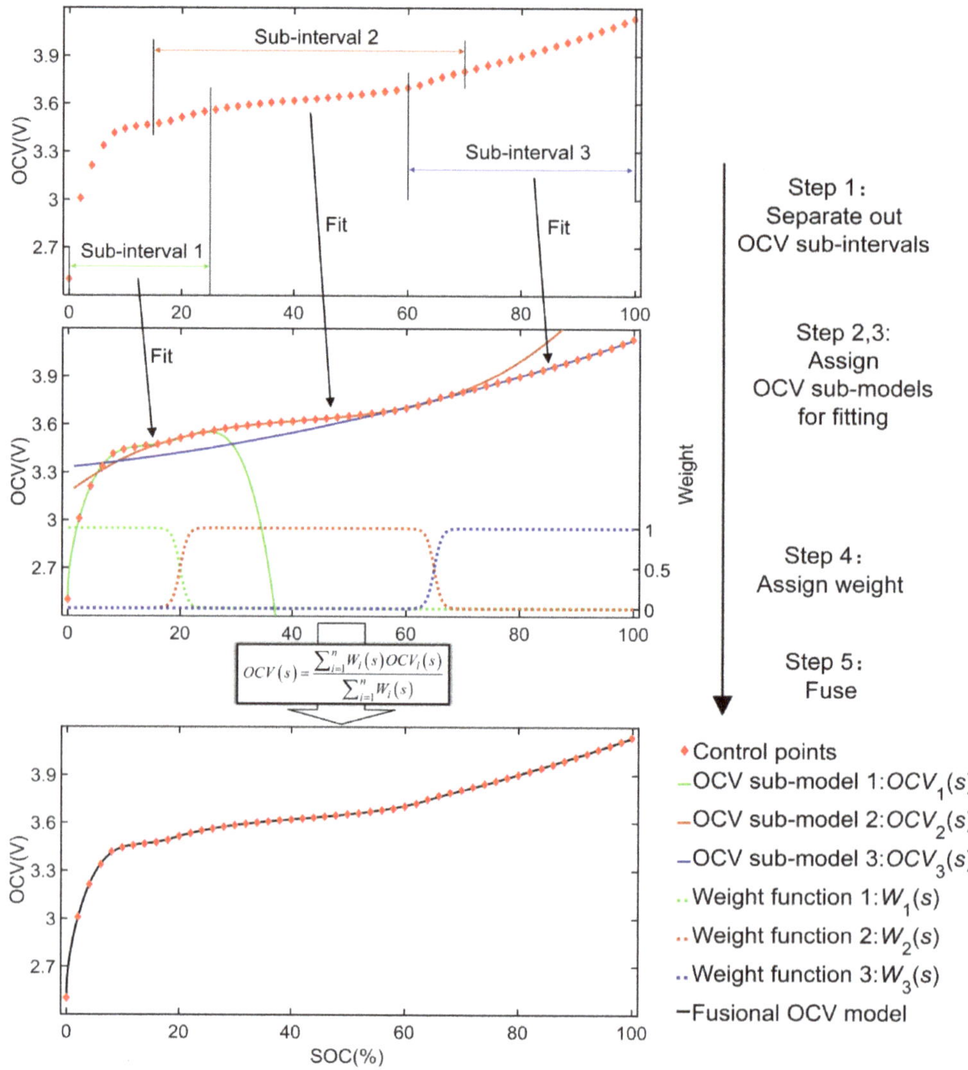

Figure 3. Flowchart of OCV model fusion method.

2.2.1. Method for $LiNiMnCoO_2$ (NMC) Battery Cell

An OCV curve of NMC battery with changing rate of OCV over SOC is presented in Figure 4. Due to the discrete form of OCV curve, the changing rate of OCV is approximately calculated through the following equation:

$$OCV'(s) \approx \frac{OCV(s+0.1\%) - OCV(s-0.1\%)}{0.2\%} \quad (6)$$

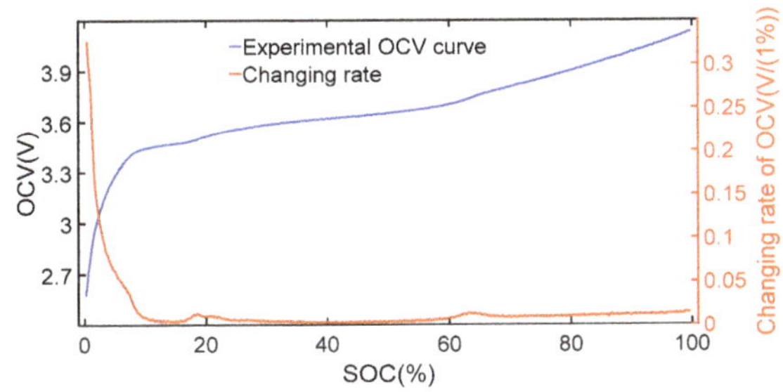

Figure 4. An OCV curve of a NMC battery with changing rate of OCV over SOC.

By analyzing the OCV curve of NMC battery, the following characteristics can be summarized:

- The OCV curve is clearly monotonous, and OCV changes dramatically when SOC drops to 0%.
- By approximately calculating the changing rate of the OCV with SOC, it is obvious that the changing rate of OCV curve has bumps around 20% SOC and 65% SOC.

The OCV models in Figure 2 are adopted to fit global OCV curve, and the results show that the accuracy fluctuates when the SOC is around 20% and 65%. Intuitively, it is easy to obtain an accurate OCV sub-model by fitting an OCV curve segment whose approximate range of SOC is (20%, 65%). By setting interval (15%, 25%) and interval (60, 70%) as conversion region, the OCV curve is divided into three parts. The corresponding sub-intervals are (0, 25%), (15%, 70%) and (60%, 100%).

In order to verify the effect of fusion method, the OCV models in Figure 2 are alternative choice of OCV sub-models. Due to the strongly non-linear variation of the OCV curve when SOC drops to 0%, the exponent, logarithm and power function are alternative choices of OCV sub-model for sub-interval (0, 25%). The OCV model in Figure 2c shows the best accuracy in sub-interval (0, 25%), so it is adopted. The variation of the OCV curve is nearly linear in sub-intervals (15%, 70%) and (60%, 100%), so corresponding sub-models with polynomial would be adequate for fitting. OCV sub-models adopted for a NMC battery are presented in Table 1.

Table 1. Sub-models of NMC battery.

Sub-Interval	Sub-Model
(0, 25%)	$OCV_{NMC,1}(s) = k_0 + k_1 s + k_2(1 - exp(-\alpha s)) + k_3(1 - exp(-\beta/(1-s)))$
(15, 70%)	$OCV_{NMC,2}(s) = k_0 + k_1 s + k_2 s^2 + k_3 s^3 + k_4 s^4$
(60, 100%)	$OCV_{NMC,3}(s) = k_0 + k_1 s + k_2 s^2 + k_3 s^3 + k_4 s^4$

After collecting enough control points in three sub-intervals respectively, the fitting results can be obtained. Weight function of each sub-interval is defined as:

$$\begin{cases} W_{NMC,1}(s) = \frac{1}{1+exp(r(s-0.2))} \\ W_{NMC,2}(s) = \begin{cases} \frac{1}{1+exp(-r(s-0.2))}, & s \leq 0.425 \\ \frac{1}{1+exp(r(s-0.65))}, & s > 0.425 \end{cases} \\ W_{NMC,3}(s) = \frac{1}{1+exp(-r(s-0.65))} \end{cases} \quad (7)$$

where r denotes the shape parameter which determines the degree of weight conversion. In this paper, r is configured as 150.

2.2.2. Method for $LiFePO_4$ (LFP) Battery Cell

An OCV curve of a LFP battery with changing rate of OCV over SOC is presented in Figure 5. Similarly, following the OCV characteristics of LFP battery can be summarized:

- The OCV curve is monotonous, and the OCV changes dramatically when SOC drops to 0% and rises to 100%. Moreover, the OCV curve has flat regions where the changing rate of OCV is close to zero.
- By approximately calculating the changing rate of OCV with SOC, it is obvious that the changing rate of the OCV curve has bumps around 20% SOC and 80% SOC.

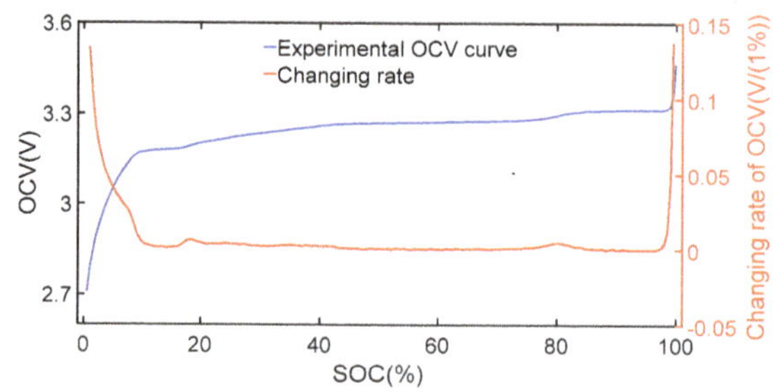

Figure 5. An OCV curve of $LiFePO_4$ (LFP) battery with changing rate of OCV over SOC.

Different OCV models in Figure 6 are adopted to fit OCV curve, and the results shows that accuracy fluctuates when SOC is around 10%, 20% and 80%. By setting interval (15%, 25%) and interval (75%, 85%) as the conversion region, OCV curve is divided into three parts. The corresponding sub-intervals are (0, 25%), (15%, 85%) and (75%, 100%). The OCV model in Figure 6c is adaptable to the strongly non-linearity of the LFP battery, which makes it become the OCV sub-model of sub-intervals (0%, 25%) and (75%, 100%). The OCV model in Figure 6b retains great precision in sub-interval (15%, 85%). The OCV sub-models adopted for the LFP battery are presented in Table 2.

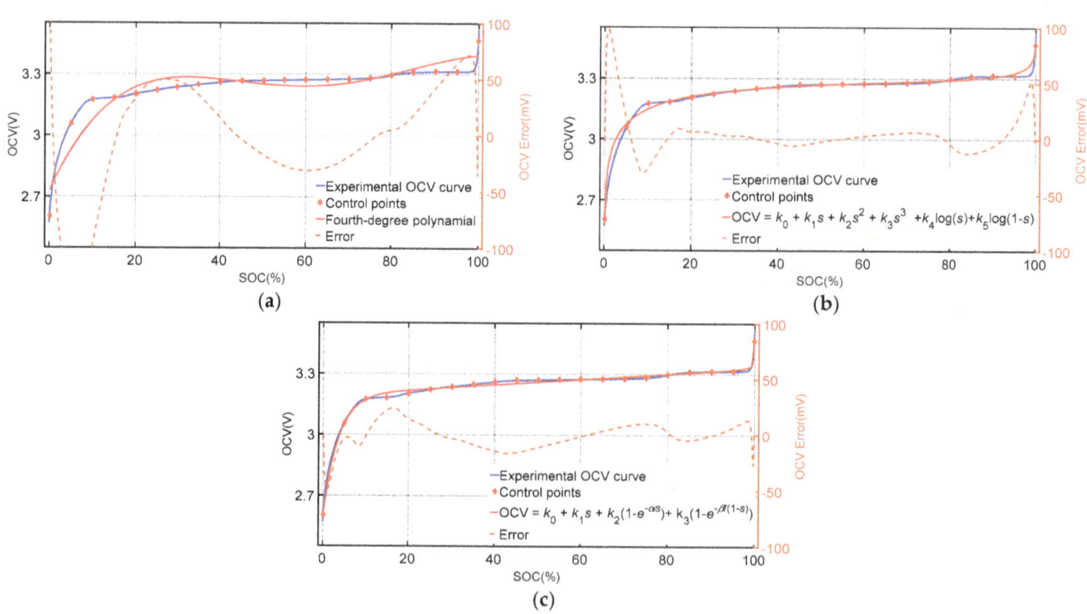

Figure 6. Conventional OCV models for LFP battery (21 control points, s denotes SOC, $k_{0,\,1,\,2,\,3,\,4}$, α, β denote fit coefficients). (**a**) Fourth-degree polynomial. (**b**) Polynomial and logarithm. (**c**) Exponent and linear function.

Table 2. Sub-models of LFP battery.

Sub-Interval	Sub-Model
(0, 25%)	$OCV_{LFP,1}(s) = k_0 + k_1 s + k_2(1 - exp(-\alpha s)) + k_3(1 - exp(-\beta/(1-s)))$
(15, 85%)	$OCV_{LFP,2}(s) = k_0 + k_1 s + k_2 s^2 + k_3 s^3 + k_4 \log(s) + k_5(1-s)$
(75, 100%)	$OCV_{LFP,1}(s) = k_0 + k_1 s + k_2(1 - exp(-\alpha s)) + k_3(1 - exp(-\beta/(1-s)))$

After collecting enough control points in three sub-intervals respectively, the fitting results can be obtained. The weight function of each sub-interval is defined as:

$$\begin{cases} W_1(s) = \frac{1}{1+exp(r(s-0.2))} \\ W_2(s) = \begin{cases} \frac{1}{1+exp(-r(s-0.2))}, & s \le 0.5 \\ \frac{1}{1+exp(r(s-0.8))}, & s > 0.5 \end{cases} \\ W_3(s) = \frac{1}{1+exp(-r(s-0.8))} \end{cases} \quad (8)$$

So far, the fusional OCV models of both batteries can be obtained by Equation (5) respectively. There are few points that need to be discussed:

- Although the fusional results are deduced from two examples, the steps of fusion method are generalized.
- According to practical condition, parameters like sub-intervals, sub-models and weight function can be explored freely.
- It is not suitable to select a sub-interval with too short a length, otherwise the number of control points need to be increased.

3. State of Charge (SOC) and Capacity Estimation Algorithm

3.1. Adaptive Cubature Kalman Filter

The discrete state space equation of nonlinear system with additive noise is:

$$\begin{cases} x_k = f(x_{k-1}, u_{k-1}) + w_{k-1} \\ z_k = h(x_k, u_k) + v_k \end{cases} \quad (9)$$

where x_k denotes the state vector at step k; u_k is the control input; $f(\cdot)$ and $h(\cdot)$ represent the process function and measurement function respectively; w_{k-1} and v_k are independent Gaussian noise with zero mean, corresponding covariance are Q_{k-1} and R_k respectively. For the Thevenin model, the parameters and vectors in Equation (9) are defined as:

$$\begin{cases} f(x_k, u_k) = Ax_{k-1} + Bu_{k-1} \\ h(x_k, u_k) = OCV_k - U_{p,k} - R_0 i_{L,k-1} \\ A = diag\left(1, exp(-\frac{\Delta t}{R_p C_p})\right) \\ B = \left[-\frac{\eta \Delta t}{C_a} R_p(1 - exp(-\frac{\Delta t}{R_p C_p}))\right]^T \\ x_k = \left[SOC_k U_{p,k}\right]^T \\ z_k = U_{t,k} \\ u_k = i_{L,k} \end{cases} \quad (10)$$

In order to cope with the nonlinearity, the distribution of random state vector is approximated by cubature points with uniform weight. Based on the spherical-radial cubature rule [33], the cubature points are generated through the following parameters:

$$\begin{cases} \xi_i = \sqrt{n}[1]_i & (i = 1, 2, \cdots, 2n) \\ \omega_i = 1/(2n) & (i = 1, 2, \cdots, 2n) \end{cases} \quad (11)$$

where n denotes dimension of the state vector, ξ_i represents the ith cubature point, ω_i denotes the weight of the ith points, [1] denotes the following set of points:

$$[1] = \left\{ \begin{bmatrix} 1 \\ 0 \\ \vdots \\ 0 \end{bmatrix}, \ldots, \begin{bmatrix} 0 \\ \vdots \\ 0 \\ 1 \end{bmatrix}, \begin{bmatrix} -1 \\ 0 \\ \vdots \\ 0 \end{bmatrix}, \ldots, \begin{bmatrix} 0 \\ \vdots \\ 0 \\ -1 \end{bmatrix} \right\} \tag{12}$$

3.2. *Process of SOC Estimation*

As one component of x_k, SOC_k can be estimated online by the recurring following process:

3.2.1. Initialization

The mathematical expectation and covariance of state vector x_0 need to be initialized as $\hat{x}_{0|0}$ and $P_{0\,|\,0}$ respectively, and covariance Q_0 and R_0 are preset.

3.2.2. Time Update

The complete CKF generates the cubature points at step $k-1$, and the cubature points are propagated to the state vector at step k by the process function. After averaging, the priori estimation at step k can be obtained. Due to the linearity of the Thevenin model process equation, the classic Kalman filter algorithm can be applied in the time update part, so that algorithm can be more concise and efficient. A hat over a letter denotes the estimation of corresponding parameter.

$$\hat{x}_{k|k-1} = f\left(\hat{x}_{k-1|k-1}, u_{k-1}\right) = A\hat{x}_{k-1|k-1} + Bu_{k-1} \tag{13}$$

$$P_{k|k-1} = AP_{k-1|k-1}A^T + Q_{k-1} \tag{14}$$

3.2.3. Measurement Update

1. Generate cubature points:

$$x_{i,k|k-1} = S_{k|k-1}\xi_i + \hat{x}_{k|k-1}, (i = 1, 2, \cdots, 2n) \tag{15}$$

$$P_{k|k-1} = S_{k|k-1}S^T_{k|k-1} \tag{16}$$

where $S_{k\,|\,k-1}$ is the Cholesky decomposition result of $P_{k\,|\,k-1}$.

2. Calculate propagated cubature points in observation space:

$$z_{i,k|k-1} = h\left(x_{i,k|k-1}, u_k\right) \tag{17}$$

3. Calculate the predicted measurement:

$$\hat{z}_{k|k-1} = \sum_{i=1}^{2n} \omega_i z_{i,k|k-1} \tag{18}$$

4. Calculate the measurement innovation covariance:

$$P_{zz,k|k-1} = \sum_{i=1}^{2n} \omega_i z_{i,k|k-1} z^T_{i,k|k-1} - \hat{z}_{k|k-1}\hat{z}^T_{k|k-1} + R_{k-1} \tag{19}$$

5. Calculate the cross-covariance:

$$P_{zx,k|k-1} = \sum_{i=1}^{2n} \omega_i x_{i,k|k-1} z^T_{i,k|k-1} - \hat{x}_{k|k-1}\hat{z}^T_{k|k-1} \tag{20}$$

6. Calculate the Kalman gain:

$$K_k = P_{zx,k|k-1} P_{zz,k|k-1}^{-1} \tag{21}$$

7. Calculate the updated state:

$$\hat{x}_{k|k} = \hat{x}_{k|k-1} + K_k \left(z_k - \hat{z}_{k|k-1} \right) \tag{22}$$

8. Calculate the updated covariance:

$$P_{k|k} = P_{k|k-1} - K_k P_{zz,k|k-1} K_k^T \tag{23}$$

3.2.4. Adaptive Update of Noise

According to the innovation sequence of terminal voltage, the process noise and measurement noise are adjusted adaptively.

1. The innovation covariance matrix:

$$H_k = \frac{1}{M} \sum_{i=k-M+1}^{k} e_i e_i^T \tag{24}$$

where M denotes the window size which is defaulted as 60, e_i denotes residual which is calculated by:

$$e_k = z_k - \hat{z}_{k|k-1} \tag{25}$$

2. The process noise covariance Q_k is updated as follows:

$$Q_k = K_k H_k K_k^T \tag{26}$$

3. The measurement noise covariance R_k is updated as follows:

$$R_k = H_k + \sum_{i=1}^{2n} \omega_i \left(z_{i,k|k-1} - z_k \right) \left(z_{i,k|k-1} - z_k \right)^T \tag{27}$$

3.3. Capacity Estimation Based on Estimated SOC

Based on the estimated SOC, the available capacity of battery can be estimated on-line. The capacity obtained through the capacity test is defaulted as the initial capacity value when SOC estimation starts to be performed. The change of capacity ΔC_k and the change of SOC ΔSOC_k are calculated as follows:

$$\Delta C_k = \sum_{i=Ls}^{k} \eta_i i_{L,i} \Delta t \tag{28}$$

$$\Delta SOC_k = SOC_k - SOC_{Ls} \tag{29}$$

where Ls denotes the step at which the capacity estimation starts. That is, the capacity estimation starts only after the SOC estimation has passed Ls step so that fluctuation of capacity estimation is reduced during the initial period. The capacity is estimated by using following equation:

$$\hat{C}_k = \frac{\Delta C_k}{\Delta SOC_k} \tag{30}$$

In order to obtain a steady value of available capacity estimation, the change rate of capacity estimation is limited by using the following equation:

$$\left| \frac{\hat{C}_k - \hat{C}_{k-Lc}}{\hat{C}_{k-Lc}} \right| \leq \varepsilon \tag{31}$$

where ε denotes the acceptable change rate of capacity estimation, *Lc* denotes the step size of capacity estimation. That is, the capacity is estimated every *Lc* step. Once Equation (31) is not satisfied, the change between new estimated capacity and last estimated capacity will be compulsively set as ε.

Figure 7 illustrates the flowchart of SOC estimation and capacity estimation.

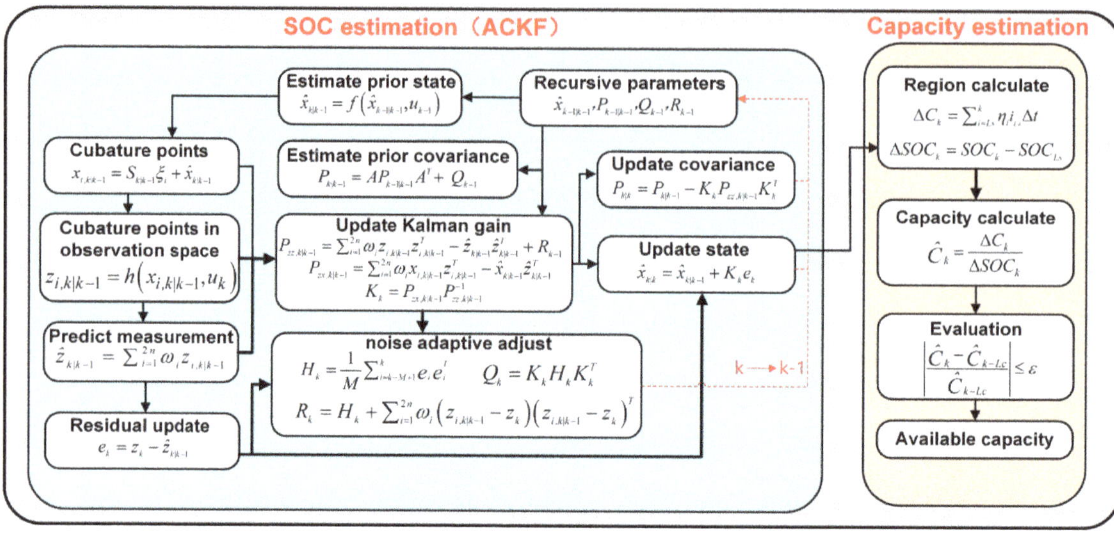

Figure 7. Flowchart of state of charge (SOC) and capacity estimation.

4. Experiment and Discussion

4.1. Experiment

The battery experimental system consisted of an Arbin BT2000 battery test machine, a thermal chamber and a computer with Arbin software. Tested NMC batteries and LFP batteries were manufactured by MGL. The capacity tests, OCV tests and dynamic stress tests (DST) were performed at the temperature of 10 °C, 25 °C, and 40 °C. Table 3 presents basic information of both batteries.

Table 3. Basis information of tested batteries.

Material	Type	Nominal Capacity (Ah)	Available Capacity (Ah)		
			10 °C	**25 °C**	**40 °C**
NMC	cylinder	25.00	28.30	28.75	29.02
LFP	pouch	20.00	19.72	19.85	19.94

This study adopted a low-current OCV test, which stimulated the battery with 0.05 C, to obtain SOC–OCV data points with 1 Hz. SOC and capacity were estimated under DST which can emulate the actual driving cycles of EVs for batteries. The test profile of DST and low-current OCV test is illustrated in Figure 8.

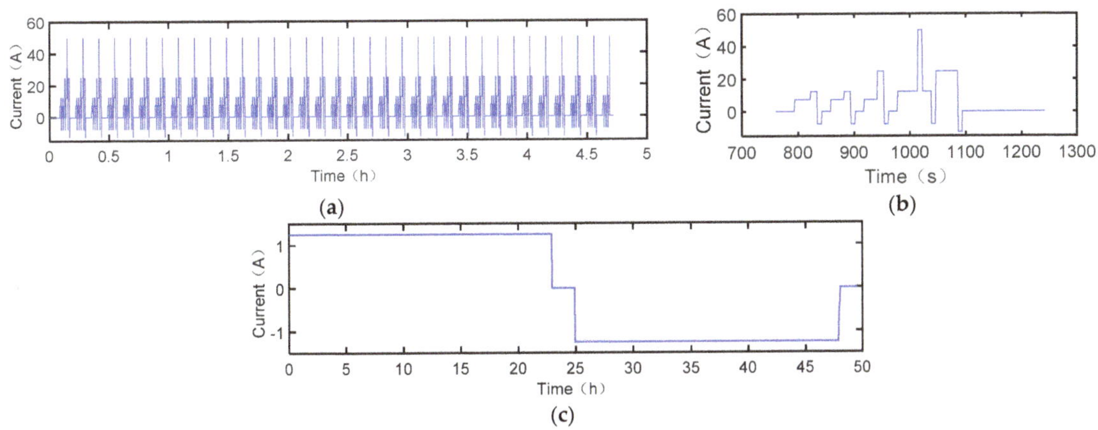

Figure 8. Current profiles of: (**a**) DST. (**b**) Cycle window of DST. (**c**) low-current OCV test.

Notably, OCV curves are regarded as OCV reference models whose OCV can be looked up in OCV curve table. In this way, the closeness between fusional OCV model and OCV curve model can be evaluated from the perspective of SOC estimation and capacity estimation.

4.2. The Fusional OCV Model

4.2.1. Fusional OCV Model of NMC Battery

The fitting results of fusional OCV models of the NMC battery at 10 °C, 25 °C and 40 °C are shown in Figure 9a,c,e. The fitting results of conventional OCV models in Table 4, which are same as the adopted sub-models, are shown in Figure 9b,d,e. All the OCV models fit the OCV curve based on 21 evenly distributed control points.

Table 4. Conventional OCV models.

Label	Model
1	$OCV_1(s) = k_0 + k_1 s + k_2 s^2 + k_3 s^3 + k_4 s^4$
2	$OCV_2(s) = k_0 + k_1 s + k_2 s^2 + k_3 s^3 + k_4 \log(s) + k_5 (1 - s)$
3	$OCV_3(s) = k_0 + k_1 s + k_2 (1 - exp(-\alpha s)) + k_3 (1 - exp(-\beta/(1 - s)))$

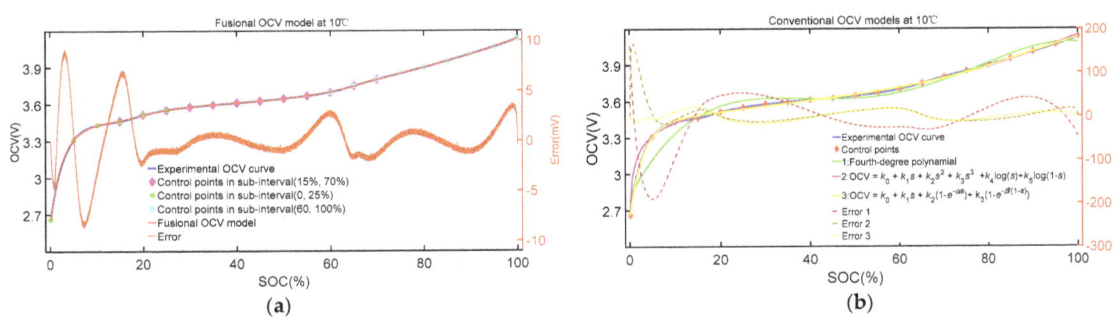

Figure 9. *Cont.*

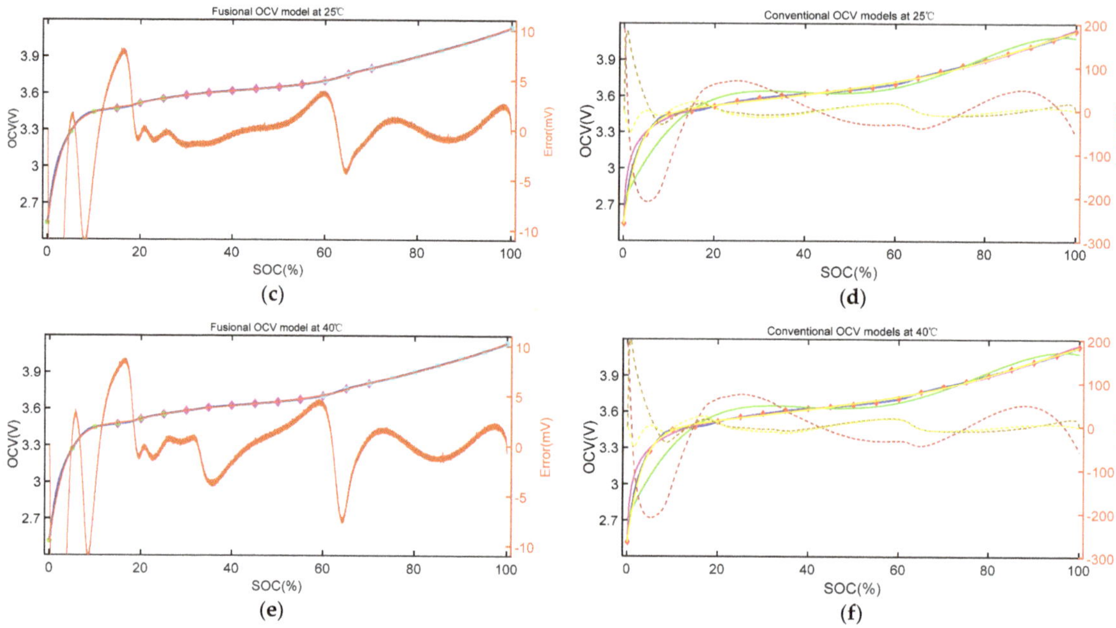

Figure 9. OCV models of NMC battery at 10 °C, 25 °C, and 40 °C. (**a**) Fusional OCV model at 10 °C. (**b**) Conventional OCV models at 10 °C. (**c**) Fusional OCV model at 25 °C. (**d**) Conventional OCV models at 25 °C. (**e**) Fusional OCV model at 40 °C. (**f**) Conventional OCV models at 40 °C.

Table 5 shows the root mean squared error (RMSE) of all models. The fourth-degree polynomial OCV model performs worst, and what make it unbearable is that all the OCV fitting curves fourth-degree polynomial OCV model are non-monotonic. The OCV models 2, 3 can capture the changing trend of the OCV, but lost the precision in several local regions. Based on the local high-precision of sub-models, the fusional OCV model highly fits the experimental curve and corresponding RMSE is reduced dramatically. Compared with using OCV sub-models to fit the global OCV curve, the average accuracy of fusional OCV models increases about 2 times. In terms of temperature effect, the performance of fusional OCV model is not influenced by ambient temperature. Notably, in order to exclude the large error of OCV model when SOC drops to 0%, the SOC range of RMSE is between 5% and 100%.

Table 5. RMSE of fusional OCV models and other models for NMC battery (V).

	10 °C	**25 °C**	**40 °C**
Fusional model	0.0022	0.0027	0.0031
Model 1	0.0473	0.0575	0.0596
Model 2	0.0106	0.0115	0.0117
Model 3	0.0109	0.0105	0.0101

4.2.2. Fusional OCV Model of LFP Battery

Similarly, the fusional OCV models of the LFP battery at 10 °C, 25 °C, and 40 °C are given in Figure 10a,c,e and the fitting results of OCV models in Table 4 are also given in Figure 10b,d,e. All the OCV models fit the OCV curve based on 21 evenly distributed control points. According to the RMSE of OCV models in Table 6, the fusional OCV models still perform with great precision, especially in the region of middle SOC. The fourth-degree polynomial OCV model is the most inaccurate and non-monotonic. Compared with using

OCV sub-models to fit the global OCV curve, the average accuracy of fusional OCV model increases about 2 times. The OCV models 2, 3 can capture the changing trend of OCV. In terms of temperature effect, the performance of the fusional OCV model is not influenced by ambient temperature. Notably, in order to exclude the large error of the OCV model when SOC drops to 0% and rises to 100%, the SOC range of RMSE is between 5% and 99%.

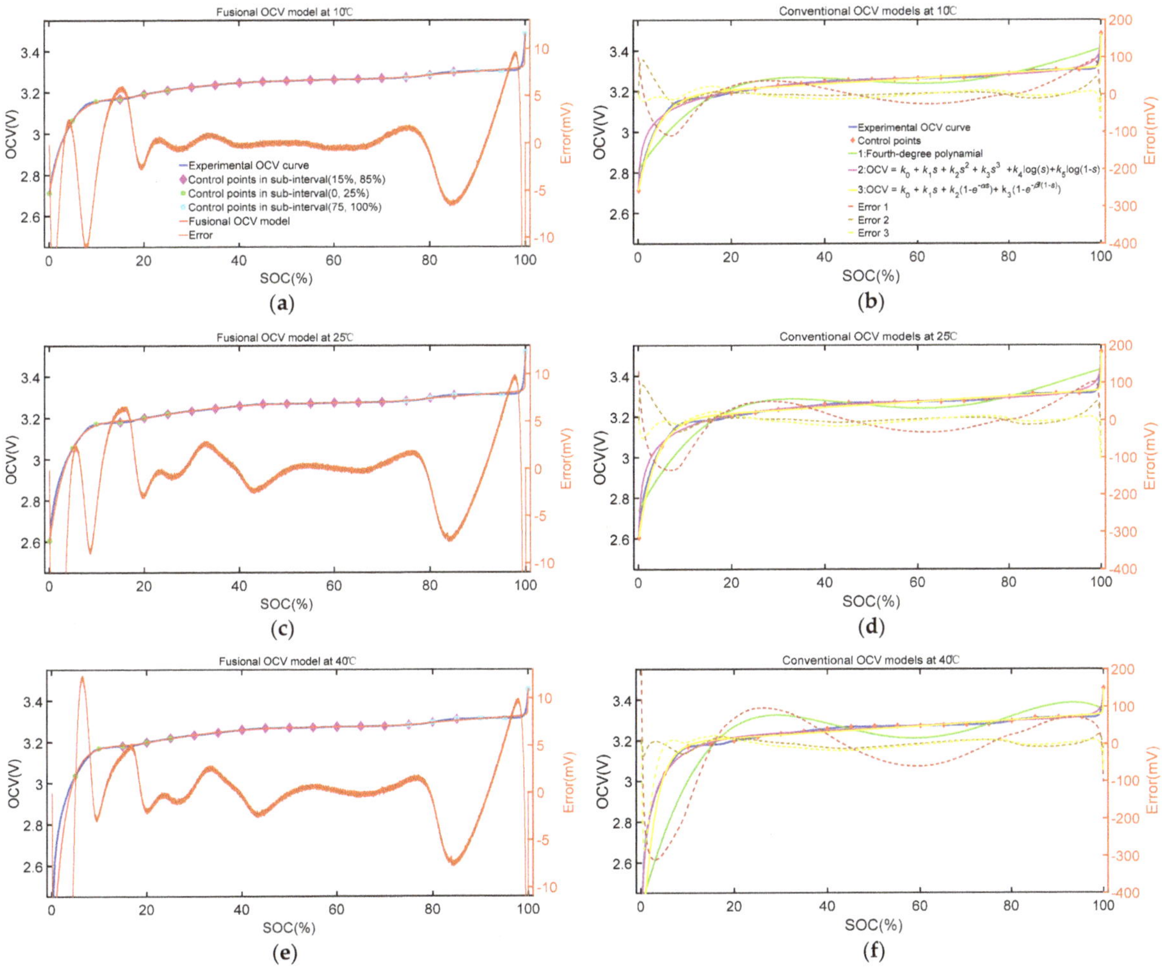

Figure 10. OCV models of LFP battery at 10 °C, 25 °C, and 40 °C. (**a**) Fusional OCV model at 10 °C. (**b**) Conventional OCV models at 10 °C. (**c**) Fusional OCV model at 25 °C. (**d**) Conventional OCV models at 25 °C. € Fusional OCV model at 40 °C. (**f**) Conventional OCV models at 40 °C.

Table 6. RMSE of fusional OCV models and other models for LFP battery (V).

	10 °C	25 °C	40 °C
Fusional model	0.0032	0.0033	0.0033
Model 1	0.0386	0.0482	0.0782
Model 2	0.0088	0.0101	0.0100
Model 3	0.0090	0.0096	0.0106

4.3. The Result of SOC Estimation with Different OCV Models

The proposed OCV model fusion method is further used to estimate SOC for verification of OCV model accuracy. By using CKF and ACKF to estimate SOC under DST at 10 °C, 25 °C and 40 °C, the estimation results of the NMC battery are shown in Figure 11 with corresponding RMSE in Table 7 and the estimation results of the LFP battery are given in Figure 12 with corresponding RMSE in Table 8. Due to the non-monotonicity and large error of fourth-degree polynomial OCV models, the corresponding SOC estimation results are invalid. The SOC estimation results of models 2 and 3 are considered.

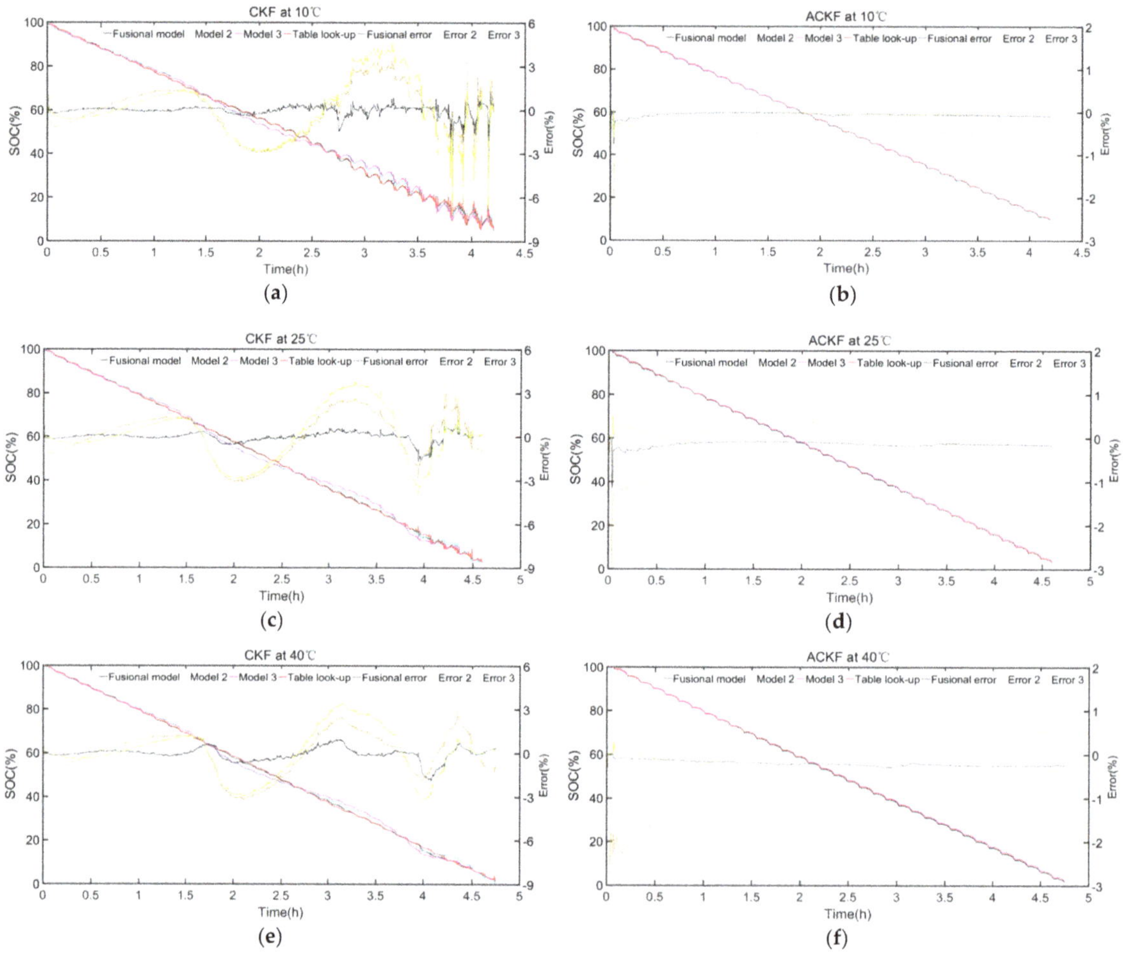

Figure 11. SOC estimation results for NMC battery by using: (**a**) CKF at 10 °C. (**b**) ACKF at 10 °C. (**c**) CKF at 25 °C. (**d**) ACKF at 25 °C. (**e**) CKF at 40 °C. (**f**) ACKF at 40 °C.

Table 7. RMSE of SOC estimation for NMC battery (%).

	10 °C		25 °C		40 °C	
	CKF	**ACKF**	**CKF**	**ACKF**	**CKF**	**ACKF**
Fusional model	0.3277	0.0725	0.3385	0.1555	0.4465	0.2125
Model 2	1.6378	0.6027	1.5454	0.5056	1.4508	0.9889
Model 3	1.9407	0.1843	1.7782	0.3828	1.6261	0.3358

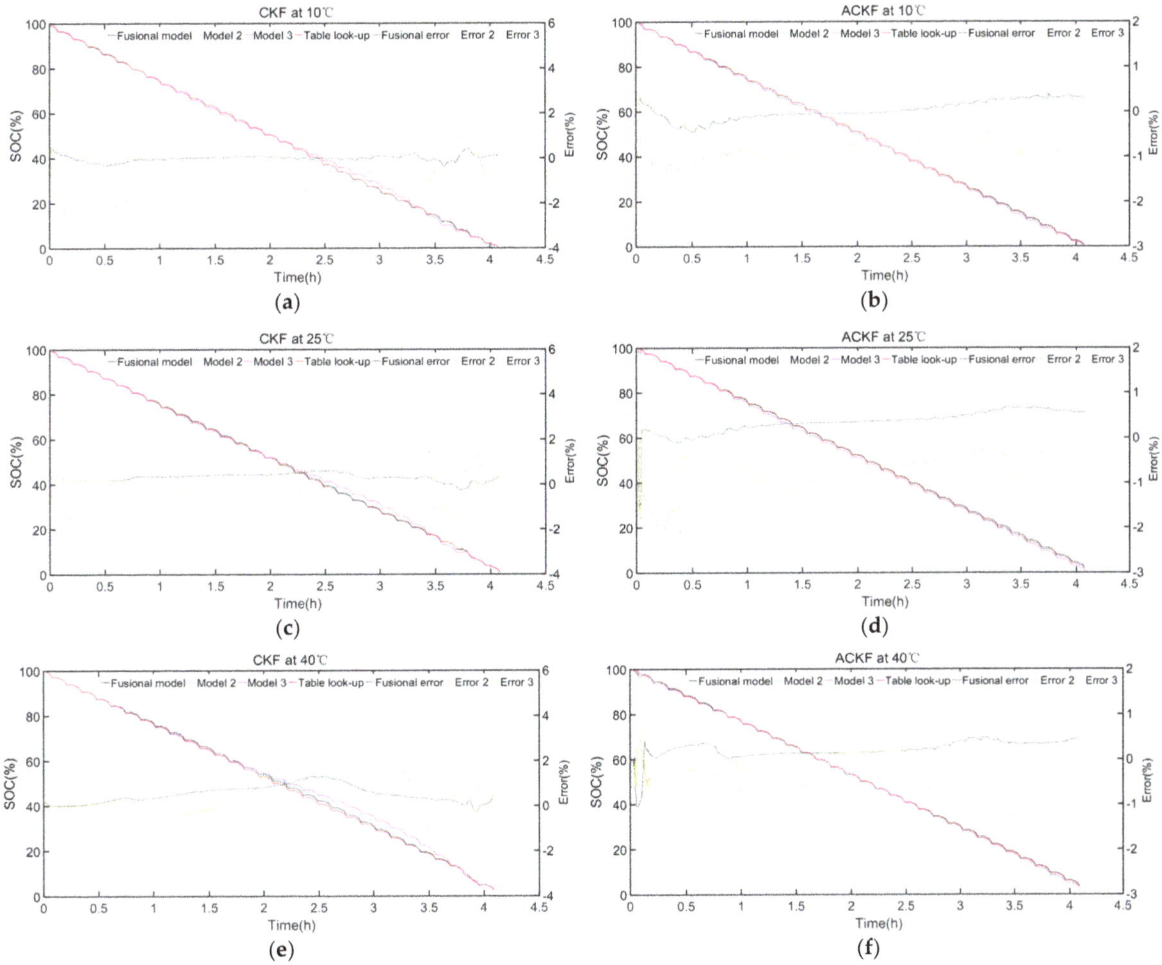

Figure 12. SOC estimation results for LFP battery by using: (**a**) CKF at 10 °C. (**b**) ACKF at 10 °C. (**c**) CKF at 25 °C. (**d**) ACKF at 25 °C. (**e**) CKF at 40 °C. (**f**) ACKF at 40 °C.

Table 8. RMSE of SOC estimation for LFP battery (%).

	10 °C		25 °C		40 °C	
	CKF	**ACKF**	**CKF**	**ACKF**	**CKF**	**ACKF**
Fusional model	0.6462	0.2049	0.3530	0.4179	0.6506	0.2905
Model 2	1.6222	0.7526	0.8740	0.9157	1.4714	0.7377
Model 3	1.2080	0.7656	1.3779	0.8355	2.4103	0.3655

When applying the CKF algorithm, the difference between the OCV models is obvious. Since the OCV curve of the NMC battery does not have the flat platform effect like LFP battery, the advantages of the fusion method are more beneficial for NMC battery. Compared with other models in the NMC battery, the SOC estimation error based on the fusional OCV model is dramatically reduced. For LFP battery, the SOC estimation error of fusional OCV models can still retain minimal, but error may fluctuate with temperature change. Overall, temperature has no significant effect on the fusion method. The addition of adaptive noise effectively weakens the influence of OCV model errors and stabilizes the estimation results. According to experimental result, the effect of the OCV model fusion method is verified.

4.4. The Result of Capacity Estimation with Different OCV Models

In order to reduce the impact of the fluctuation during the initial period, capacity estimation starts to be performed after three DST cycles have finished. In consideration of the time cost of one DST cycle is 480 s in the experiment, start step L_S is set as 1440 s. Cycle step L_c is set as 360 s and acceptable change rate ε is set as 2%. As shown in Figures 13 and 14, it can be seen that fluctuation of capacity estimation by using the CKF algorithm is more obvious than capacity estimation by using the ACKF algorithm. What is more important, as shown in Tables 9 and 10, is that fusional OCV models still perform best in tracking capacity estimation results of OCV curve models. In terms of temperature effect, ambient temperature has no significant effect on the OCV fusion method.

Table 9. RMSE of capacity estimation for NMC battery (Ah).

	10 °C		25 °C		40 °C	
	CKF	**ACKF**	**CKF**	**ACKF**	**CKF**	**ACKF**
Fusional model	0.1833	0.0138	0.1718	0.1556	0.2226	0.1427
Model 2	0.9395	0.3558	0.8801	0.4317	0.8161	0.9693
Model 3	0.9877	0.1973	0.9414	0.1940	0.8499	0.1866

Table 10. RMSE of capacity estimation for LFP battery (Ah).

	10 °C		25 °C		40 °C	
	CKF	**ACKF**	**CKF**	**ACKF**	**CKF**	**ACKF**
Fusional model	0.1523	0.1866	0.1073	0.1791	0.2808	0.1240
Model 2	0.6295	0.4062	0.4753	0.2461	0.5442	0.3240
Model 3	0.3775	0.4850	0.4307	0.4102	0.8465	0.1283

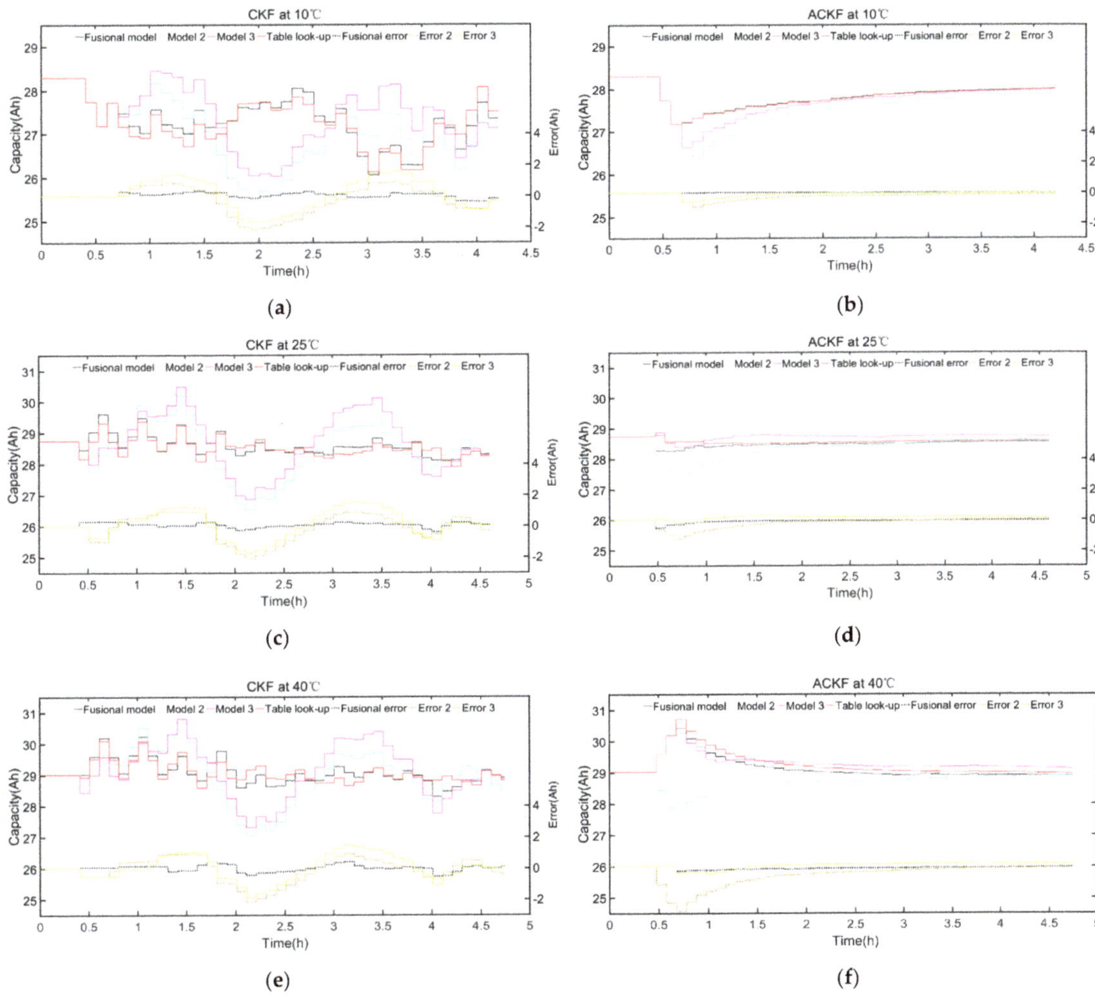

Figure 13. Capacity estimation results for NMC battery by using: (**a**) CKF at 10 °C. (**b**) ACKF at 10 °C. (**c**) CKF at 25 °C. (**d**) ACKF at 25 °C. (**e**) CKF at 40 °C. (**f**) ACKF at 40 °C.

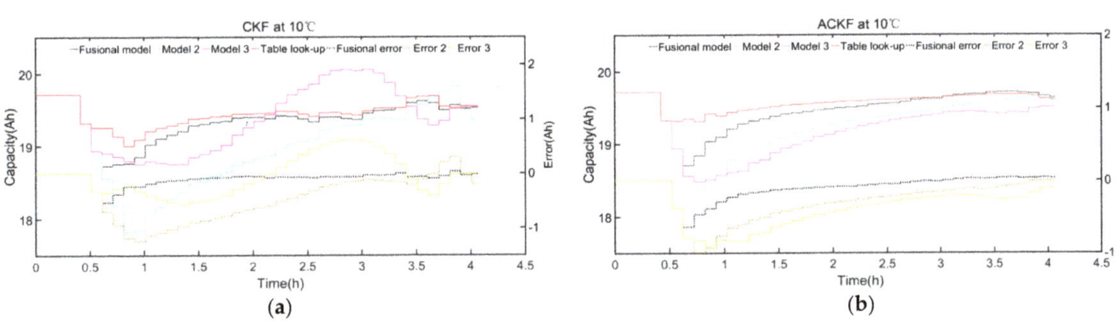

Figure 14. *Cont.*

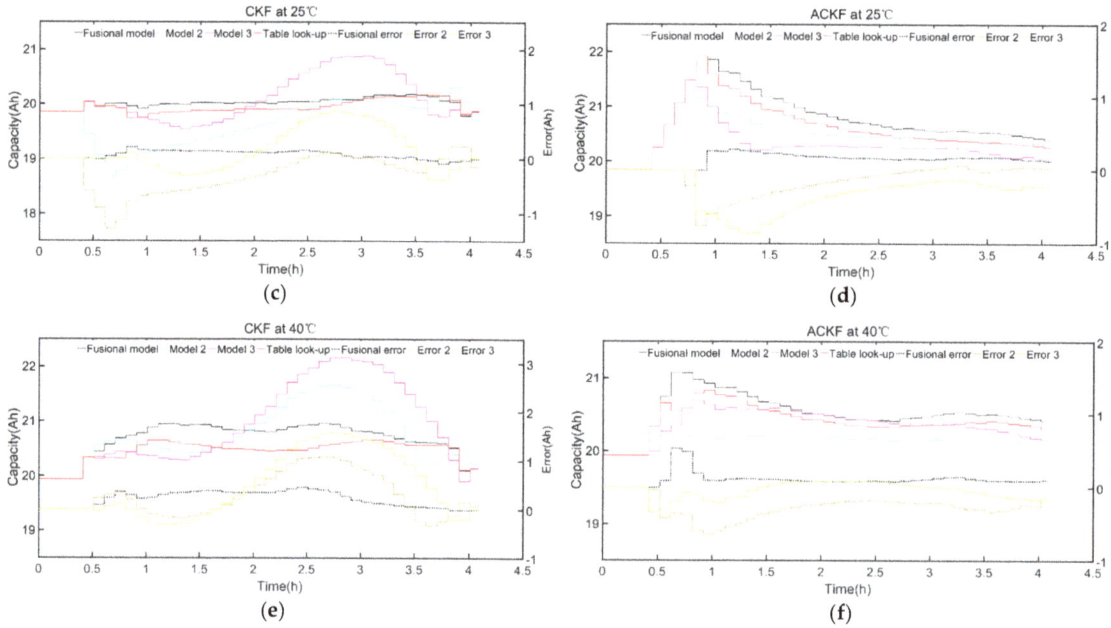

Figure 14. Capacity estimation results for LFP battery by using: (**a**) CKF at 10 °C. (**b**) ACKF at 10 °C. (**c**) CKF at 25 °C. (**d**) ACKF at 25 °C. (**e**) CKF at 40 °C. (**f**) ACKF at 40 °C.

5. Conclusions

At present, the research on battery SOC and SOH estimation mainly focuses on improving the structure of an equivalent circuit model and proposing new estimation algorithms, ignoring the influence of the OCV model on the accuracy of state estimation. Using a single OCV model cannot have a good fitting effect in the entire battery discharge interval. For this reason, an OCV model fusion method is proposed that can effectively obtain a high-fidelity OCV model. The method makes it possible to focus on the fitting accuracy in a certain SOC interval, so as to capture the changing trend of the OCV in a specific SOC region as much as possible. Furthermore, it has good adaptability for fitting complex OCV curves, and provides a good solution for OCV curves that are difficult to fit globally. The generalization fitting performance of the method is verified by the application on NMC battery and LFP battery. Experimental results at three ambient temperatures showed that the performance of the fusional OCV model was not sensitive to temperature. Regarding the influence of the OCV fusion model on the accuracy of state estimation, this paper used CKF and ACKF to estimate the SOC and capacity under three ambient temperatures, the results indicating that the fusional OCV model can effectively track the performance of OCV curve model in terms of supporting algorithm.

The fusion method has an important implication for reconstructing a global OCV characteristic curve based on curve segments. It is not only suitable for offline low-current OCV experiments, and we will carry out research work on applying it to offline incremental OCV experiments and online OCV reconstruction processes.

Author Contributions: Conceptualization, C.W. and X.Z.; methodology, Q.Y.; software, L.E.; validation, T.L., and J.L.; writing—original draft preparation, Q.Y.; writing—review and editing, Y.H. and T.L.; All authors have read and agreed to the published version of the manuscript.

Funding: This work has been supported by the Shandong Provincial Natural Science Foundation Project (ZR2020ME209).

Institutional Review Board Statement: Not applicable.

Informed Consent Statement: Not applicable.

Data Availability Statement: Data sharing not applicable.

Acknowledgments: The systemic experiments of the lithium-ion batteries were performed at the Advanced Energy Storage and Application (AESA) Group, Beijing Institute of Technology.

Conflicts of Interest: The authors declare no conflict of interest.

References

1. Abada, S.; Marlair, G.; Lecocq, A.; Petit, M.; Sauvant-Moynot, V.; Huet, F. Safety focused modeling of lithium-ion batteries: A review. *J. Power Sources* **2016**, *306*, 178–192. [CrossRef]
2. Lai, X.; Zheng, Y.J.; Sun, T. A comparative study of different equivalent circuit models for estimating state-of-charge of lithium-ion batteries. *Electrochim. Acta* **2018**, *259*, 566–577. [CrossRef]
3. Yang, R.; Xiong, R.; Shen, W.; Lin, X. Extreme Learning Machine Based Thermal Model for Lithium-ion Batteries of Electric Vehicles under External Short Circuit. *Engineering* **2020**. [CrossRef]
4. Yu, Q.; Xiong, R.; Yang, R.; Pecht, M. Online capacity estimation for lithium-ion batteries through joint estimation method. *Appl. Energy* **2019**, *255*, 113817. [CrossRef]
5. Xing, Y.J.; He, W.; Pecht, M.; Tsui, K.L. State of charge estimation of lithium-ion batteries using the open-circuit voltage at various ambient temperatures. *Appl. Energy* **2014**, *113*, 106–115. [CrossRef]
6. Hu, Y.; Yurkovich, S. Battery cell state-of-charge estimation using linear parameter varying system techniques. *J Power Sources* **2012**, *198*, 338–350. [CrossRef]
7. Chemali, E.; Kollmeyer, P.J.; Preindl, M.; Emadi, A. State-of-charge estimation of Li-ion batteries using deep neural networks: A machine learning approach. *J. Power Sources* **2018**, *400*, 242–255. [CrossRef]
8. Yu, Q.Q.; Xiong, R.; Wang, L.Y.; Lin, C. A Comparative Study on Open Circuit Voltage Models for Lithium-ion Batteries. *Chin. J. Mech. Eng.* **2018**, *31*, 64. [CrossRef]
9. Fang, Y.; Xiong, R.; Wang, J. Estimation of Lithium-Ion Battery State of Charge for Electric Vehicles Based on Dual Extended Kalman Filter. *Energy Procedia* **2018**, *152*, 574–579. [CrossRef]
10. Plett, G.L. Extended Kalman filtering for battery management systems of LiPB-based HEV battery packs: Part 3. State and parameter estimation. *J. Power Source* **2004**, *134*, 277–292. [CrossRef]
11. Xiong, R.; Yu, Q.; Wang, L.Y. Open circuit voltage and state of charge online estimation for lithium ion batteries. *Energy Procedia* **2017**, *142*, 1902–1907. [CrossRef]
12. Arasaratnam, I.; Haykin, S. Cubature Kalman filters. *IEEE Trans. Autom. Control* **2009**, *54*, 1254–1269. [CrossRef]
13. Xiong, R.; Sun, F.C.; He, H.W. Data-driven state-of-charge estimator for electric vehicles battery using robust extended Kalman filter. *Int. J. Automot. Technol.* **2014**, *15*, 89–96. [CrossRef]
14. Mohamed, A.H.; Schwarz, K.P. Adaptive Kalman filtering for INS/GPS. *J. Geodesy* **1999**, *73*, 193–203. [CrossRef]
15. Han, J.; Kim, D.; Sunwoo, M. State-of-charge estimation of lead-acid batteries using an adaptive extended Kalman filter. *J. Power Sources* **2009**, *188*, 606–612. [CrossRef]
16. Sun, F.C.; Hu, X.S.; Zou, Y.; Li, S.G. Adaptive unscented Kalman filtering for state of charge estimation of a lithium-ion battery for electric vehicles. *J. Power Sources* **2011**, *36*, 3531–3540. [CrossRef]
17. Zheng, F.D.; Xing, Y.J.; Jiang, J.C.; Sun, B.X.; Kim, J.; Pecht, M. Influence of different open circuit voltage tests on state of charge online estimation for lithium-ion batteries. *Appl. Energy* **2016**, *183*, 513–525. [CrossRef]
18. Petzl, M.; Danzer, M. Advancements in OCV Measurement and Analysis for Lithium-Ion Batteries. *IEEE Trans. Energy Convers.* **2013**, *28*, 675–681. [CrossRef]
19. Zhu, J.G.; Knapp, M.; Darma, M.S.D.; Fang, Q.H.; Wang, X.Y.; Dai, H.F.; Wei, X.Z.; Ehrenberg, H. An improved electro-thermal battery model complemented by current dependent parameters for vehicular low temperature application. *Appl. Energy* **2019**, *248*, 149–161. [CrossRef]
20. Lavigne, L.; Sabatier, J.; Francisco, M.; Guillemard, F.; Noury, A. Lithium-ion Open Circuit Voltage (OCV) curve modelling and its ageing adjustment. *J. Power Sources* **2016**, *324*, 694–703. [CrossRef]
21. Plett, G.L. Extended Kalman filtering for battery management systems of LiPB-based HEV battery packs: Part 2. Modeling and identification. *J. Power Sources* **2004**, *134*, 262–276. [CrossRef]
22. Xiong, R.; Li, L.; Yu, Q.; Jin, Q.; Yang, R. A set membership theory based parameter and state of charge co-estimation method for all-climate batteries. *J. Clean. Prod.* **2020**, *249*, 11389. [CrossRef]
23. Xiong, R.; Sun, F.C.; Chen, C.; He, H.W. A data-driven multi-scale extended Kalman filtering based parameter and state estimation approach of lithium-ion polymer battery in electric vehicles. *Appl. Energy* **2014**, *113*, 463–476. [CrossRef]
24. Dong, G.Z.; Wei, J.W.; Zhang, C.B.; Chen, Z.H. Online state of charge estimation and open circuit voltage hysteresis modeling of $LiFePO_4$ battery using invariant imbedding method. *Appl. Energy* **2016**, *162*, 163–171. [CrossRef]
25. Zhang, C.P.; Jiang, J.C.; Zhang, L.J.; Liu, S.J. A generalized SOC-OCV model for lithium-ion batteries and the SOC estimation for LNMCO battery. *Energies* **2016**, *9*, 900. [CrossRef]

26. Hu, X.S.; Li, S.B.; Peng, H.; Sun, F.C. Robustness analysis of state-of-charge estimation methods for two types of Li-ion batteries. *J. Power Sources* **2012**, *217*, 209–219. [CrossRef]
27. Ganesan, N.; Basu, S.; Hariharan, K.S.; Kolake, S.M.; Song, T.; Yeo, T.; Sohn, D.K.; Doo, S. Physics based modeling of a series parallel battery pack for asymmetry analysis, predictive control and life extension. *J. Power Sources* **2016**, *322*, 57–67. [CrossRef]
28. Widanage, W.D.; Barai, A.; Chouchelamane, G.H.; Uddin, K.; McGordon, A.; Marco, J.; Jennings, P. Design and use of multisine signals for Li-ion battery equivalent circuit modelling. Part 2: Model estimation. *J. Power Sources* **2016**, *324*, 61–69. [CrossRef]
29. Yu, Q.; Xiong, R.; Lin, C.; Shen, W.X.; Deng, J. Lithium-ion battery parameters and state-of-charge joint estimation based on H-infinity and unscented Kalman filters. *IEEE Trans. Veh. Technol.* **2017**, *66*, 8693–8701. [CrossRef]
30. Hu, X.S.; Li, S.B.; Peng, H. A comparative study of equivalent circuit models for Li-ion batteries. *J. Power Sources* **2012**, *198*, 359–367. [CrossRef]
31. Xiong, R.; Yu, Q.; Wang, L.; Lin, C. A novel method to obtain the open circuit voltage for the state of charge of lithium ion batteries in electric vehicles by using H infinity filter. *Appl. Energy* **2017**, *207*, 346–353. [CrossRef]
32. Xiong, R. Estimation of Battery Pack State for Electric Vehicles Using Model-Data Fusion Approach. Ph.D. Thesis, Beijing Institute of Technology, Beijing, China, June 2014.
33. Liu, M.; Lai, J.Z.; Li, Z.M.; Liu, J.Y. An adaptive cubature Kalman filter algorithm for inertial and land-based navigation system. *Aerosp. Sci. Technol.* **2016**, *51*, 52–60. [CrossRef]

energies

MDPI

Article

A Multi-Model Probability Based Two-Layer Fusion Modeling Approach of Supercapacitor for Electric Vehicles

Bo Huang [1,2], Yuting Ma [2,3,*], Chun Wang [1,2,*], Yongzhi Chen [1] and Quanqing Yu [4]

1 School of Mechanical Engineering, Sichuan University of Science and Engineering, Zigong 643000, China; huangbojx@suse.edu.cn (B.H.); 18011060104@stu.suse.edu.cn (Y.C.)
2 Artificial Intelligence Key Laboratory of Sichuan Province, Sichuan University of Science and Engineering, Zigong 643000, China
3 School of Mathematics and Statistics, Sichuan University of Science and Engineering, Zigong 643000, China
4 School of Automotive Engineering, Harbin Institute of Technology, Weihai 264209, China; qqyu@hit.edu.cn
* Correspondence: 18341010329@stu.suse.edu.cn (Y.M.); wangchun@suse.edu.cn (C.W.)

Abstract: The improvement of the supercapacitor model redundancy is a significant method to guarantee the reliability of the power system in electric vehicle application. In order to enhance the accuracy of the supercapacitor model, eight conventional supercapacitor models were selected for parameter identification by genetic algorithm, and the model accuracies based on standard diving cycle are further discussed. Then, three fusion modeling approaches including Bayesian fusion, residual normalization fusion, and state of charge (SOC) fragment fusion are presented and compared. In order to further improve the accuracy of these models, a two-layer fusion model based on SOC fragments is proposed in this paper. Compared with other fusion models, the root mean square error (RMSE), maximum error, and mean error of the two-layer fusion model can be reduced by at least 23.04%, 8.70%, and 30.13%, respectively. Moreover, the two-layer fusion model is further verified at 10, 25, and 40 °C, and the RMSE can be correspondingly reduced by 60.41%, 47.26%, 23.04%. The results indicate that the two-layer fusion model proposed in this paper achieves better robustness and accuracy.

Keywords: supercapacitor; parameter identification; genetic algorithm; fusion model

Citation: Huang, B.; Ma, Y.; Wang, C.; Chen, Y.; Yu, Q. A Multi-Model Probability Based Two-Layer Fusion Modeling Approach of Supercapacitor for Electric Vehicles. *Energies* **2021**, *14*, 4644. https://doi.org/10.3390/en14154644

Academic Editor: Stefano Bracco

Received: 15 June 2021
Accepted: 26 July 2021
Published: 30 July 2021

1. Introduction

In recent decades, a new energy technology, which has been rapidly developed and applied in the field of electric vehicles (EV) has attracted the attention of many countries such as China, the United States, Germany, the United Kingdom and Japan [1]. Lithium-ion batteries are widely used in EV power systems due to their high energy and power density and low self-discharge rate [2,3]. The high rate of charge and discharge current will seriously affect the life of the battery, which can generally only be controlled within 2C rate. However, supercapacitors can not only achieve high-rate charge and discharge, but also have unparalleled advantages in terms of power density and long cycle life. Therefore, in the EVs application, supercapacitors are often combined with lithium-ion batteries to serve as a hybrid energy storage system for EV energy supply [4–6]. In view of the prominent characteristic of the high power density, supercapacitors can not only provide the peak current urgently needed by electric vehicles and absorb excessive braking current, but also extend the cycle life of the power system and alleviate the impact of surge current on bus voltage. Supercapacitor models are strongly related to the optimal control of EV power systems. Therefore, inaccurate model parameters easily degrade the monitoring function of the power system, and may also lead to potential problems such as low efficiency, fires, and explosion of electric vehicles [7,8]. Therefore, the increased redundancy of the supercapacitor model is the significant approach to enhance the accuracy and guarantee the reliability of power systems for EVs.

At present, the most commonly used models of supercapacitor include the black box model, electrochemical model, and equivalent circuit model (ECM) [9,10].

(1) The black box model can describe the relationship between specific parameters and external characteristics with good flexibility and model precision. Optimization algorithms, including neural networks, fuzzy control, and machine learning, are employed to train the model on a large number of experimental data [11]. Wu et al. established an equivalent circuit neural network nonlinear dynamics model with temperature and voltage as input variables. Although the results are reliable, a large amount of data is needed for simulation [12]. Zhang et al. constructed a residual capacity estimation model based on an artificial neural network to represent the dynamic performances of supercapacitors, considering various currents and uncertain temperatures [13]. The experimental results show that the proposed model is feasible and effective, which can provide accurate prediction of residual capacity. Nevertheless, the black box model requires a large amount of data for training to improve the prediction accuracy.

(2) In order to accurately describe the internal parameters and external characteristics of supercapacitors, the electrochemical model, including many partial differential equations, is widely used in supercapacitor modeling [14]. Drummond et al. studied two electrochemical models to simulate the nonlinear partial differentiation of supercapacitors and found that the spectral discrete model can improve computational efficiency while ensuring accuracy. [15] Wang et al. proposed a three-dimensional model, which not only makes it possible to simulate the dynamic performances of electric double layer capacitors (EDLCs), but also provides standard rules for achieving the maximum charging performance of EDLCs [16]. Drummond et al. presented an absolute voltage stability method, which combines the electrochemical parameters with electrical properties of the supercapacitor. The method can obtain a stable voltage with less experimental data [17]. Tian et al. conducted a comparative study on five fractional models and found that the composition and structure of the models would affect the voltage simulation and state of charge (SOC) estimation [18]. In fact, the model accuracy is not directly proportional to the parameter complexity. Although the electrochemical model has many advantages, it is not conducive to practical application due to its complex structure and huge computation [19].

(3) An equivalent circuit model is a circuit network composed of a capacitor, inductor, resistor, and other circuit elements to represent the voltage response characteristics of supercapacitors. At present, equivalent circuit models mainly include the internal resistance model, RC model, and PNGV model, etc. [20]. Since the equivalent circuit model has fewer parameters and can balance the accuracy and complexity of dynamic simulation, it has been extensively used in the model construction of automotive supercapacitors [21–23]. Spyker et al. proposed a classical equivalent circuit, which consists of an equivalent series resistance, equivalent parallel resistance, and main capacitor, but it only describes the dynamic performance of supercapacitors in a short time [24]. From ref [22], a variable resistance equivalent circuit model for supercapacitors is presented to accurately simulate the charging, redistribution, and self-discharge processes of supercapacitors. Compared with the energy recursive model, it can provide a more accurate terminal voltage estimation of the supercapacitor. Liu et al. described the relationship between model parameters and temperature variation based on different functions. In this way, an equivalent circuit model considering temperature uncertainty is introduced for enhancing model fitness at various temperatures [25–27].

There is a lot of research that has discussed the supercapacitor models, but most of this has focused on the methods to improve the accuracy under a single model. However, each model has the particular advantage under different SOC ranges. Therefore, the improvement of model accuracy and the offset of the single model drawback are the key problems in the supercapacitor modeling field. The fusion model modeling method is an effective and popular solution to the problem. The fusion method consists of a physical fusion method and a data fusion method. Additionally, the combination of physical fusion method and data fusion method can theoretically further improve the model accuracy.

Liu et al. established a model on the basis of a composite model [28] and presented a model combination method, which can be used to construct data fusion by a multi-model combination. Moreover, many data fusion methods including SOC fusion estimation and weight allocation optimization fusion are introduced in the literature [29]. Li et al. proposed a fusion estimation method of SOC based on Gaussian process regression (GPR), which significantly improved the accuracy of the model [30]. However, a single Gaussian distribution is difficult to resist external interference, and the results are uncertain. Wei et al. used the normalized weights of multiple Gaussian distributions to calculate the weight of Gaussian components, and proposed a SOC estimation method based on Gaussian mixture model (GMM). The simulation results show that this method can effectively resist external interference and improve the accuracy of the model [31]. Meanwhile, Lyu et al. proposed a data fusion model method to estimate battery capacity by local charging curve using Gaussian regression, and smoothing incremental capacity curve by local weighted scatter smoothing can effectively improve the model accuracy [32]. Researchers can also use a data fusion method when constructing a terminal voltage fusion model. However, because the different fusion methods will affect the accuracy of the supercapacitor model, how to choose the fusion method effectively is a problem worth studying.

In this paper, a two-layer fusion model based on a multi-model supercapacitor is proposed. This fusion method adopts physical data fusion, including three fusion models: a fusion model based on SOC fragments, a fusion model based on a Bayesian algorithm [33], and fusion model based on residual normalization.

This paper is organized as follows. Section 2 introduces the feature experiments for supercapacitors. In addition, eight popular equivalent circuit models of supercapacitors are presented in Section 3. Section 4 discusses the parameter identification method for supercapacitor models. In Section 5, a two-layer fusion model is proposed, and the conclusion is in Section 6.

2. Characteristics of Supercapacitors

The characteristic experiments of the supercapacitors mainly include hybrid pulse power characterization (HPPC) and urban dynamometer driving schedule (UDDS). The HPPC experiment can reflect the relationship between voltage characteristics, depth of discharge (DoD), and charge–discharge rate under different SOCs, which aims to provide experimental data supporting the parameter identification of the supercapacitor model. The UDDS character experiments are mainly used to test the performance of the supercapacitor under actual driving conditions.

The capacitance of the experimental supercapacitor is 1500 F, and the upper cut-off voltage is 2.7 V (SOC = 100%) and the lower cut-off voltage is set as 0.5 V (SOC = 0%). The HPPC test procedure is shown follows:

Step 1: Charge the supercapacitor to 2.7 V with constant current 1 A;

Step 2: Hold the supercapacitor for 10 h to reach a stable state;

Step 3: Discharge the supercapacitor with low current to 0.5 V;

Step 4: Hold for 5 min;

Step 5: Then, charge with 1 A constant current constant voltage (CC-CV) up to the cut-off voltage of 2.7 V, until the current is less than 0.05 A (define: $t = 0$).

Step 6: Complete charging and discharging tests in different rates of current, $t = t + 1$. Firstly, discharge for 5 s and charge for 5 s at a current of 1A; then, discharge for 5 s and charge for 5 s at a current of 5A; finally, discharge at 10A for 5 s and charge for 5 s. It is worth noting the need to stand for a period of time after each charge and discharge.

Step 7: Keep 10% of the rated capacity of 1 A constant discharge. If $t \leq 10$, return to Step 6.

The pulse step, rest step, and discharge step of the test are shown in Figure 1. Positive denotes current discharge and negative denotes current charge.

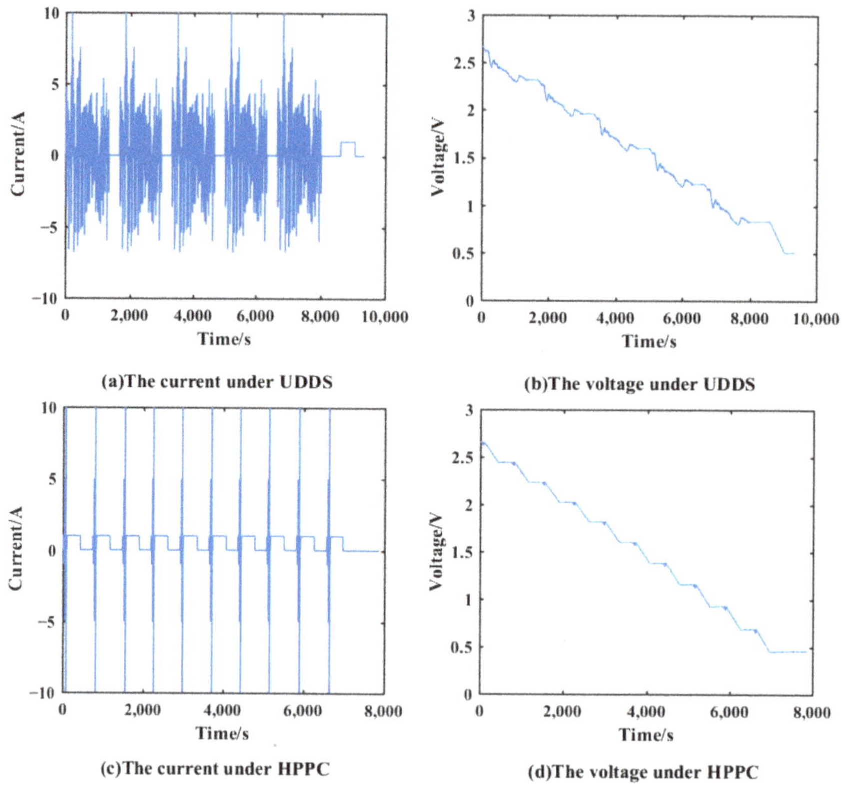

Figure 1. HPPC current and voltage diagram and UDDS current and voltage diagram. (**a**) Current under UDDS. (**b**) Voltage under UDDS. (**c**) Current under HPPC. (**d**) Voltage under UDDS.

3. Introduction to the Supercapacitor Model

Since there are many models of supercapacitors, the accuracy of the model and its complexity should be taken into consideration when selecting the model.

The accuracy and complexity of supercapacitor models are the significant factors of concern in EV application. Moreover, there are many common supercapacitor models reported in previous literatures. Consequently, eight popular equivalent circuit models including the Rint model [6], Thevenin model [6], dual-polarization model [6], PNGV model [34], GNL model [35], dynamic model [24], first-order *RC* model with one-state hysteresis [36], and second-order *RC* model with one-state hysteresis [36], were comprehensively considered and selected in this paper.

The Rint model consists of a power module and an internal resistance module.

The Thevenin model considers the polarization characteristics of the supercapacitor. In the model, the ideal voltage source, U_{oc}, describes the open-circuit voltage, and R_D and C are the polarization internal resistance and polarization capacitance, respectively. U_D is the voltage drop of *RC* parallel link, which is used to simulate the polarization voltage of the supercapacitor.

In the dual-polarization model, two *RC* modules in series are added on the basis of the Rint model to describe the supercapacitor polarization characteristics.

In the PNGV model, U_{oc} is the ideal voltage source and represents the open-circuit voltage. R_0 is the ohmic resistance of the battery. R_p is the battery polarization resistance. C_p is the parallel capacitance beside R_p. C_p describes the change in open-circuit voltage as the load current accumulates over time.

GNL model considers the effects of ohmic polarization, electrochemical polarization, concentration polarization, and self-discharge. In the model, the open-circuit voltage is represented as U_{oc}. R_1 and C_1 are concentration polarization resistance capacitance parameters, respectively. R_2 and C_2 are the resistance-capacitance parameters of the electrochemical polarization of the power source. R_e is ohmic internal resistance and R_s is the internal resistance of self-discharge.

The dynamic model is composed of a series resistor, a series capacitor, and two *RC* networks. In the model, U_{oc} is the ideal pressure source. u0, u1, and u2 correspondingly represent the main capacity and the terminal voltages of the two *RC* networks.

The model with one-state hysteresis considers changes in y (dependent variable) behind changes in x (independent variable). In the supercapacitor energy storage system, the voltage changes behind the current changes. Therefore, the lag level *h* is added in the calculation of the first-order *RC* model with one-state hysteresis, and second-order *RC* model with one-state hysteresis.

4. Model Parameter Identification

4.1. Genetic Algorithm (GA)

GA is a kind of adaptive global optimization probabilistic search algorithm, which has good adaptability and optimization ability in parameter identification. It starts with a randomly generated population. After the initial population is generated, the principle of survival of the fittest needs to be implemented in order to finally approach the optimal solution. In each generation, individuals are selected according to the fitness in the problem field, and the filial generation representing the new solution set is generated by genetic operators. This process is the same as natural selection to make offspring better adapted to the environment. After decoding, the optimal individual in the last generation population may be used as the optimal solution of the problem. In this paper, the fitness function of genetic algorithm is the square sum of the error between the terminal voltage of the equivalent circuit model and the actual measured terminal voltage.

The genetic algorithm flow diagram is shown in Figure 2. The specific steps of genetic algorithm are listed as follows:

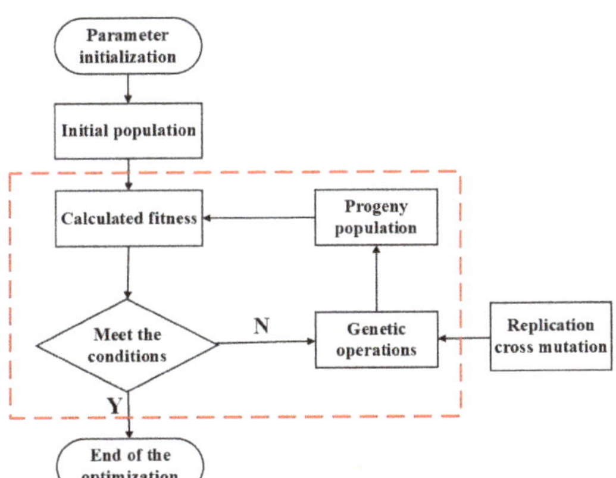

Figure 2. Flow chart of genetic algorithm. (The cyclic process of optimizing parameters is reflected in the red dotted box in the figure).

Step 1. Set the boundary conditions of the parameters.

Step 2. Generate the initial population.

Step 3. Calculate the fitness of individuals in the population and judge whether the requirements are met. If satisfied, identification is over; otherwise, proceed the next step.

Step 4. Carry out inheritance, crossover, and mutation of the population to obtain offspring.

Step 5. Return to Step 3.

4.2. Parameter Identification

This section takes the Thevenin model as an example to describe, and the rest of the models are similar. Circuit diagrams of the Thevenin model of the supercapacitor is shown in Table 1, where i_L is the load current; R_D and C are the polarization internal resistance and polarization capacitance, respectively; and U_D is the voltage drop of RC parallel link, which is used to simulate the polarization voltage of the supercapacitor [37]. The circuit equation of this circuit model is Equation (1).

$$\begin{cases} \dot{U}_D = \frac{i_L}{C} - \frac{U_D}{R_D C} \\ U_t = U_{oc} - U_D - i_L R_i \end{cases} \tag{1}$$

Table 1. Circuit diagram and equation of each model.

Model	Circuit Diagram	Equation
Rint model	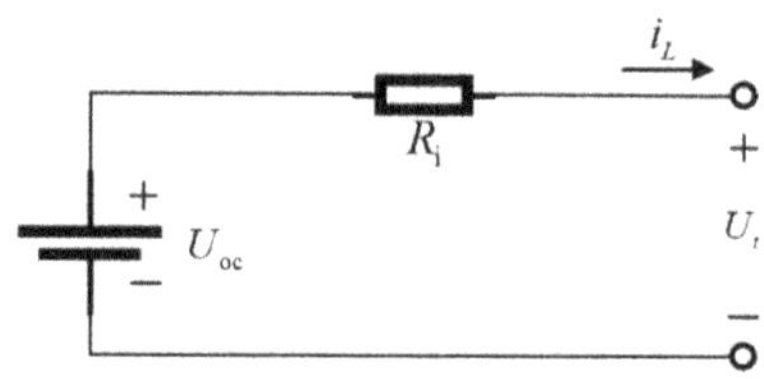	$U_t = U_{oc} - i_L R_i$
Thevenin model	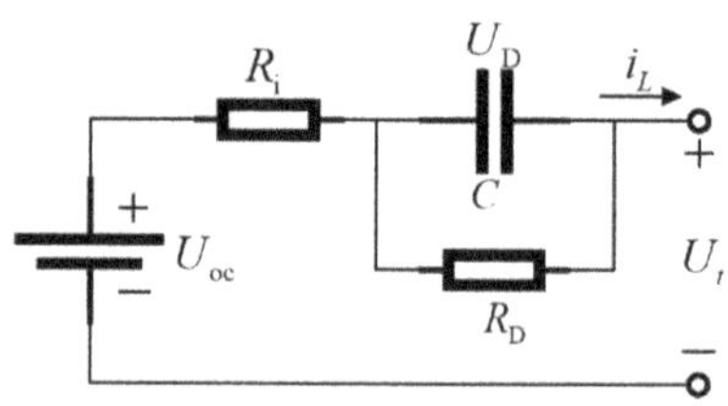	$\begin{cases} \dot{U}_D = \frac{i_L}{C} - \frac{U_D}{R_D C} \\ U_t = U_{oc} - U_D - i_L R_i \end{cases}$
Dual-polarization model	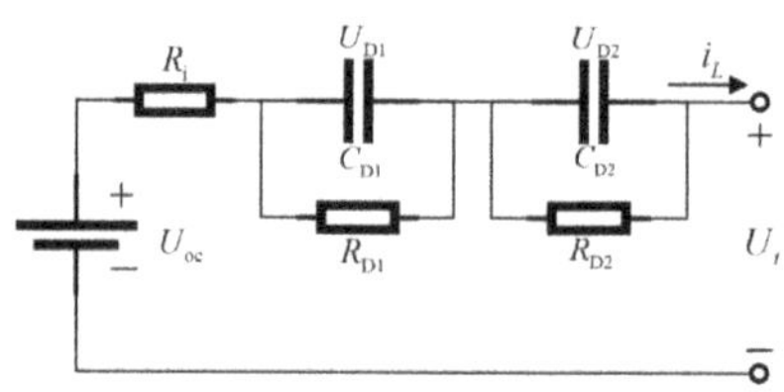	$\begin{cases} U_{D1} = \frac{i_L}{C_{D1}} - \frac{U_{D1}}{R_{D1} C_{D1}} \\ U_{D2} = \frac{i_L}{C_{D2}} - \frac{U_{D2}}{R_{D2} C_{D2}} \\ U_t = U_{oc} - U_{D1} - U_{D2} - I_L R_i \end{cases}$
PNGV model	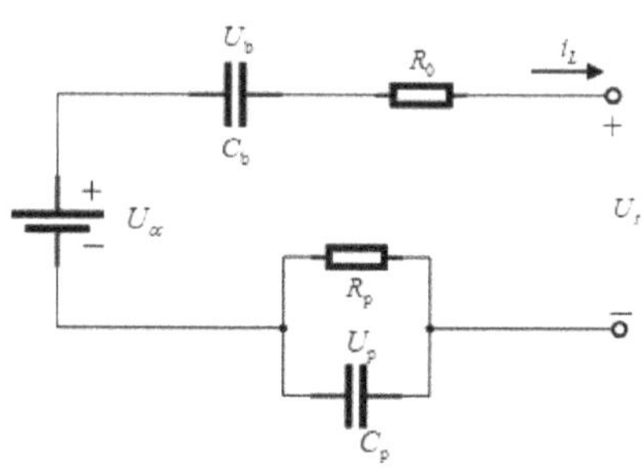	$\begin{cases} \begin{bmatrix} \frac{dU_p}{dt} \\ \frac{dU_b}{dt} \end{bmatrix} = \begin{bmatrix} -\frac{1}{C_p R_p} & 0 \\ 0 & 0 \end{bmatrix} \begin{bmatrix} U_p \\ U_b \end{bmatrix} + \begin{bmatrix} \frac{1}{C_p} \\ \frac{1}{C_b} \end{bmatrix} [I] \\ [U_L] = [U_{oc}] + \begin{bmatrix} -1 & -1 \end{bmatrix} \begin{bmatrix} U_p \\ U_b \end{bmatrix} - R_0 [I] \end{cases}$

Table 1. *Cont.*

Model	Circuit Diagram	Equation
GNL model	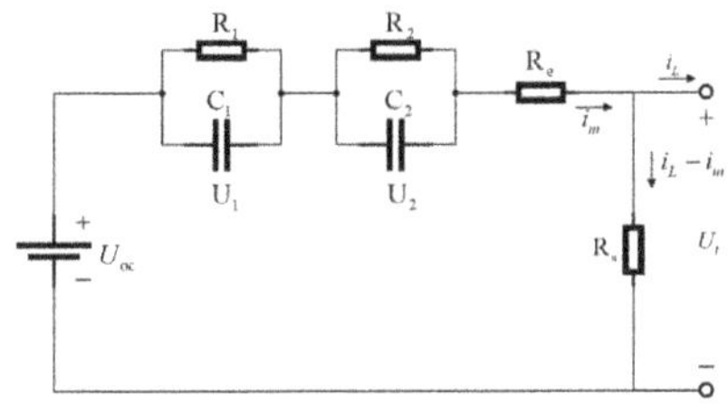	$\begin{cases} I_m = \frac{U_1}{R_1} + C_1\frac{dU_1}{dt} \\ I_m = \frac{U_2}{R_2} + C_2\frac{dU_2}{dt} \\ U_L = U_{oc} - I_m R_e - U_1 - U_2 \\ U_L = (I - I_m)R_s \end{cases}$
Dynamic model	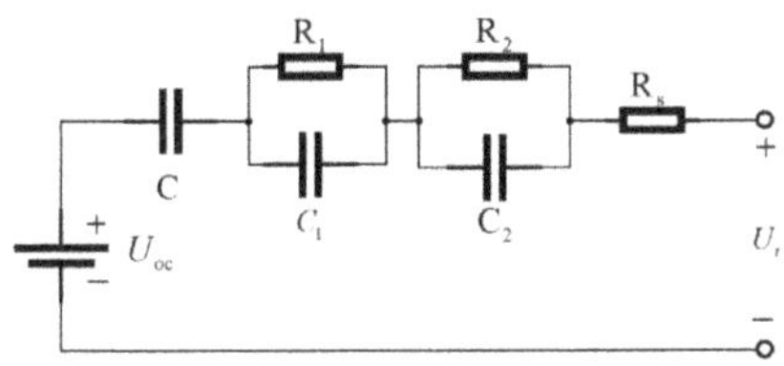	$\begin{cases} \frac{u_1}{R_1} + C_1\frac{du_1}{dt} = i_c \\ \frac{u_2}{R_2} + C_2\frac{du_2}{dt} = i_c \\ u_1 + u_2 + R_0 i = u_c \end{cases}$
First-order RC model with one-state hysteresis	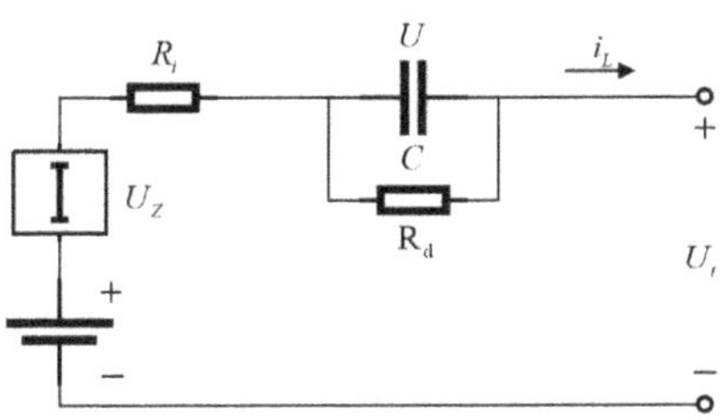	$\begin{cases} h = -\lvert\kappa i_L\rvert h + \lvert\kappa i_L\rvert \text{sgn}(i_L) \\ U = -\frac{1}{CR_d}U + \frac{1}{C}i_L \\ U_L = U_{oc} + hM - U - i_L R_0 \end{cases}$
Second-order RC model with one-state hysteresis.	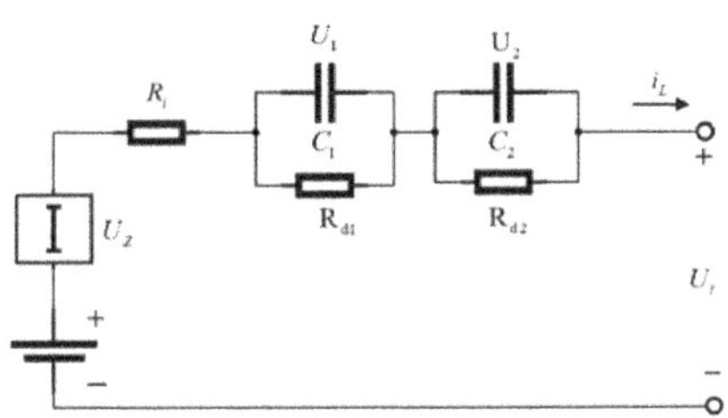	$\begin{cases} h = -\lvert\kappa i_L\rvert h + \lvert\kappa i_L\rvert \text{sgn}(i_L) \\ U_1 = -\frac{1}{C_1 R_{d1}}U_1 + \frac{1}{C_1}i_L \\ U_2 = -\frac{1}{C_2 R_{d2}}U_2 + \frac{1}{C_2}i_L \\ U_L = U_{oc} + hM - U_1 - U_2 - i_L R_0 \end{cases}$

The model is discretized before parameter identification. The polarization voltage of the supercapacitor model is obtained as Equation (2).

$$U_D[(k+1)\Delta t] = e^{-\Delta t/\tau}U_D(k\Delta t) + R_D i_L[(k+1)\Delta t][1 - e^{-\Delta t/\tau}] \tag{2}$$

Among them, $\tau = R_D \times C$. Meanwhile, the discretization calculation equation of the supercapacitor SOC can be obtained as shown in Equation (3).

$$z_k = z_{k-1} - \eta_i i_L \Delta t / C_{max} \tag{3}$$

Z_k represents the SOC value at time k; Δt represents the segment time of current acquisition; η_i represents the coulomb efficiency; and C_{max} represents the rated capacity of the supercapacitor.

The parameters to be identified in the Thevenin model of the supercapacitor include R_i, R_D, and τ. Since the parameters will change under different SOC state estimations, it is necessary to identify the three parameters in each SOC segment. In order to balance the identification accuracy and efficiency, the discharge segment of the supercapacitor was divided into ten segments ranging from 100% to 0%. The identification results and errors are shown in Table 2:

Table 2. Parameter identification results of Thevenin model for supercapacitors.

SOC	R_i (mΩ)	R_D (mΩ)	τ
100%–90%	0.5760	4.079	318.778
90%–80%	0.5874	4.956	159.593
80%–70%	0.5790	4.289	124.776
70%–60%	0.5679	9.450	263.936
60%–50%	0.5557	3.965	134.347
50%–40%	0.5437	7.719	254.746
40%–30%	0.5488	9.117	309.871
30%–20%	0.5494	6.066	180.197
20%–10%	0.5530	6.383	225.290
10%–0%	0.4853	0.681	11.2813

Figure 3 shows the error diagram of the terminal voltage simulation results and experimental values under UDDS conditions. In general, the identification error can be divided into two ranges, namely the SOC range of [50%–100%] and [0%–50%]. As can be seen from the Figure 3, the Thevenin model has a relatively excellent simulation accuracy in the segment of [50%–100%], and the error in this segment can be controlled within 10 mV without large fluctuations. Similar results were obtained by verifying the other seven models. When the SOC drops to [0%–50%], the accuracy of all models decreases. When the SOC is above 50%, a better precision can be obtained. The results show that the genetic algorithm can effectively identify the relevant parameters in each model.

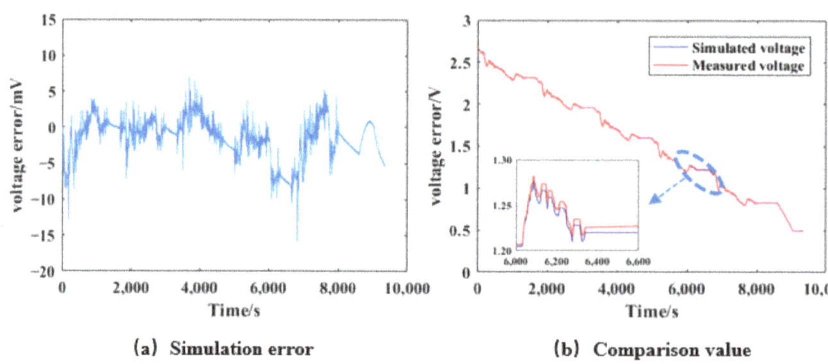

Figure 3. UDDS-based Thevenin model validation diagram. (**a**) Simulation error variation figure (**b**) Comparison of simulation terminal voltage and measured value.

After parameter identification based on the data obtained from the HPPC, the maximum error, mean error, and root mean square error (RMSE) of the eight models were, respectively, calculated under the UDDS, as shown in Table 3.

By analyzing the terminal voltage error model of supercapacitors, it can be seen that more complex models can not necessarily achieve higher accuracy of training data sets. In fact, if the nature of the model is too complex, it will be more susceptible to uncertainty. Then, models with overly complex features are not suitable for model validation datasets.

From Table 3, the Rint model has better simulation accuracy compared with other models. Due to the difference in the simulation accuracy of the SOC segment, the variation of the SOC segment should be taken into consideration in the comparison. For example, in the SOC range of (90%–100%), the RMSE of the Rint model is 0.0285 mV, which is the model with the highest accuracy in this SOC range. However, in the SOC range of (50%–60%), the accuracy of the dynamic model is the highest, and the root mean square error is 1.7946 mV.

Table 3. Errors of the eight models.

Model	Maximum Error (mV)	Mean Error (mV)	Root Mean Square Error (mV)
Rint model	15.062	2.2248	3.0864
Thevenin model	15.953	2.2727	3.1445
Dual-polarization model	18.411	3.9966	4.5970
PNGV model	17.110	3.9035	5.3282
GNL model	15.317	2.2960	3.1851
First-order *RC* model with one-state hysteresis	20.038	5.5645	6.8858
Second-order *RC* model with one-state hysteresis.	26.105	12.2051	12.948
Dynamic model	18.775	3.6325	4.4737

On account of model differences affecting the simulation accuracy of terminal voltage, none of the models can maintain the optimal simulation accuracy of terminal voltage at different times. It is difficult for a single model to maintain the optimal accuracy in a changing external environment. Therefore, the supercapacitor fusion model based on multi-model probabilistic is proposed in this paper.

5. Multi-Model Probabilistic Fusion Model

It has been verified that the model with the best accuracy is different in the varying SOC interval. Therefore, results based on a single model are not guaranteed to be optimal in the entire SOC segment. It is worth proposing an optimization algorithm based on multiple models to further optimize accuracy. Consequently, four kinds of multi-model voltage residuals are presented to scientifically determine the model switching objective function.

Fusion model based on SOC fragments: The objective function was established to find the minimum root mean square error model in different SOC intervals, and the fusion model was established by combining them. The operation is to divide the SOC into 10 segments, and then calculate the RMSE of the different models in each segment. The model with the smallest RMSE value is used as the fusion model of the current SoC segment.

Fusion model based on Bayesian algorithm: The advantages of different models are combined by giving weight to the eight models, respectively, for fusion. In order to determine the weight of each model, the probability is adopted in this paper to describe the degree of closeness between the predicted terminal voltage and the real voltage. When selecting weights, the statistical characteristics of the residual are added, and a Bayesian algorithm is used to obtain the conditional distribution probability of terminal voltage. A Bayesian algorithm is the estimation of the prior knowledge to the posterior knowledge in the inspection process. By using the discrete Bayesian algorithm, the probability of the previous moment is considered as a deterministic probability, and then the probability of the later moment can be estimated. In this way, the weight of each model at the next moment is obtained.

Fusion model based on residual normalization: The fusion result is the weighted sum of each model, which is taken as the initial value of the state estimation at the next moment, so as to obtain the prior estimate. When the weight is selected, the instantaneous terminal voltage residuals represent the estimation accuracy. The specific operation is to normalize the terminal voltage residuals of the eight models, and the obtained probability based on the normalization of the residual is the model weight.

5.1. Fusion Model Based on SOC Fragments

RMSEs of different models are calculated under the SOC segment, and the model corresponding to the minimum RMSE is determined. The model terminal voltage is taken as the terminal voltage of the segment under the fusion model. According to the minimum RMSE of each model in 10 SOC segments, the models selected for each SOC segment are shown in Table 4.

Table 4. Selection of different SOC fragment models.

SOC	The Target Model
100%–90%	second-order RC model with one-state
90%–80%	GNL model
80%–70%	Rint model
70%–60%	Rint model
60%–50%	PNGV model
50%–40%	Rint model
40%–30%	Rint model
30%–20%	Rint model
20%–10%	Rint model
10%–0%	Thevenin model

It can be seen from the selection of models in different SOC segments that the Rint model has the highest probability of being defined as the target model. There is a 60% probability that the Rint model will be selected, which is in line with the overall optimal result of the Rint model in Table 3. The fusion model method based on the SOC segment incorporates 75% of the target model through optimization selection. This method eliminated some models with poor accuracy in each SOC segment and retained the models with better accuracy, which greatly reduced the operation time and improved the fusion efficiency.

5.2. Fusion Model Based on Bayesian Algorithm

The Bayesian estimation process is simple and fast, and considering the influence of the previous moment on the next moment, the predicted value of terminal voltage is set as:

$$\hat{U}(k) = \sum_{i=1}^{8} w_i \widetilde{U}_i(k) \tag{4}$$

Among them, $\widetilde{U}_i(k)$ is the predicted value of terminal voltage of a single model, w_i is the weight coefficient of each model, and w_i satisfies $\sum_{i=1}^{8} w_i = 1$.

According to Bayes' theorem:

$$p(\widetilde{U}(k)|p(U(k))) = \frac{p(U(k)\Big|\widetilde{U}(k))p(w(k))}{p(U(k))} \tag{5}$$

$U(k)$ is the terminal voltage to be evaluated at k, and p is the probability. The fusion probability of each target model is calculated as follows:

$$p(s_i(k)|U(k)) = \frac{p(s_i(k), U(k))}{p(U(k))} \tag{6}$$

where $s_i(k)$ is the parameter set of the ith model under the SOC basis at time k.

The predicted value of terminal voltage can be rewritten as:

$$\hat{U}(k) = \sum_{i=1}^{8} p_i \widetilde{U}_i(k) \tag{7}$$

Residuals for: $e_i(k) = U_i(k) - \widetilde{U}_i(k)$

Then:

$$\Psi_i(k) = \frac{1}{(2\pi)^{1/2} Q_i^{1/2}(k)} \exp(-e_i(k)^2 Q_i(k)^{-1}/2) \tag{8}$$

Among them, $Q_i(k) = \sum_{k=1}^{n} e_i(k)^2/n$. This is the variance of the residuals of each model. Therefore, the weight coefficient is:

$$w_i(k) = \frac{\Psi_i(k) w_i(k-1)}{\sum_{i=1}^{8} \Psi_i(k) w_i(k-1)} \tag{9}$$

5.3. Fusion Model Based on Residual Normalization

Without considering the influence of weights at the previous moment on the weights at the later moment, the probability is adopted to describe the approximation degree between the predicted terminal voltage and the real value. The predicted value of terminal voltage is:

$$\hat{U}(k) = \sum_{i=1}^{8} w_i z_k \tag{10}$$

Z_k is the SOC value. w_i is the weight coefficient of each model, and w_i satisfies $\sum_{i=1}^{8} w_i = 1$.

Residuals for: $e_i(k) = U_i(k) - \widetilde{U}_i(k)$

Then,$w_i(k) = \frac{S(k) - e_i{}^2(k)}{(N-1)S(k)}$, S(k) is the sum of squares of the residuals of each model. In other words, $S(k) = \sum_{i=1}^{8} e_i{}^2(k)$.

5.4. Two-Layer Fusion Model

In order to further improve the accuracy and adaptability of the model, the two-layer fusion model is proposed. The optimal root mean square error (RMSE) is taken as the decision variable, and the corresponding data to be fused is taken as the result of the two-layer fusion model. The algorithm process of the double-layer fusion model is shown in Figure 4.

The specific steps are as follows:

Step 1: Input the error matrices E_1, E_2 and E_3 under UDDS of each model.

Step 2: Calculate RMSE with SOC segment to obtain matrix RE_{ij}, i = 3, j = 10.

Step 3: Search the position information of the minimum RMSE of each segment, respectively, to obtain M_{ij} and N_{ij}.

Step 4: Assign the position information M_{ij} and N_{ij} to the corresponding objective function. That is, get the objective function of different SOC segments.

Under different SOC segments, the selected target models are shown in Table 5.

In the selection of two-layer fusion models of different SOC fragments, it can be seen that the fusion model based on residual normalization has the highest probability to be defined as the target model, up to 80%. However, the fusion model based on SOC fragments and the fusion model based on a Bayesian algorithm only have a 10% probability to be defined as the target model. The results show that the fusion model based on residual normalization is more advantageous in general. However, it is not reliable to explain the accuracy of the model only according to the probability of the model being selected. Therefore, it is more necessary to compare and analyze each model in depth.

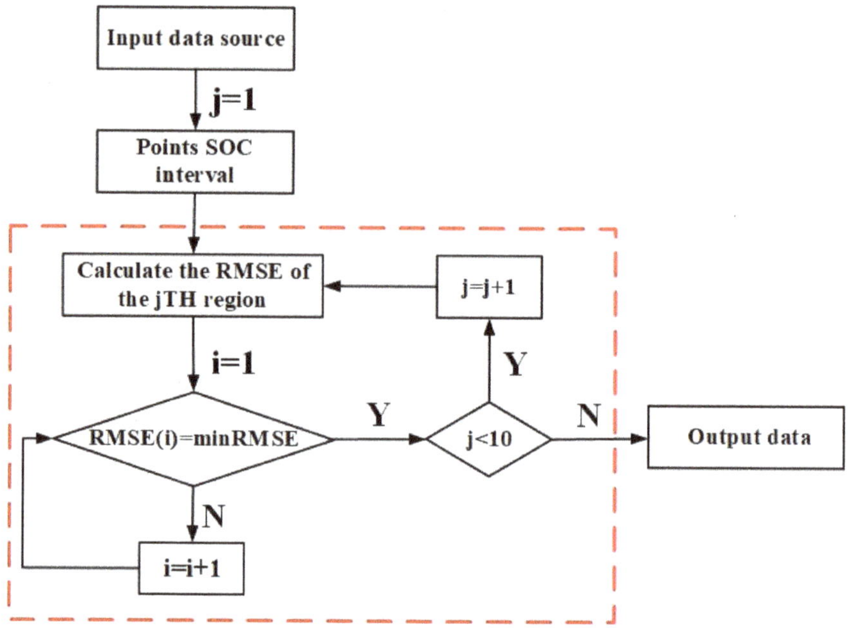

Figure 4. Flow chart of two-layer fusion model. (The red dashed box in the figure shows the algorithm cycle process).

Table 5. Selection of *two-layer fusion model* of different SOC fragments.

SOC	The Target Model
100%−90%	Fusion model based on residual normalization
90%−80%	Fusion model based on residual normalization
80%−70%	Fusion model based on Bayesian algorithm:
70%−60%	Fusion model based on residual normalization
60%−50%	Fusion model based on SOC fragments
50%−40%	Fusion model based on residual normalization
40%−30%	Fusion model based on residual normalization
30%−20%	Fusion model based on residual normalization
20%−10%	Fusion model based on residual normalization
10%−0%	Fusion model based on residual normalization

5.5. Results and Analysis of Different Fusion Models

Based on the data obtained from the fusion model, it is divided into 10 segments according to per 10% SOC under the UDDS to verify its errors. On the basis of the errors, the maximum error, mean error, and RMSE of the four fusion models are calculated, respectively. The comparison between the simulated values of the four fusion models and the measured terminal voltages is shown in Figure 5.

It can be concluded that, due to the good convergence and small error, the two-layer fusion model does not show obvious advantages to improve the accuracy in high SOC segments. In the middle and low SOC segments, the initial data is more volatile than that of high SOC segment, while the two-layer fusion model obviously shows the advantage of fast convergence. Combined with Table 6, the mean error and RMSE of the two-layer fusion model under the three fusion models are reduced by at least 2.08% and 1.36%, respectively. Moreover, compared with the other three fusion models, the two-layer fusion model can make different SOC segments retain the optimal simulation voltage. The two-layer fusion model can reduce the errors of the single fusion model in different SOC segments to further improve the model precision.

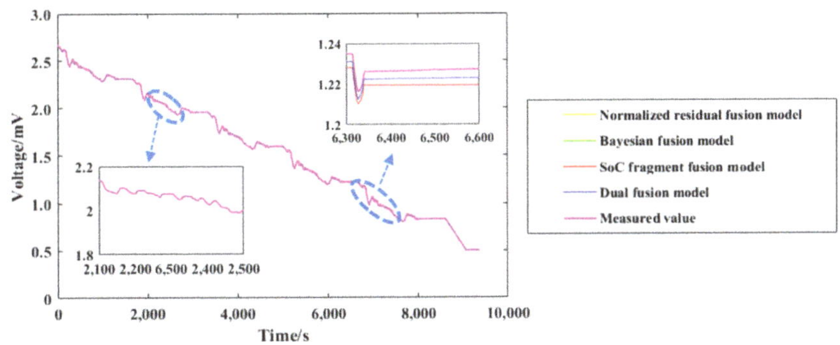

Figure 5. Simulation error comparison of models under the same UDDS.

Table 6. Errors of the four fusion models.

Model	Maximum Error (mV)	Mean Error (mV)	Root Mean Square Error (mV)
Fusion model based on SOC fragments	15.062	2.1179	3.0023
Fusion model based on Bayesian algorithm	15.062	2.1973	3.0567
Fusion model based on residual normalization	12.565	1.5112	2.1043
two-layer fusion model	12.565	1.4798	2.0756

5.6. Validation of Fusion Models at Different Temperatures

The supercapacitor model should adapt to the complex and changeable operation environment. It is well known that different temperatures will make a difference in the performance of supercapacitor [38]. In order to take into account the temperature, this paper selected test data under the experimental environment of 10, 25, and 40 °C, respectively, and repeated the parameter identification processes of GA to obtain the eight parameter sets. Similarly, the obtained parameter set is substituted into the UDDS to verify and calculate the simulation value of the terminal voltage. Then, the maximum error, mean error, and RMSE of the four fusion models are obtained.

From Figure 6, it is found that the two-layer fusion model can ensure the minimum root mean square error and average error in the optimal section at different temperatures. Figure 6 shows that, under the experimental conditions of 10, 25, and 40 °C, the RMSE can be reduced by 60.41%, 47.26%, and 23.04%, the maximum error can be decreased by 9.51%, 19.87%, and 8.70%, and the mean error can be declined by 68.21%, 48.48%, and 30.13%, respectively. The improvement effect of the maximum error is not as obvious as that of the mean error and the RMSE. In fact, the two-layer fusion model takes the minimum RMSE as the decision variable and imports the fusion data again for settlement based on the SOC segment. Therefore, for the two-layer fusion model, the interference of other fusion models in the period of large estimation error can be greatly avoided in terms of RMSE and mean error, so as to improve the accuracy of the model again. It is demonstrated that the two-layer fusion model has strong adaptability against temperature and can combine the advantages of each fusion model at different temperatures to achieve the optimal results.

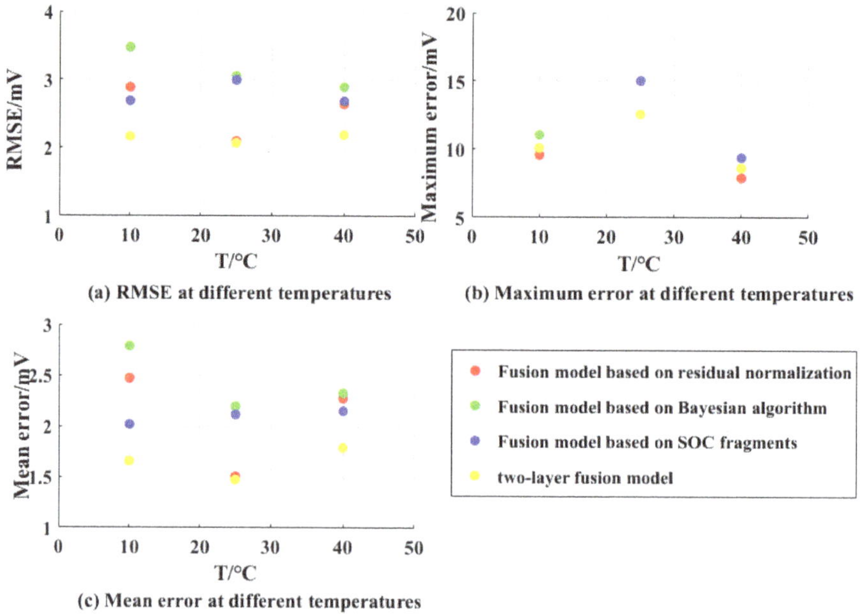

Figure 6. Comparison of the fusion model errors at different temperatures. (**a**) RMSE. (**b**) Maximum error. (**c**) Mean error.

6. Conclusions

A two-layer fusion model is proposed in this paper based on three fusion models. In the two-layer fusion model, the terminal voltages at different times can quickly converge to the true values, and RMSE can reduce by 23.04%, which indicates that it significantly improves the accuracy of the model. The two-layer fusion model was validated at ambient temperatures of 10, 25, and 40 °C, respectively. Compared with the previous three fusion models, the RMSE, maximum error, and mean error of the two-layer fusion model are all reduced. For RMSE, the two-layer fusion model correspondingly reduced by 60.41%, 47.26%, and 23.04%, which indicates that it has reliability redundancy. Finally, the two-layer fusion model proposed in this paper can effectively play the advantages of physical fusion and data fusion with rapid convergence. It can significantly avoid the interference of the larger model errors in different SOC intervals, improves the accuracy of the estimation results, and has high applicability.

Author Contributions: Conceptualization, B.H. and C.W.; methodology, Y.M. and C.W.; software, Y.M.; validation, Y.M.; formal analysis, Y.M. and Y.C.; writing—original draft preparation, Y.M. and Y.C.; writing—review and editing, B.H., C.W. and Q.Y.; visualization, Y.M.; supervision, B.H.; project administration, C.W.; funding acquisition, C.W. All authors have read and agreed to the published version of the manuscript.

Funding: This research was funded by the National Natural Science Foundation of China (Grant No. 51907136), Zigong Key Science and Technology Project (Grant No. 2019YYJC14), Talent-Introduction Project of Sichuan University of Science & Engineering (Grant No. 2019RC15), and Foundation of Artificial Intelligence Key Laboratory of Sichuan Province (Grant No. 2020RYY01). The systemic experiments were performed at the Advanced Energy Storage and Application (AESA) Group, Beijing Institute of Technology.

Institutional Review Board Statement: Not applicable.

Informed Consent Statement: Not applicable.

Conflicts of Interest: The authors declare no conflict of interest.

References

1. Chen, X.K.; Sun, D.; Chen, X.H. Modelingand State of Charge Robust Estimation for Lithium-ion Batteries. *Trans. China Electrotech. Soc.* **2015**, *30*, 141–147.
2. Xiong, R.; Sun, W.; Yu, Q.Q.; Sun, F. Research Progress, Challenges and Prospects of Fault Diagnosis on Battery System of Electric Vehicles. *Appl. Energy* **2020**, *279*, 115855. [CrossRef]
3. Xiong, R.; Li, L.; Yu, Q.Q.; Jin, Q.; Yang, R. A set membership theory based parameter and state of charge co-estimation method for all-climate batteries. *J. Clean. Prod.* **2020**, *249*, 11389. [CrossRef]
4. Hu, X.S.; Xu, L.; Lin, X.K.; Pecht, M. Battery Lifetime Prognostics. *Joule* **2020**, *4*, 310–346. [CrossRef]
5. Zhang, L.; Hu, X.S.; Wang, Z.P.; Sun, F.C.; Dorrell, D.G. Experimental impedance investigation of a supercapacitor at different conditions for electric vehicle applications. *J. Power Sources* **2015**, *287*, 129–138. [CrossRef]
6. Zhang, L.; Hu, X.S.; Wang, Z.P.; Sun, F.C.; Dorrell, D.G. A review of supercapacitor modeling, estimation, and applications: A control/management perspective. *Renew. Sustain. Energy Rev.* **2018**, *81*, 1868–1878. [CrossRef]
7. Wang, B.; Xu, J.; Cao, B.G.; Ning, B. Adaptive mode switch strategy based on simulated annealing optimization of a multi-mode hybrid energy storage system for electric vehicles. *Appl. Energy* **2017**, *194*, 596–608. [CrossRef]
8. Herrera, V.; Milo, A.; Haizea, G.; Ion, E.O.; Igor, V.; Haritza, C. Adaptive energy management strategy and optimal sizing applied on a battery-supercapacitor based tramway. *Appl. Energy* **2016**, *169*, 831–845. [CrossRef]
9. Lai, X.; Wang, S.Y.; Ma, S.D.; Xie, J.Y.; Zheng, Y.J. Parameter sensitivity analysis and simplification of equivalent circuit model for the state of charge of lithium-ion batteries. *Electrochim. Acta* **2020**, *330*, 135239. [CrossRef]
10. Li, J.; Adewuyi, K.; Lotfi, N.; Landers, R.G.; Park, J. A single particle model with chemical/mechanical degradation physics for lithiumion battery State of Health (SOH) estimation. *Appl. Energy* **2018**, *212*, 1178–1190. [CrossRef]
11. Hu, X.S.; Tang, X.L. Review of Modeling Techniques for Lithium-ion Traction Batteries in Electric Vehicles. *Trans. China Electrotech. Soc.* **2017**, *53*, 20–31. [CrossRef]
12. Wu, C.H.; Hung, Y.H.; Hong, C.W. On-line supercapacitor Dynamic models for energy conversion and managemen. *Energy Convers. Manag.* **2011**, *53*, 337–345. [CrossRef]
13. Zhang, L.; Wang, Z.P.; Hu, X.S.; Dorrell, D.G. Residual Capacity Estimation for Supercapacitors in Electric Vehicles Using Artificial Neural Network. *IFAC Proc. Vol.* **2014**, *47*, 3899–3904.
14. Gabano, J.D.; Poinot, T.; Kanoun, H. LPV continuous fractional modeling applied to supercapacitor impedance identification. *Control Eng. Pract.* **2015**, *45*, 86–97. [CrossRef]
15. Drummond, R.; Valmorbida, G.; Stephen, R.D. Equivalent Circuits for Electrochemical Supercapacitor Models. *IFAC PapersOnLine* **2017**, *50*, 2671–2676. [CrossRef]
16. Wang, H.N.; Pilon, L. Mesoscale modeling of electric double layer capacitors with three-dimensional ordered structures. *J. Power Sources* **2013**, *221*, 252–260. [CrossRef]
17. Drummond, R.; David, A.H.; Stephen, R.D. Low-order mathematical modelling of electric double layer supercapacitors using spectral methods. *J. Power Sources* **2015**, *277*, 317–328. [CrossRef]
18. Tian, J.P.; Xiong, R.; Shen, W.X.; Wang, J. A Comparative Study of Fractional Order Models on State of Charge Estimation for Lithium Ion Batteries. *Chin. J. Mech. Eng.* **2020**, *33*, 106–120. [CrossRef]
19. Alexander, P.; Schmidt, M.B.; Árpád, W.I.; Lino, G. Experiment-driven electrochemical modeling and systematic parameterization for a lithium-ion battery cell. *J. Power Sources* **2010**, *195*, 5071–5080.
20. Vlasta, S.; Josef, S.; Majzner, J.; Sedlak, P.; Kuparowitz, T.; Brandon, B.; Vasina, P. Supercapacitor equivalent electrical circuit model based on charges redistribution by diffusion. *J. Power Sources* **2015**, *286*, 58–65.
21. Akram, E.; Mohamed, A.; Olivier, B.; Vinassa, J.M. Online parameter identification for real-time supercapacitor performance estimation in automotive applications. *Int. J. Electr. Power Energy Syst.* **2013**, *51*, 162–167.
22. Zhang, Y.; Yang, H.Z. Modeling and characterization of supercapacitors for wireless sensor network applications. *J. Power Sources* **2011**, *196*, 4128–4135. [CrossRef]
23. Xu, D.; Zhang, L.; Wang, B.; Ma, G.L. Estimation of supercapacitor energy based on particle swarm optimization algorithm for its equivalent circuit model. *Energy Procedia* **2019**, *158*, 4974–4979. [CrossRef]
24. Zhang, L.; Wang, Z.P.; Hu, X.S.; Sun, F.C.; David, G.D. A comparative study of equivalent circuit models of supercapacitors for electric vehicles. *J. Power Sources* **2015**, *274*, 899–906. [CrossRef]
25. Ivan, Ž.; Viktor, Š.; Ban, Ž.; Dubravko, K. Algorithm with temperature dependent maximum charging current of a supercapacitor module in a tram regenerative braking system. *J. Energy Storage* **2021**, *36*, 102378.
26. Yang, X.G.; Zhang, G.S.; Wang, C.Y. Computational design and refinement of self-heating lithium ion batteries. *J. Power Sources* **2016**, *328*, 203–211. [CrossRef]
27. Yang, X.G.; Leng, Y.J.; Zhang, G.S.; Ge, S.H.; Wang, C.Y. Modeling of lithium plating induced aging of lithium-ion batteries: Transition from linear to nonlinear aging. *J. Power Sources* **2017**, *360*, 28–40. [CrossRef]
28. Liu, X.T.; Chen, Z.H.; Zhang, C.B.; Wu, J. A novel temperature-compensated model for power Li-ion batteries with dual-particle-filter state of charge estimation. *Appl. Energy* **2014**, *123*, 263–272. [CrossRef]
29. Zhang, L.; Xu, D.; Wang, B.; Ma, G.L.; Li, Q.Y.; Zhou, H. Parameter Identification of the Equivalent Circuit Model of Supercapacitor Weight Distribution Optimization. *J. Xi'an Jiao Tong Univ.* **2018**, *52*, 96–102.

30. Li, J.B.; Wei, M.; Ye, M.; Jiao, S.J.; Xu, X.X. SOC estimation of lithium-ion batteries based on Gauss process regression. *Energy Storage Sci. Technol.* **2020**, *9*, 131–137.
31. Wei, M.; Li, J.B.; Ye, M.; Gao, K.P.; Xu, X.X. SOC estimation of Li-ion battery based on gaussian mixture regression. *Energy Storage Sci. Technol.* **2020**, *9*, 958–963.
32. Lyu, Z.Q.; Gao, R.J.; Li, X.Y. A partial charging curve-based data-fusion-model method for capacity estimation of Li-Ion battery. *J. Power Sources* **2021**, *483*, 229131. [CrossRef]
33. Wang, Y.Q.; Lu, F.C.; Li, H.M. Synthetic Fault Diagnosis Method of Power Transformer Based on Rough Set Theory and Bayesian Network. In *Advances in Neural Networks—ISNN 2008*; Springer: Berlin/Heidelberg, Germany, 2008; pp. 498–505.
34. Gan, Y.; Li, Y.; Yao, J. Study on Parameter Identification and SOC Estimation Based on PNGV Model for Energy Storage Lithium-ion battery. *Energy Res. Inf.* **2017**, *33*, 194–199.
35. Yan, X.W.; Guo, Y.W.; Wang, Y.W.; Deng, H.R.; Guo, Q. Electric Vehicle Battery State of Charge Estimation Based on GNL Model Adaptive Kalman Filter. *Sci. Technol. Eng.* **2018**, *18*, 94–100.
36. Chen, C.; Sun, F.C.; Xiong, R.; He, H.W. A Novel Dual H Infinity Filters Based Battery Parameter and State Estimation Approach for Electric Vehicles Application. *Energy Procedia* **2008**, *103*, 154–196. [CrossRef]
37. Yang, R.X.; Xiong, R.; Shen, W.X. On-board soft short circuit fault diagnosis of lithium-ion battery packs for electric vehicles using extended Kalman filter. *CSEE J. Power Energy Syst.* **2020**. [CrossRef]
38. Yang, R.X.; Xiong, R.; Shen, W.X.; Lin, X.F. Extreme Learning Machine-Based Thermal Model for Lithium-Ion Batteries of Electric Vehicles under External Short Circuit. *Engineering* **2021**, *7*, 395–405. [CrossRef]

energies

Article

Electrification of LPT in Algeciras Bay: A New Methodology to Assess the Consumption of an Equivalent E-Bus

Carola Leone [1], Giorgio Piazza [2], Michela Longo [1,*] and Stefano Bracco [2]

1 Department of Energy, Politecnico di Milano, 20156 Milano, Italy; carola.leone@polimi.it
2 Electrical, Electronics and Telecommunication Engineering and Naval Architecture Department, University of Genoa, 16126 Genoa, Italy; giorgio.piazza@edu.unige.it (G.P.); stefano.bracco@unige.it (S.B.)
* Correspondence: michela.longo@polimi.it

Abstract: The present paper proposes a new methodology to aid the electrification process of local public transport (LPT). In more detail, real drive cycles of traditional buses currently in use are evaluated together with other data to simulate the consumption of equivalent e-buses (electric buses) with similar characteristics. The results are then used in order to design the best charging infrastructure. The proposed methodology is applied to the case study of Algeciras Bay, where a specific line of LPT is considered. Real measurements are used as data for the simulation model, and the average consumption of an equivalent e-bus is obtained for different operating conditions. Based on these results, different sizes and locations for fast-charging infrastructure are proposed, and the size of the depot charging system is defined trying to maintain the current buses timetable. Finally, some future developments of the present work are presented by considering other bus lines that may benefit from the introduction of the defined charging systems.

Keywords: electric heavy-duty vehicles; local public transport; charging infrastructure; electric bus consumption; simulation

Citation: Leone, C.; Piazza, G.; Longo, M.; Bracco, S. Electrification of LPT in Algeciras Bay: A New Methodology to Assess the Consumption of an Equivalent E-Bus. *Energies* **2021**, *14*, 5117. https://doi.org/10.3390/en14165117

Academic Editor: Thanikanti Sudhakar Babu

Received: 16 July 2021
Accepted: 11 August 2021
Published: 19 August 2021

1. Introduction

With the Paris Agreement in 2015, the members of the UNFCCC agreed to undertake ambitious efforts to keep the global average temperature rise well below 2 °C above pre-industrial levels and to strive for an increase of less than 1.5 °C within this century [1]. In order to reach this ambitious long-term goal, nations have to take action and drastically reduce their overall carbon dioxide (CO_2) emissions. Regarding the EU's total emissions, the road transportation sector makes a share of 21% [2], whereas road buses in combination with other heavy-duty vehicles contribute to 5.6% of the total emitted CO_2 [3]. Following the conversion of conventional-powered buses to purely electric-drive ones fed by green energy would have a significant impact on the overall determined goal, which is the reason why the EU started the European Clean Bus deployment initiative in 2016 [4], which is based on three pillars. The first initiative is a public declaration from cities and regions, manufacturers, and transport organizations endorsing a common ambition to accelerate the roll out of clean buses. The second one is the creation of a deployment platform to exchange information, create coalitions, leverage potential investment action, and issue recommendations on specific policies. Finally, the third pillar is the creation of an expert group bringing together actors from the demand and the supply side. In this context, e-buses can play an important role, by completely avoiding local emissions in the city and reducing the overall emissions proportionally to the increase in renewables in each national energy mix. Nevertheless, the adoption of electric buses is currently slow, as depicted in Figure 1, where the electric bus registration in thousands of units is presented for the different regions on the left vertical axis, and the percentage deviation in respect to the previous year for the same regions is reported with straight lines on the right vertical axis.

As can be seen, China is leading the market with a registration of 78,000 new vehicles in 2020, up in respect to the previous year by 9% and a total sales share of 27%.

Europe is steadily growing at a relatively low pace, registering about 7000 e-buses in 2020 making up 4% of all new bus registration in Europe. On the other hand, considering the number of announced available models of e-buses for the years to come, it is possible to notice how in China and Europe, the numbers are constant, whereas in the USA, an increase is foreseen over the years.

The reasons behind this slow introduction of electric buses into the market are mainly due to the high upfront capital investment in comparison to traditional buses [5] and the need for proper designing and realization of charging facilities [6,7]. Moreover, the performances of an electric bus are highly influenced by driving distances, road orography, weather conditions, traffic congestions, and different passenger demands [8], thus being difficult for local public transport companies to assess whether a specific e-bus model is capable to substitute the actual traditional vehicle and if the substitution determines different timetables due to charging necessities. In the literature, the estimation of the energy consumption along a route of electric buses is usually obtained considering driving cycles, which are standardized, as reported in [9], where a tool for assessing the energy consumption of city transit electric buses is proposed. In other works, driving cycles are acquired from real operation data of electric vehicles, as reported in [10], where standard cycles and real data from Beijing electric vehicles are compared and assessed, or as in [11], where a real-world dataset from more than 16,000 electric buses in Shenzhen has been used to understand operating and charging patterns. Moreover, in [12], GPS devices are used to collect locations and times of all the bus lines in Cuenca, Ecuador in order to define which line to prioritize for the electrification according to emissions criteria. Finally, in [7], the energy consumption value to be used to size the charging infrastructure of an electric bus fleet in Finland is obtained by testing a 12 m prototype. Table 1 summarizes the review of the studies presented above. In the present work, real data are collected from traditional internal combustion engine buses that are currently in operation to derive a specific drive cycle that accurately resembles the actual operating conditions to which an equivalent e-bus would be subjected. These data are provided to a simulation tool that evaluates the consumption of the equivalent e-bus. In more detail, in Section 2, the methodology proposed to study the electrification of a line of local public transport service is presented. Then, in Section 3, the methodology is applied to a real test-case scenario constituted by the electrification of one line of the local public transport in Algeciras Bay to evaluate the overall energy consumption of an equivalent e-bus along the specific line considered. Finally, in Section 4, several considerations on the best charging system infrastructure are reported, while in Section 5, conclusions are presented.

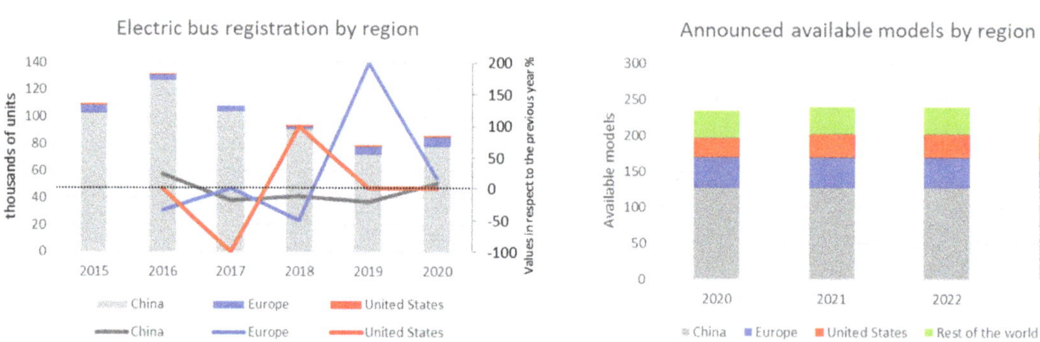

Figure 1. Electric bus registration and announced available models by region. Elaboration of data from IEA [13].

Table 1. Review of data collection methods to assess the energy consumption of e-buses.

	Collection Data Method	Case Study
[10]	Standard cycles and real operation data	Beijing, China
[11]	Real-world dataset	Shenzen, China
[12]	GPS devices	Cuenca, Ecuador
[7]	Consumption values of a test e-bus	Tampere city/Finland
This work	The data are collected with GPS devices from the real-world route driven by conventional diesel buses.	Algeciras, Spain

2. Methodology

The present work aims to propose a methodology to aid the transition towards electrification of public transport service companies. In more detail, as depicted in Figure 2, the idea is to use real data coming from actual buses currently deployed by the public transport service companies (such as the bus speed profile along a specific line, the corresponding altitude profile, and the operating conditions of the bus, i.e., the auxiliary power absorption, the number of passengers along the route, and the weather conditions) to run a Simulink simulation model developed to evaluate the consumption of an e-bus. This model simulates the corresponding consumption for an equivalent electric bus in terms of size and seat capacity subjected to the same operating conditions, speed, and altitude profile of the traditional propelled bus. The e-bus considered is characterized inside the model by considering its motor, battery, brake system, and driveline specifications. The main outcome of the model is the simulated consumption of the equivalent e-bus, which is very useful information to make several other considerations in case of the electrification of the line, such as the need for modification of the actual bus schedule, the choice of the best charging system in terms of typology (depot or opportunity charging system), geographical location, and size (capacity of charge).

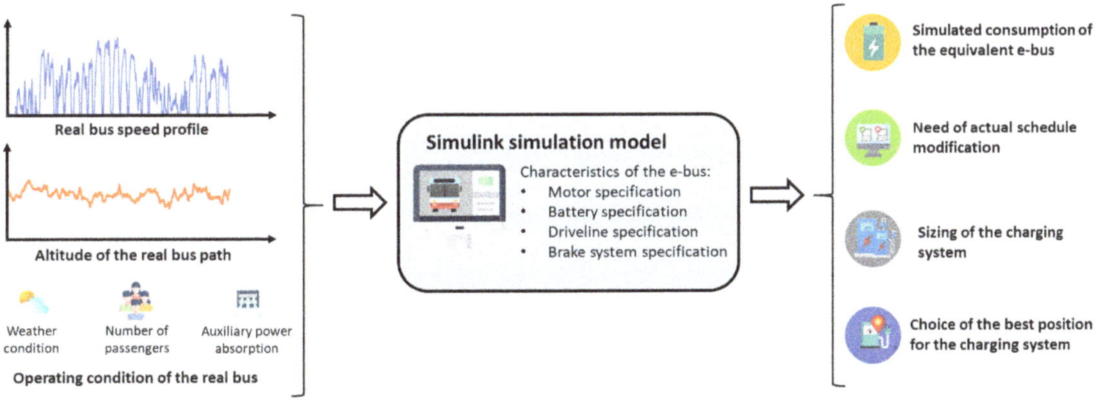

Figure 2. Graphic representation of the implemented methodology.

2.1. Simulink Simulation Model

The Simulink simulation model considered in the present work takes inspiration from [14], where an e-car simulation model is presented. More specifically, in the present work, the driving cycle that is fed to the model is not a standardized driving cycle such as the New European Drive Cycle (NEDC), but derives from real data related to a real bus during its scheduled working hours. In this way, the driving cycle takes into account not only the path conditions but also the impact generated by traffic conditions. Moreover, the Simulink model has been modified in order to be capable of accepting an altitude profile. Indeed, from the altitude variation, it is possible to calculate the slope of the road for each single time frame. This parameter is fundamental while considering the motion equation as described in the following. Furthermore, since the Simulink model in [15] was considered for an electric car, thus, most of the parameters of the model have been changed to represent a bus scenario.

In Figure 3, the graphic representation of the Simulink model used in the present work is reported. The model can be divided into six main blocks: the driver, the brake system, the electric motor, the battery, the driveline, and the motion equation block. The real bus speed profile is compared with the simulated speed, and the driver block works as a proportional integral derivative (PID) controller and adapts the acceleration and brake pedal position in order to follow the real bus speed profile. According to the brake pedal position the brake system block gives as an output the braking force and the regenerative braking command, respectively, to the driveline and motor blocks. As a function of technical parameters of the considered motor (peak torque, rated power) and of variables of the simulator (angular speed of the motor, position of the accelerator, regenerative braking command), the motor block calculates the motor torque and power, which are, respectively, fed to the driveline and the battery. The driveline block calculates the traction force acting on the vehicle as a function of the input motor torque, the torque spin loss, the friction braking force, and the ratio between the gear and wheel radius. As will be presented shortly, the motion equation block (according to the traction force and all of the opposing forces acting on the vehicle) calculates the simulated speed of the vehicle, which is given in a closed loop to the driver block. Finally, the battery block estimates the energy that is used by the bus considering the motor power and the auxiliary power that are needed. In the following, the blocks will be described in more detail. The main inputs of the model can be summarized into three categories: real bus operating data, assumed operating data, and rated data of the equivalent considered e-bus. Concerning the inputs relative to the real bus, the speed profile and the contextual altitude profile have been obtained by using the GPS of a smartphone Google Pixel 4a by means of the free-to-use application Phyphox [16]. The data have been filtered and elaborated before being fed to the Simulink simulation model. Regarding the operating conditions of the real bus, several assessments were made considering weather data for the location in terms of temperature and pressure (which are used to calculate the density of the air) and the average number of passengers along the line. The main assumed operating data are the tire pressure, the auxiliary power, and the initial state of charge of the battery. On the other hand, the main rated data of the equivalent considered e-bus are the width and height (useful to determine the frontal area of the bus), the curb weight of the vehicle, the gear ratio, the tire nominal pressure and size, the drag coefficient, the maximum power and torque of the electric motor, the maximum braking force of the braking system, and the regenerative braking threshold in terms of velocity. Furthermore, rated parameters of the battery are also taken into account such as the open-circuit voltage, the rated capacity, and the internal resistance of the battery.

2.2. Brake System Block

This block represents the braking system of the e-bus. It receives as inputs the pedal position of the brake, which is transformed into a desired brake force by scaling linearly the pedal position in respect to the maximum brake force, calculated as a function of the road adhesion coefficient. The desired brake force is then split into the regenerative brake force and the friction brake force according to a specific braking factor. Moreover, the regenerative brake force can assume non-null values only for vehicle speed above a certain threshold usually in the range 10–25 km/h [17]. Finally, the regenerative brake force is converted into a corresponding torque at the crankshaft by means of a parameter taking into account the gear ratio and the tire radius.

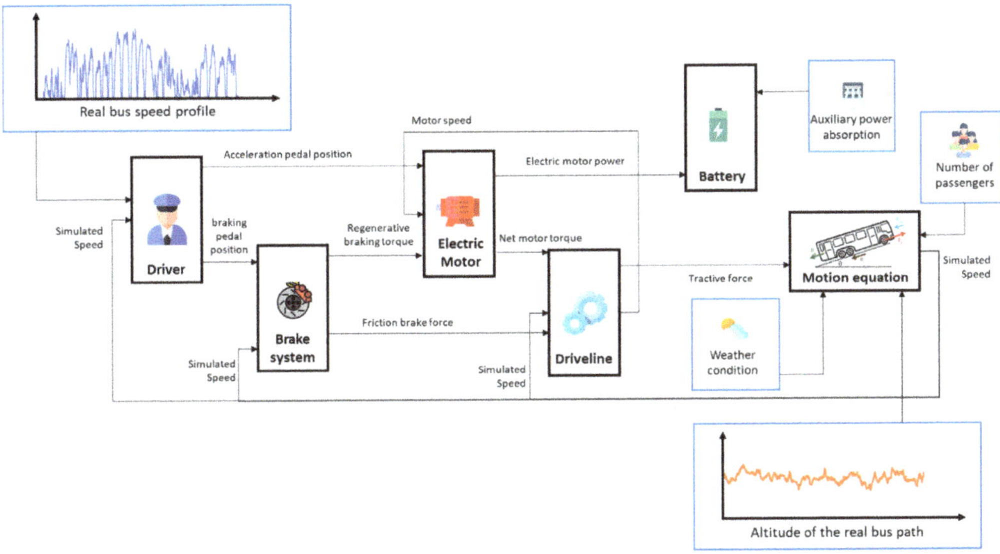

Figure 3. Simulink model information diagram.

2.3. Electric Motor Block

The electric motor block receives as input from the driver block the accelerator pedal position. This signal is converted into a corresponding positive torque by linearly scaling the pedal position in respect to the minimum in each instant between the rated torque of the electric motor and the torque coming from the ratio between the rated power and the actual angular motor speed, respectively, if it is working in the constant torque region or in the constant power region, as depicted in Figure 4.

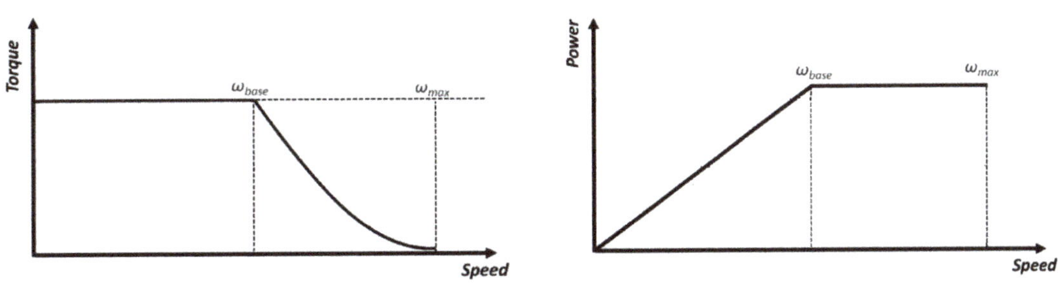

Figure 4. Typical characteristic torque/power curves of in function of angular speed [14].

Moreover, the allowable regenerative torque is defined in respect to the previously cited minimum according to a specific factor and then used as a lower bound to convert the regenerative brake torque at the crankshaft coming from the brake block into the actual regenerative torque, which can be developed. The next step is adding with opposite signs the accelerating torque and the regenerative torque in order to obtain for each time interval the net motor torque T_{net}^{mot}, which is used to calculate the mechanical output power of the motor P_{mech}^{mot} by multiplying it by the angular speed of the motor. On the other hand, the electric input power of the motor P_{el}^{mot} supplied by the battery is defined as a function of losses as reported in (1), where the losses are defined as in (2) as a function of the net torque of the motor, the angular speed of the motor ω^{mot}, and several parameters, which take into account the copper losses k_c, the iron losses k_i, the windage losses k_ω, and constant power losses c.

$$P^{mot}_{mech} = P^{mot}_{el} - P^{mot}_{losses} \tag{1}$$

$$P^{motor}_{losses} = k_c \cdot \left(T^{mot}_{net}\right)^2 + k_i \cdot \omega^{mot} + k_\omega \cdot \left(\omega^{mot}\right)^3 + c \tag{2}$$

During the regenerative braking phase, the mechanical power coming from the driveline is converted into electric power by the electric motor acting as a generator.

Finally, the electric motor block outputs are the electric motor power needed from the battery and the net motor torque supplied to the driveline.

2.4. Driveline Block

The driveline block receives as inputs the net motor torque from the electric motor, the friction brake force from the braking system, and the simulated speed of the vehicle. The simulated speed of the bus is converted into the angular speed of the crankshaft through the ratio between the tire radius and the gear ratio. The angular speed is then multiplied by the torque spin losses to obtain the driveline power losses. The torque spin losses are evaluated as a linear function of the rated torque of the motor and of the spin loss coefficient, which is assumed to be constant. The torque spin losses are also subtracted from the net motor torque coming from the electric motor block in order to determine the driveline torque output. The latter is then converted into the positive tractive force supplied by the driveline to the tires by means of the ratio between the tire radius and the gear ratio. The friction brake force coming from the brake system block is then subtracted from the positive tractive force to obtain the net tractive force acting on the vehicle, which is delivered to the motion equation block.

2.5. Motion Equation Block

The motion equation block receives as an input the net tractive force acting on the vehicle and compares it with the sum of resistances acting in the opposite direction on the bus. Indeed, in every second of the trip of the bus, its motion mainly depends on the following four forces (Figure 5): the inertial force, the aerodynamic force (R_{air}), the resistance force due to the grade (R_g), and the rolling resistance (R_r) [18].

$$R_r = K_r \cdot mgcos\theta \tag{3}$$

$$R_{air} = \frac{1}{2}\rho_{air} \cdot A \cdot C_{air} \cdot v^2 \tag{4}$$

$$R_g = m \cdot g \cdot sin\theta \tag{5}$$

where K_r is a rolling coefficient function of the pressure of the tires and of the simulated speed of the vehicle v. The rolling resistance is given by the product of the rolling resistance coefficient times the normal component of the weight force, the force perpendicular to the road on which the vehicle's wheels are rolling; therefore, if the vehicle is running on an inclined plane, the force vertical component is defined by the product of the vehicle's mass m, the acceleration of gravity g times the cosine of the road slope θ. Concerning the aerodynamic force, it is a function of the density of the air ρ_{air} (which is calculated as a function of the ambient conditions), of the frontal area of the bus A, of the drag coefficient C_{air}, and of the simulated speed of the vehicle. The grade resistance is a function of the nominal mass of the vehicle plus the mass of all the onboard passengers m and of the slope of the road profile θ, which is derived from the variation of the altitude profile along the line.

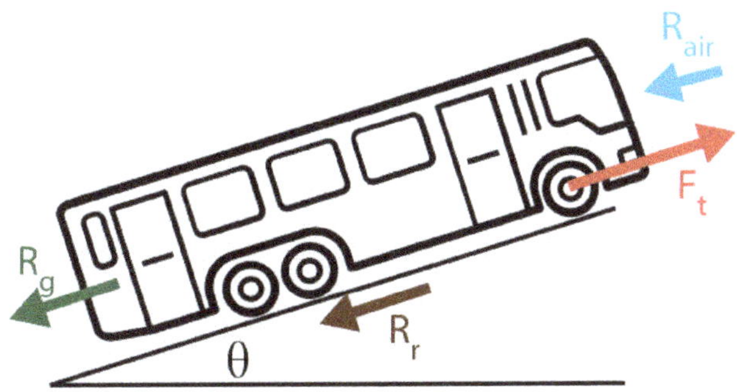

Figure 5. Forces acting on a moving bus.

The inertial force related to all the rotating components inside the vehicle is instead taken into account by considering the concept of equivalent mass expressed in (6). To obtain the equivalent mass m_e, the static mass m (which is a function of the vehicle and the number of passengers onboard) is increased by a factor β, which varies from vehicle to vehicle.

$$m_e = m(1 + \beta) \quad (6)$$

As already mentioned, to accelerate the vehicle, the net tractive force acting on the vehicle F_t must exceed the sum of all the resistances, as expressed by the motion equation in (7).

$$F_t - \sum_{i=1}^{3} R_i = m_e \cdot a \quad (7)$$

The net tractive force F_t can assume positive values, hence the vehicle is said to be in the powering mode, negative values, which means the vehicle is braking (braking mode), and finally, a nil value of the F_t causing the vehicle's natural deceleration (coasting mode). The motion of the vehicle is an alternation of these three phases. Exploiting (7), it is possible to calculate the acceleration of the vehicle and, consequently, its simulated speed. The latter is then supplied back to the driver, brake system, and driveline block in a closed loop.

2.6. Battery Block

The battery block receives as inputs the electric motor power needed from the electric motor and the power absorbed by the auxiliary systems. The two are summed to evaluate the total power of discharge of the battery. For the battery, the Thevenin equivalent circuit shown in Figure 6 is considered. Therefore, the battery is represented by an ideal voltage source defined as the open-circuit voltage V_{oc}^{bat} in series with the resistance R_{int}^{bat}, which represents the internal resistance of the battery. According to the Thevenin equivalent circuit, assuming constant no-load voltage, battery losses will vary with the current depending on the total power consumed by the electric motor. Therefore, battery losses can be expected to be lower if the total power consumed by the engine is reduced.

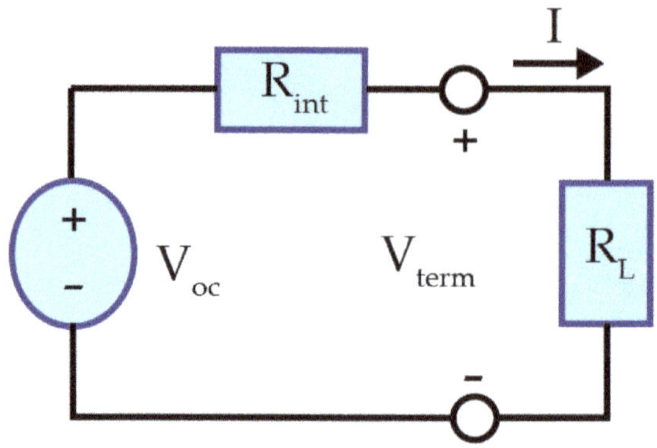

Figure 6. Battery's Thevenin equivalent circuit.

According to (8), the real discharge power of the battery is equal to the ideal discharge power net of the losses linked to the internal resistance of the battery.

$$P_{real}^{dch,bat} = P_{ideal}^{dch,bat} - P_{losses}^{dch,bat} = V_{oc}^{bat} I^{dch,bat} - R_{int}^{dch,bat} I^{dch,bat2} \quad (8)$$

from which it is possible to retrieve the discharge current (9).

$$I^{dch,bat} = \frac{V_{oc}^{bat} - \sqrt{\left(V_{oc}^{bat}\right)^2 - 4R_{int}^{bat} P_{real}^{dch,bat}}}{2R_{int}^{bat}} \quad (9)$$

Knowing the discharge current against time, it is possible to evaluate the state of charge (SOC) of the battery (10) as a function of the rated energy content of the battery E^{bat} measured in kWh. The SOC provides the current battery status. It is expressed in percentage values, where 0% indicates that the battery is completely discharged and 100% indicates a full charge.

$$SOC^{bat} = \frac{-\int_0^t V_{oc}^{bat} I^{dch,bat} dt}{E^{bat}} \cdot 100 \quad (10)$$

At this point, the energy average consumption of the e-bus $\mathbb{C}$ measured in kWh/km can be calculated according to (11) as a function of the initial amount of energy inside the battery E_{in}^{bat} and the overall traveled distance in km d.

$$\mathbb{C} = \frac{E_{in}^{bat} - \frac{SOC^{bat}}{100} \cdot E^{bat}}{d} \quad (11)$$

As a consequence, the consumption of the e-bus can be simulated along its route, giving the possibility to draw several significant considerations on the more proper charging system infrastructure for each considered case study, as will be shown in more detail in Sections 3 and 4.

3. Case Study Scenario

The methodology described in Section 2 is applied to an existing bus line that connects the city of Algeciras to La Línea de la Concepción, both located in the Autonomous Province of Andalucía, in the south of Spain. The bus and coach services in Algeciras and, in general, in Andalucía are operated by different private bus companies. Most of them operate from the main bus terminals in the towns. Some, however, do use their bus station. The main bus terminals from which the considered bus line starts and ends its route are the so-called San Bernardo Station in Algeciras and La Linea bus station in La Línea de la Concepción, depicted in Figure 7a,b, respectively.

(a)

(b)

Figure 7. Bus stations (**a**) San Bernardo and (**b**) La Linea.

3.1. Bus Data

Currently, the bus employed to cover this route is the 12 m diesel bus Volvo B12B, shown in Figure 8, which is a rear-engine coach and intercity bus chassis with 52 seats built by Volvo for the European market starting from 2001.

Figure 8. Current bus employed for the analyzed line.

The characteristics of the corresponding electric bus model chosen as a substitute are listed in Table 2 and its layout is depicted in Figure 9. The chosen electric model is a 15 m long bus, which is completely electric and equipped with onboard lithium-ion battery packs; it is certified both for class I vehicles (as a city bus) and for class II vehicles (as a bus for inter-city transport as in this case). The bus is equipped with six battery packs each one of about 78 kWh for a total energy capacity of about 470 kWh. In particular, four out of six packs are installed in the rear of the bus, the remaining two are mounted on the roof [19]. The nominal maximum power the asynchronous motor can provide is 300 kW, while the supply battery voltage is 400 V.

Table 2. Electric bus technical details.

Parameter	Value
Length [m]	14.89
Width [m]	2.550
Height [m]	3.465
Mass [kg]	19 000
P_{aux} [kW]	[3.5/16/27.5] *
Tire pressure [bar]	8
Seats	55
Payload mass [kg]	[700/1540/2800] *
Inertial factor β	0.15
Equivalent mass [kg]	15 250
Max. motor power [kW]	300
Efficiency	0.85
% Regenerative braking	67
Battery energy capacity [kWh]	6 × 78

* in accordance with scenarios described in the text.

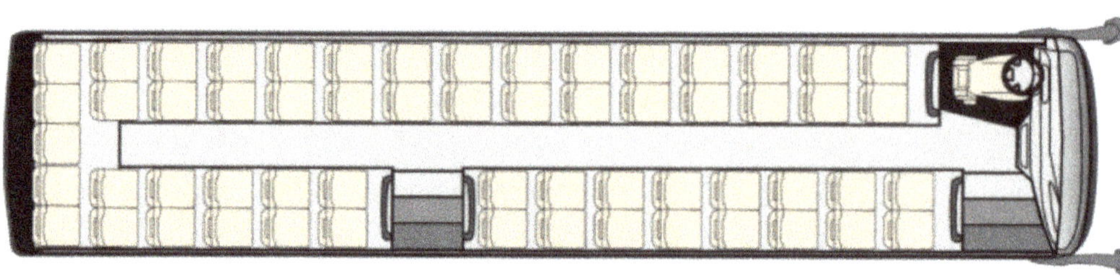

Figure 9. Considered bus layout.

The traction battery must also provide the energy to supply the auxiliary services (i.e., rear and front lights, HVAC, doors automatic system, etc.). In electric buses, and electric vehicles in general, the most energy-consuming auxiliary service is the heating, ventilating, and air conditioning (HVAC) unit, and hence, particular attention must be given to this system in the energy consumption estimation. The air conditioning system of conventional buses mainly consists of a rooftop mounted evaporator and condenser and of a compressor assembly mounted on the side of the engine. The air conditioning compressor in diesel buses is directly driven by the internal combustion engine (ICE) instead, but in electric buses, it is powered by a dedicated electric motor always supplied by the battery [20]. For the heating system, the question is a little more complicated. In ICE buses, the heating system employs the heat coming from the engine's coolant. About 30% of the heat generated during combustion is transferred to the coolant, giving an easy and fast source of heat. The incoming air is, hence, warmed as it passes through a radiator that holds the hot engine coolant. In an electric bus, none of these components exist. For electric buses, three main heating solutions can be used: electric resistance heating, electric heat pump, and fuel heating.

In [21], the authors carried out a cost analysis for the different HVAC systems that can be installed on a city electric bus, and their results reveal that the heat pump system helps to achieve great energy savings compared to the heating resistance system. Nevertheless, pondering a lifetime of 12 years, the current high initial capital cost of a heat pump system is not compensated when only considering direct costs.

The thermal load required by the bus to achieve internal thermal comfort for passengers varies at different outdoor conditions. According to [22], with a resistance heating system, a constant power of 24 kW is necessary for warming the cabin of a 12 m electric bus with a difference from outside ambient temperate of 27 °C, leading in this way to an increase in the vehicle consumption between 1.3–2 kWh/km depending on the average forward speed. Instead, running the compressor to cool the cabin on a hot summer day with 35 °C outside requires 12.5 kW, which results in a consumption increase of about 0.7–1 kWh/km. Therefore, the heating system represents the worst operating consumption case scenario, and in the case of extreme outdoor conditions, it can reduce the overall driving range by 50% [23,24].

Other auxiliary services presented in an electric bus, as previously mentioned, are the illumination system, doors automatic system, battery cooling, and pump steering. Again, according to the results in [22], the overall power demand of these auxiliaries in a 12 m electric bus is lower compared to the HVAC system, and it is about $3 \div 5$ kW.

To analyze the impact of the different auxiliary services load on the bus energy consumption, in this paper, three different scenarios are considered. In the first case, we assume that the HVAC system is turned off, since the internal temperature is already in the comfort range of 19 °C–23 °C. Therefore, the overall power absorbed by the auxiliaries is set at 3.5 kW. The second scenario instead foresees the use of the air conditioning system in order to cool the cabin up to a temperature in the comfort range with a temperature greater than 30 °C on the outside; therefore, a value equal to 16 kW has been set as the power of the auxiliaries. Lastly, the third case represents the worst-case scenario with the heating system working at the maximum power of 24 kW and, hence, an overall auxiliary power of about 27.5 kW.

With all the six battery modules, the electric bus weighs around 20 t, which means it is 14% heavier than the conventional gas model; this increment is mainly due to the presence of the lithium-ion battery and its lower energy density (only 0.10–0.27 kWh/kg or 0.25–0.70 kWh/L) with respect to diesel fuel (11.6 kWh/kg or 9.7 kWh/L). Choosing an inertial factor β of 0.15 [25], the value for the equivalent mass results 15.25 t. To this value, the mass of the passenger must be added, which is computed by multiplying the number of considered passengers times the average weight of European people of 70 kg.

One of the most significant advantages of electric vehicles is the possibility to harvest energy during the braking phase, employing the so-called regenerative braking. This feature is particularly important in electric buses, since they have heavy mass, fixed routes, and many stop-and-go events. However, assessing the amount of regenerative braking in an electric vehicle is not a trivial task, since it depends on many factors such as the initial and final braking speed, the mass of the vehicle, the braking rate, the vehicle structure, the power-train layout, electric motor, and battery characteristics [26]. Based on papers that analyze electric bus regenerative braking, the amount of the recovered energy thanks to the regenerative braking in this study has been set equal to 67% [27,28]. In the model, regenerative braking is not considered when the bus is driving lower than 10 km/h.

Another advantage of an electric bus over a conventional one is its higher efficiency both in the well-to-tank and tank-to-wheels analysis [29]. In this study, the charging system efficiency has been set equal to 85% [30].

One more benefit of choosing an electric bus fleet instead of a gasoline one is the reduction in pollution and fuel costs. Furthermore, e-buses are less loud, and since the electric motors produce far fewer vibrations, e-buses are more comfortable for onboard passengers and need less maintenance. Finally, if well displayed, the recharging process for an electric bus is more efficient and safer than the gasoline one in terms of reliability.

The feature of the regenerative braking system must not be forgotten, as it represents a great advantage compared to the conventional buses, which are not able to recover any energy in any driving phase. All these combined advantages confirm that if a modernization of the bus lines is needed, the better option is to replace the gasoline buses with electric ones. Although the initial introduction of an electric transport system and fleet can be costly, as a long-term mode of public transport they are surprisingly cost effective [31].

3.2. Route Characteristics

As depicted in Figure 10, the considered M-120 line starts its route in San Bernardo station in Algeciras and ends in the bus station in La Linea de Concepcion, for an overall trip about 22 km long. While the first and the end parts of the line pass through the urban context, the central and longer part of the line develops in highway A-7. Twice a day, at 7.00 and 15.00, direct service is performed by the line M-120D (roundtrip). The same path of line M-120 is followed but many fewer stops are made. This feature will allow us to estimate the difference between electric bus energy consumption in urban and suburban contexts. The path elevation profile found with Google Earth is reported in Figure 11.

(**a**) (**b**)

Figure 10. Analyzed routes (**a**) complete line M-120 (**b**) direct line M-120D.

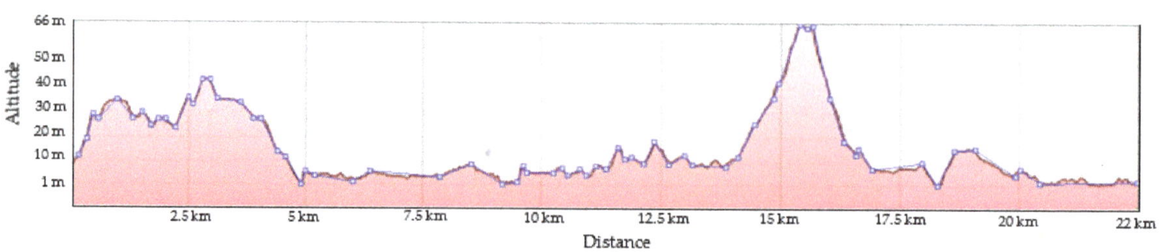

Figure 11. Elevation profile.

3.3. Results and Discussion

3.3.1. Line M-120D

In Figure 12, the driving cycles measured for the line M-120D are reported. The bus leaves on time at 3 p.m. from Algeciras bus station with approximately 25 people onboard, and it arrives at La Linea station at 3.30 p.m. to leave for the return trip at 3.45 p.m. As mentioned before, since the central part of the route is performed on the highway, the bus reaches a higher speed touching the maximum one of 100 km/h. The stopping time at the bus stops varies in the range from 20 s up to 80 s; it is strongly influenced by the number of onboarding passengers. As a matter of fact, passengers are allowed to enter the bus only from the front door, and at that moment, they can buy the ticket or validate their subscription/ticket. This policy on one side greatly decreases fair evasion; however, on the other side, it slows down the boarding procedure.

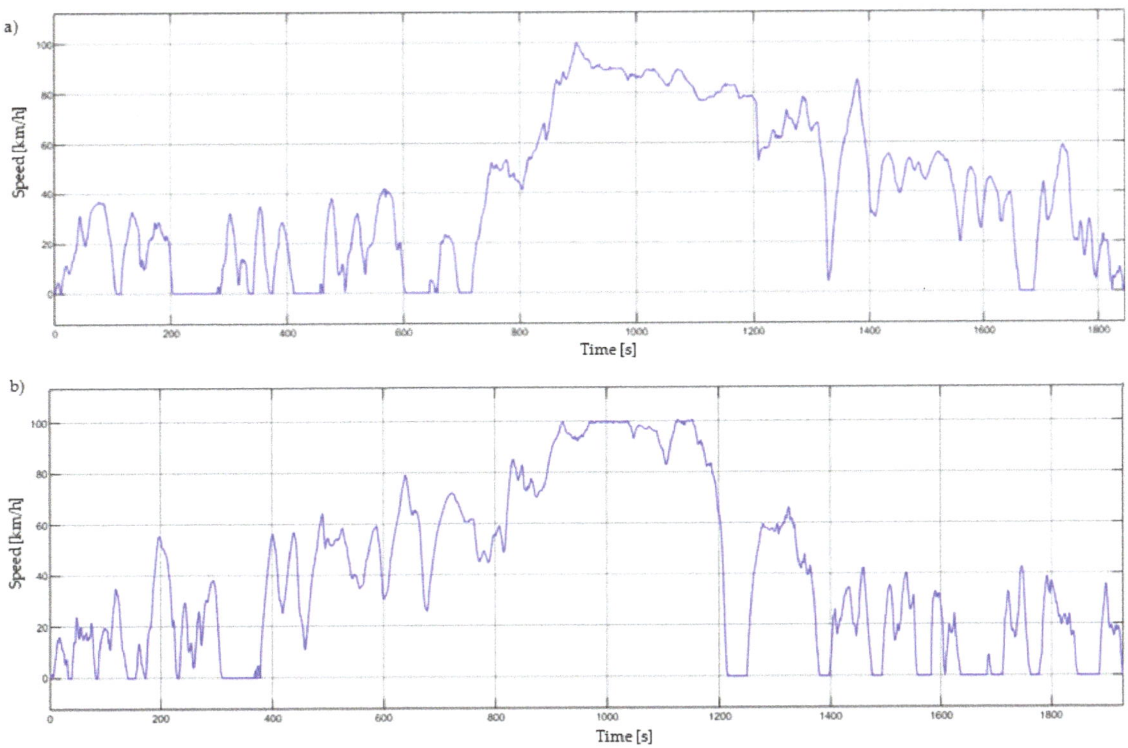

Figure 12. Measured driving cycles: (**a**) Algeciras-La Linea (**b**) La Linea-Algeciras.

In practical measurements, each trip really lasts about 30 min, as expressed in the timetable, and therefore, the declared service is guaranteed.

Introducing as input for the simulator the measured driving cycles, the average energy consumption per kilometer, also considering the charging efficiency, for a single trip is simulated in different operation conditions. The results for the line M-120D are reported in Table 3, and we can conclude that the power absorbed by the auxiliaries and, hence, the outside temperature has a major effect on the energy consumption than the number of onboard passengers. Indeed, passing from 10 up to 40 passengers causes an increase in the energy consumption between 3.5 ÷ 5%. Instead passing from an auxiliary power of 3.5 kW up to 27.5 kW will cause an increase in the energy consumed from 40% up to 50%.

Table 3. Line M-120D simulation results for electric bus.

# Passengers	Auxiliary Power [kW]	Energy Consumption Algeciras—La Linea [kWh/km]	Energy Consumption La Linea—Algeciras [kWh/km]	Energy Consumed Round Trip [kWh]
40	3.5	1.492	1.693	68.24
22		1.456	1.640	66.33
10		1.427	1.604	64.94
40	16	1.866	2.06	84.11
22		1.823	2.007	82.06
10		1.794	1.971	80.66
40	27.5	2.204	2.399	98.61
22		2.161	2.345	96.53
10		2.133	2.31	95.18

3.3.2. Line M-120

The same procedure is followed for the line M-120, which presents 19 stops for the outward trip (12 more with respect to the direct line) and 17 for the return one (10 more than M-120D). It is worth mentioning that the bus does not stop at all the stations because rarely all of them are called; however, it must pass in correspondence with all of them and decelerate so that the driver can check if any passengers are waiting at the stop. In Figure 13, the driving cycles measured for this line are depicted. As can be seen, the trends are characterized by many more stops and more speed variations with respect to those presented in Figure 12. Not all the halts correspond to bus stops; between 300 s up to 600 s, the bus runs on an always congested road segment due to the presence of an important traffic light.

In this case, as well, the maximum speed reached by bus both in the outward and return journeys is 100 km/h; however, this speed is kept just for a few seconds.

The results found for the line M-120 are shown in Table 4. Comparing these results with those of the direct line, it can be seen that the energy consumption per kilometer, in this case, is about 5–10% higher. This result could seem in contrast with the spread knowledge that EVs consume less in an urban context with frequent stop-and-go. However, by analyzing the driving cycles, it can be noticed that even if in line M-120 more stops are performed, since the deceleration rates are very high, the bus cannot fully exploit the regenerative braking [32], resulting, hence, in higher consumptions. Nevertheless, in the case of electric buses used in LPT services, since the stops are almost all planned, the driver could be easily taught to adopt an optimized driving behavior to maximize the regenerative energy.

Table 4. Line M-120 simulation results for electric bus.

# Passengers	Auxiliary Power [kW]	Energy Consumption Algeciras—La Linea [kWh/km]	Energy Consumption La Linea—Algeciras [kWh/km]	Energy Consumed Round Trip [kWh]
40	3.5	1.626	1.796	73.56
22		1.558	1.731	73.01
10		1.518	1.686	73.54
40	16	2.067	2.21	94.97
22		2.001	2.142	91.97
10		1.966	2.098	90.24
40	27.5	2.473	2.587	112.37
22		2.405	2.522	109.42
10		2.373	2.477	107.71

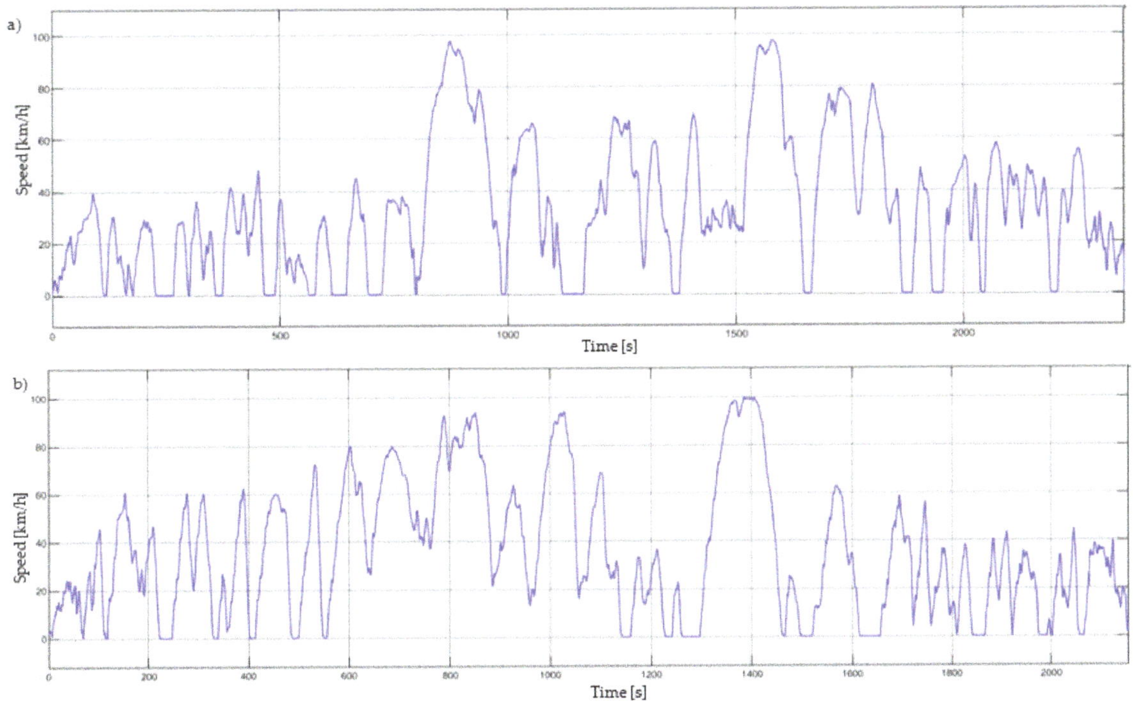

Figure 13. Measured driving cycles: (**a**) Algeciras-La Linea (**b**) La Linea-Algeciras.

4. Charging System

An electric bus fleet can rely on two main charging systems: overnight and opportunity charging. Their representation is sketched in Figure 14. Overnight charging systems, as the name suggests, aim to charge the fleets of buses during the night at the depot with plug-in connectors and charging power for each connector from 40 up to 150 kW. The output power mainly depends on whether an AC or DC charging system is installed. In the case of an AC charging system, the onboard charger of the bus is employed to perform the AC-DC conversion, and the offboard structure only includes power and communication cables, metering, and protection devices. The AC charging system (mode 3) allows maximum power up to about 43 kW (86 kW if two Type2 plugs are used). If, instead, higher charging powers are required or the electric buses do not dispose of the onboard charger, a DC charging system is necessary (mode 4), which means that the conversion stage is now performed offboard the vehicle inside the charging system, and therefore, the infrastructure capital costs are higher. In this case, the power limit is imposed by the cable and the connector; as a matter of fact, the Combo 2 (CCS2) connector allows us to reach 200 A without the need for special liquid cooling systems. Normally, in order not to increase the power required too much from the public distribution network, the output current is limited to 150 A, which corresponds to a nominal power of about 100 kW considering the typical voltage values of the batteries of the actual electric buses on the market. It may happen that, to complete the service without weighing down the vehicle too much with larger batteries, electric buses require additional daytime recharges or occasional recharges that take advantage of the halt times at the terminus and/or at the stops; this bus charging system takes the name of opportunity charging. The opportunity charging system employs overhead pantographs, which can support charging powers up to 600 ÷ 750 kW. Given the high power required, the connection to the electricity grid cannot take place directly from the low voltage distribution but must take place from the medium voltage distribution.

(**a**) (**b**)

Figure 14. Representation of (**a**) depot and (**b**) opportunity charging systems.

In this paragraph, the aim is to find out the location and the size of the two above-mentioned charging systems, which assure a correct operation of the bus line without degrading the service.

The most important data necessary in addition to the vehicle characteristics, to display the analysis, are the total energy consumption on the entire line, the number of round trips each bus of the line performs, and the stopping time at each terminal station.

Battery performances, in particular rated power and capacity, degrade over time due to multiple aging mechanisms. Battery aging can be divided into calendar and cycling aging. Calendar aging includes all the processes that lead to a degradation of the battery cell occurring while the battery is at rest. In recent lithium-ion batteries, the main aging mechanism impacting on calendar aging is the solid electrolyte interphase (SEI) formation on the negative electrode [33].

Cycling aging, instead, is related to the use of the battery and its severity mainly depends on charging/discharging current rates, cycle depths, and battery temperature. Therefore, in this field, the persistent use of high charging powers related to the opportunity charging system could significantly increase the impact of this side effect. However, given the high battery capacity of the electric buses considered, the C-rate does not exceed the common and suggested value of 1C. Therefore, the aging effect of the charging systems considered in this study is neglected.

The negative effects of the battery aging on the energy capacity have been taken into account by selecting a higher minimum SOC with respect to the one suggested by the manufacturer.

Following the indications of Line M-120 timetable [34], it can be deduced that in order to cover all the scheduled trips, three buses are necessary for the peak time slot in the weekdays (four if we also consider the bus necessary to cover the direct service of line M-120D); this number decreases to two on Saturday and to one on Sunday. At the weekend, the direct service is not provided.

4.1. Opportunity Charging

Given the fast pace of the timetable and the length of the considered line, the installation of opportunity charging systems becomes necessary.

The overall timetable of line M-120 is divided into the planning timetables of the three necessary buses as shown in Figure 15. As can be seen, the scheduled duration of the trip is 45 min, and according to the reported timetable, the bus immediately starts another trip as it arrives at one of the terminals. However, the real case is that the trip between Algeciras and La Linea always lasts less than 45 min; as a matter of fact, in all the measured runs (three round trips and, hence, six single runs) the running time was between 36 ÷ 40 min. This implies a stopping time at the two terminals, which range between 5 up to 10 min.

η^{ch} is the efficiency if the charging system. They result in 7 h, 2.5 h, and 4.6 h for Buses 1, 2, and 3, respectively.

$$t^{ch} = \frac{E^{bat} - E^{bat}.\frac{SOC^{bat}}{100}}{P^{ch}.\eta^{ch}} \tag{12}$$

Therefore, by overlapping recharging time with the stopping time of the three buses during the night, it can be concluded that the two AC charging poles of 43 kW are enough to recharge all the buses before their first run of the morning. In particular, one charging station is used by the second bus from 20 h up to about 22.30 h, then, from midnight, it will be dedicated to Bus 1. Finally, the second charging pole during the night will recharge Bus 3, and hence, all the vehicles will be able to start the operation with a full battery.

4.3. Future Expansion

The installation of the charging infrastructure necessary for the electrification of line M-120 could represent a good starting point for the electrification of other lines, which start or end the operation in the analyzed terminals. For instance, line M-121, which connects La Linea to Los Barrios could take advantage of the opportunity charger installed in La Linea bus station. This line, whose route is 23 km long and is represented in Figure 19a), is operating from Monday to Friday with the schedule reported in Figure 19b).

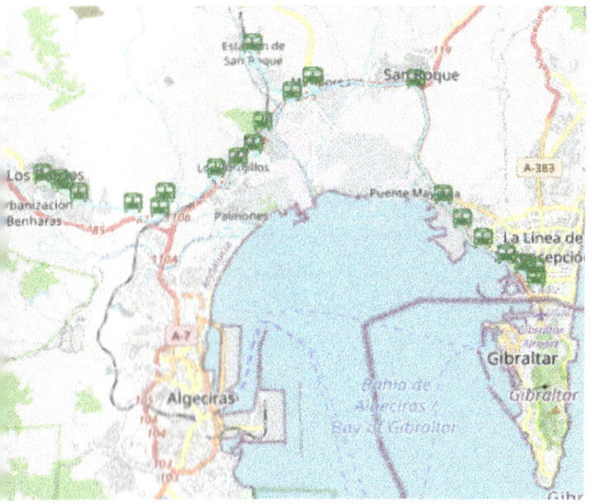

(a)

Line M-121	
La Linea-Los Barrios	La Linea-Los Barrios
6:30-7:30	7:30-8:28
8:30-9:30	9:30-10:28
15:15-16:15	16:15-17:13
19:30-20:30	20:30-21:28

(b)

Figure 19. Line M-121: (**a**) route and (**b**) timetable.

Since the covered path overlaps with that of Line M-120 for most of the travel time and the bus model currently employed is the same, the energy consumption calculated for line M-120 of 2.522 kWh/km is used also for this line. Therefore, it results that to compute a round trip, the bus will consume about 116 kWh. Given the low number of runs the bus of this line must cover, its electrification would be possible even by using only the Opp Charger installed in La Linea, just ensuring that its charging time does not overlap those of line M-120.

Another line that could effectively benefit from the charging infrastructure installed in the bus station of Algeciras is line M-150, which connects Algeciras to Tarifa for an overall length of 21 km. The covered path and the timetable of the two buses necessary to provide the service are reported in Figure 20a,b, respectively. As for Line M-120, the official duration of the trip is set at 45 min; however, the real run lasts less as measured during the real tests. In particular, it has never exceeded 40 min.

This period is indeed necessary also for the offboarding and onboarding process of the passengers and to apply the opportunity charging.

BUS 1	
Algeciras-La Linea	La Linea-Algeciras
07:00-07:45	07:45-08:30
08:30-09:15	09:15-10:00
10:00-10:45	10:45-11:30
12:00-12:45	12:45-13:30
13:30-14:15	14:15-15:00
15:00-15:45	15:45-16:30
16:30-17:15	17:15-18:00
18:00-18:45	-
19:30-20:15	20:15-21:00
21:00-21:45	21:45-22:30
22:30-23:15	23:15-00:00

BUS 2	
Algeciras-La Linea	La Linea-Algeciras
07:30-08:15	08:15-09:00
09:00-09:45	09:45-10:30
10:30-11:15	11:15-12:00
12:30-13:15	13:15-14:00
14:00-14:45	14:45-15:30
18:30-19:15	-

BUS 3	
Algeciras-La Linea	La Linea-Algeciras
08:00-08:45	08:45-09:30
09:30-10:15	10:15-11:00
11:00-11:45	11:45-12:30
14:30-15:15	15:15-16:00
16:00-16:45	16:45-17:30
17:30-18:15	18:15-19:00
19:00-19:45	19:45-20:30
20:30-21:15	21:15-22:00

Figure 15. Timetables of the three buses.

To size the charging system necessary to electrify a bus line, the worst-case scenario must be considered so that the service is correctly provided also in the most disadvantageous case. In our study, this scenario is verified when the auxiliaries absorb the maximum power of 27.5 kW, leading in this way to a consumption (considering on average 22 passengers on the bus) of 2.405 kWh per km for the trip from Algeciras to La Linea and of 2.522 for the return one (for the normal service and not the direct one).

Different solutions for the opportunity charging systems have been simulated and the obtained results are described in the following. However, in the proposals for the opportunity charging system, the technical limits mentioned by the manufacturer must be considered, and according to the datasheet of the chosen electric bus, a maximum charging power of 450kW is allowed. Moreover, a recommended maximum depth of discharge for a lithium-ion battery is usually about 80% and an additional safety margin of 10% must be considered in order to allow the bus to return to the depot at any moment in case of an emergency and to deal with the decrease in energy capacity over time due to battery aging. Therefore, a minimum SOC of 30% must be always guaranteed.

Finally, if the electric bus stops at the Algeciras terminal for a time interval long enough to be recharged with the overnight charging system, then this charging solution will be preferred. This situation occurs for both Buses 2 and 3 in the intervals 15:30–18:30 h and 12:30–14:30, respectively. Therefore, in these time intervals, the electric buses are connected to the charging system installed for the overnight charging, which is usually rated at 43 kW AC, instead of to the opportunity charger.

4.1.1. One Opportunity Charger of 450 kW

The first solution reckons on the installation of just one opportunity charger (Opp Charger) with a rated power of 450 kW installed in the San Bernardo bus station. In Figure 16, the SOC of the three buses with this charging solution is displayed. As can be noticed, even if this is the best solution in terms of cost, it does not allow for the correct operation of the service, since both the SOC of Bus 1 and Bus 3 fall below the safe limit margin of 30%.

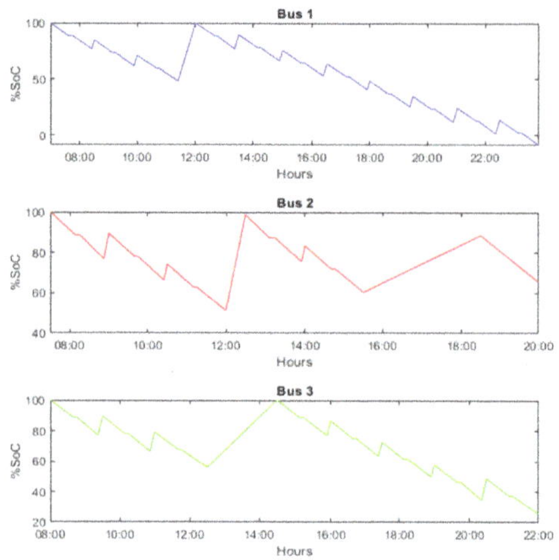

Figure 16. SOC of the three buses with one Opp Charger of 450 kW.

4.1.2. Two Opportunity Chargers of 300 kW

The second solution foresees the installation of two opportunity chargers of 300 kW, one in each of the two terminals. In this case, as shown in Figure 17, the only critical situation is that of Bus 1 whose SOC, about 21:30–22 h, decreases under the limit of 30%. However, this aspect could be solved by performing a switch of buses between Buses 1 and 2 to perform the last two round trips at 21 h and 22:30 h. Indeed, Bus 2 ends the operation at 20 h with 70% of SOC. It could be recharged in the time interval 20–21 h using the depot charging system and, then, picked back to perform the last two round trips instead of Bus 1, which, hence, will return to the depot at about 09:00 PM with a SOC of 34%.

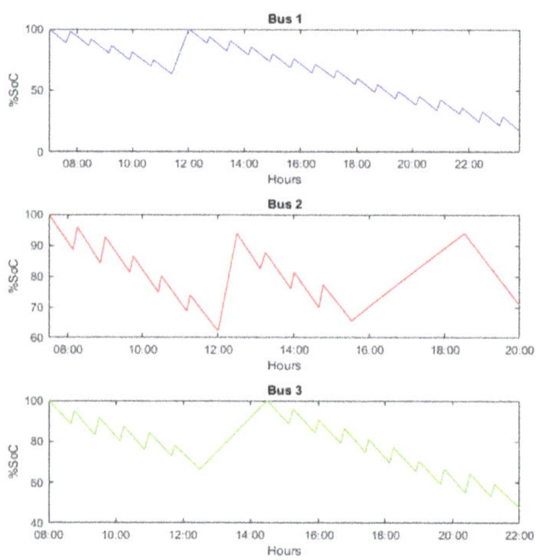

Figure 17. SOC of the three buses with two Opp Chargers of 300 kW.

With this solution, the maximum DOD is reached by Bus 1 at the end of its daily operation. Nevertheless, since the bus recharged approximately at the end of each trip, along the day, the SOC variations have small entity. This fact helps to prolong the battery life; as a matter of fact, the smaller the discharge (low DoD), the longer the battery will last. The situation is even better for Buses 2 and 3; their SOC varies throughout the day in the range 65–95% and 50–100%, respectively.

4.1.3. Two Opportunity Chargers of 350 kW

In this last case, the rating power of the two Opp Chargers is inc
The SOC of the three buses along the day, recharged according to this sol
in Figure 18. As can be seen, in all three cases, the SOC always remain
limit, and no buses switch is necessary.

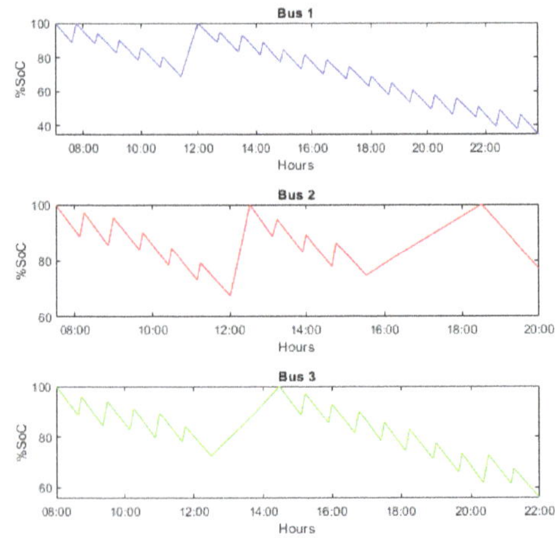

Figure 18. SOC of the three buses with two Opp Chargers of 350 kW.

The maximum DOD in this case is always achieved by Bus 1 at t
operation, and it does not differ so much from the previous case. Instead
buses, the situation is slightly meliorate with respect to the previous cas
DOD is 30% and 45%, respectively.

For the service on Sunday, currently, only one bus is employed co
the day. However, in the case of an electric bus fleet, the best solution w
of the bus with another one of the three every time its battery is almost
value. Therefore, in this way, all three (four considering that of the direct
will be used along the day, and they will not rely on the opportunity ch

4.2. Depot Charging

The buses of Line M-120 start and finish their operation at the dep
Algeciras. Therefore, the overnight charging infrastructure must be instal

To size the depot charging system, the most critical case is the
weekdays, since in this period, three buses are employed (the fourth b
runs only twice). In Table 5, the arrival times and SOC of the buses are re
the departing times on the following morning.

Table 5. Buses stopping time at night.

#Bus	End Operation	Start Operation	SOC	Nigl
Bus 1	00:00	07:00	35%	
Bus 2	20:00	07:30	77%	
Bus 3	22:00	08:00	58%	

If we assume to employ an AC 43 kW charging system, the necessar
each of the three buses are computed through (12), where E^{bat} is the ene
battery, SOC^{bat} is the final SOC with which the bus ends the daily ope
at the bus depot (this value is taken from Table 5), P^{ch} is the charging

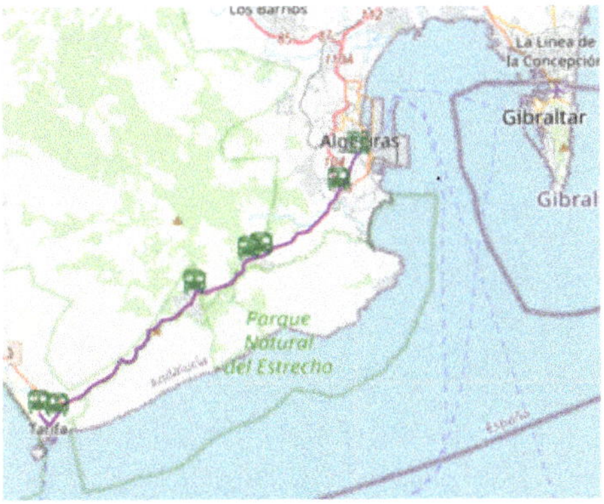

BUS 1	
Tarifa-Algeciras	Algeciras-Tarifa
6:30-7:15	8:45-9:30
9:30-10:15	10:15-11:00
11:00-11:45	11:45-12:30
12:30-13:15	13:15-14:00
14:00-14:45	14:45-15:30
15:30-16:15	18:00-18:45
18:45-19:30	19:30-20:15
20:15-21:00	21:00-21:45

BUS 2	
Tarifa-Algeciras	Algeciras-Tarifa
7:05-7:50	8:15-9:00
-	21:45-22:30
22:30-23:15	-

a) b)

Figure 20. Line M-150: (**a**) route and (**b**) timetable of the employed buses.

The energy consumption in the worst-case scenario for a roundtrip in this case is about 106 kWh. The simulated SOC profile of Bus 1 is reported in Figure 21. Instead, the SOC of Bus 2 is not shown, since for this bus, the charging time does not represent a problem given the very long stopping times. The SOC of Bus 1 after 15:30 h goes under the limit value of 30%; however, this critical situation could be avoided by using Bus 2 for the runs, which are performed in the time interval 14:45–16:15 h. Indeed, it is stopped at Algeciras depot from about 9:30 h (it returns to Algeciras at 9:00 h after the last morning run); therefore, it would have all the necessary time to be recharged up to 100% and, hence, substitute Bus 1 for the critical runs.

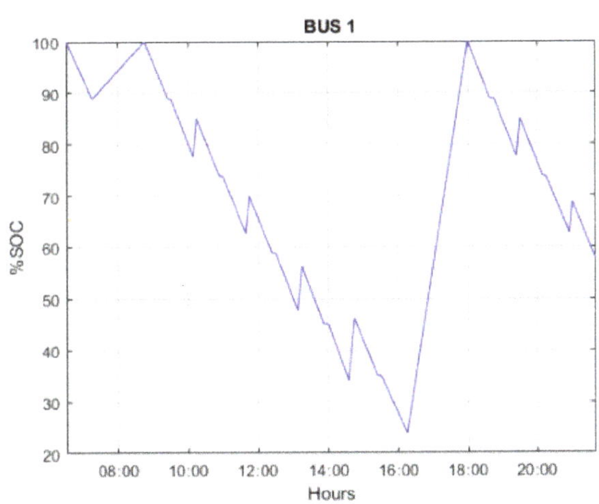

Figure 21. SOC of Bus 1, which runs Line M-150.

Nevertheless, to correctly electrify line M-150, an overnight charging system composed of one AC 43-kW charging pole is needed to be installed in Tarifa bus station.

Lastly, in Figure 22, the occupation profile of the two opportunity chargers is reported. Values of the Y axes equal to 1 imply that the Opp Charger is occupied and, hence, in operation. On the contrary, Y values equal to 0 denote that the Opp Charger is free and, hence, turned off.

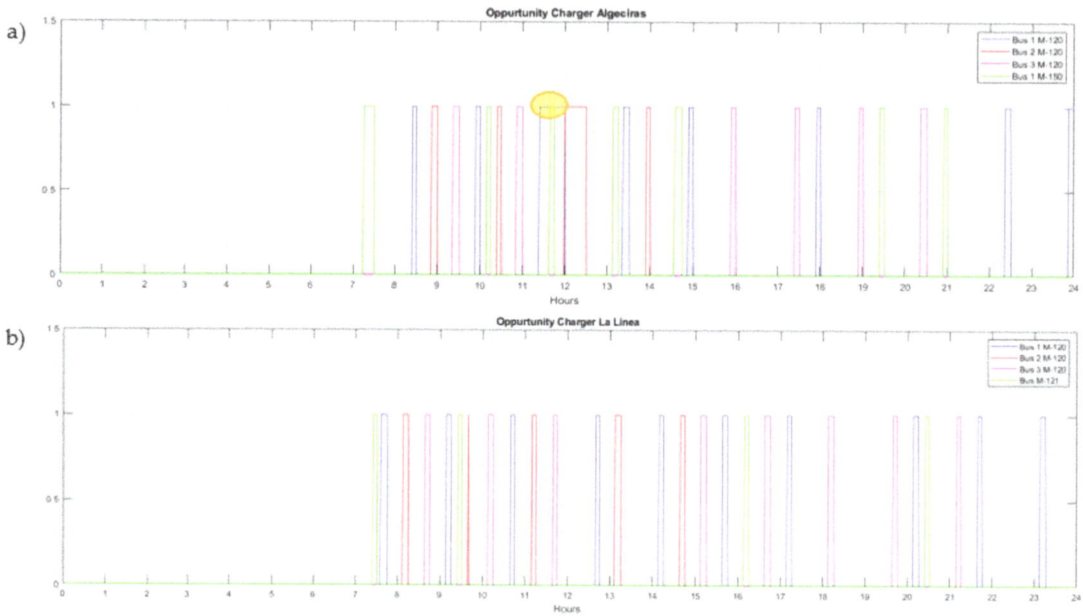

Figure 22. Opp Chargers occupation profiles installed in (**a**) Algeciras and (**b**) La Linea.

For the charger installed in Algeciras, a critical overlapping situation (highlighted by a yellow circle) occurs between 11.30–12 h (the charging period of Bus 1 line M-120), which can last 5/10 min (the charging time of the bus of line M-150). Bus 1 of line M-120 arrives some minutes before 11.30 h to Algeciras station with about 380 kWh in the battery; therefore, ideally, it has about 35 min to recharge the consumed 94 kWh and, hence, reach 100% SOC. With a charging power of 350 kW, to fill the missing energy, about 16 min are necessary, this means that the bus of line M-120 can leave the opportunity charger for the overlapped 5–10 min to the bus of line M-150 without compromising its own operation. Instead, the opportunity charger installed in La Linea does not report any critical overlap.

5. Conclusions

This paper deals with a public transport electrification scenario-based in terms of energy consumption evaluation and, consequently, charging infrastructure design and planning, by considering various operational concerns. The results confirmed the possibility to replace the conventional old buses currently employed in line M-120 between the cities of Algeciras and La Linea in Spain with new electric buses. However, the operation of an electric bus is greatly affected by various factors, such as the outside temperature, the number of stops, the elevation profile of the path, the number of onboard passengers, etc. On this aspect, the findings highlight that the main role is played by the power absorbed by the auxiliaries that, in the worst case, can lead to an increase in energy consumption by 50%.

To correctly electrify the public bus line, both overnight and opportunity charging systems are required. The slow charge during the night could be performed by two AC charging poles of 43 kW installed in Algeciras bus station. Instead, the installation of the opportunity charging system in only one of the two bus stations resulted to not be enough for the proper operation of the service; therefore, two fast chargers must be installed, one in each terminal. Finally, from the installation of this charging infrastructure, the electrification could be easily extended to two other lines of the same company.

The proposed study does not incorporate an economic and environmental analysis coming from the electrification. Therefore, the future steps of research will focus on the computation of a cost and emissions assessment also evaluating a possible integration in the charging infrastructure of renewable energy sources. Moreover, since in this work, the effect of the aging of the battery has not been computed, to assess with a higher precision the impact of this side effect, a more accurate battery model, which comprises not only the electrical sub-model but also the aging and thermal sub-models, can be included in future steps.

Author Contributions: Conceptualization, C.L., G.P., S.B. and M.L.; methodology, G.P. and S.B.; software, G.P. and S.B.; formal analysis, C.L.; investigation, C.L. and G.P.; resources, M.L. and S.B; data curation, C.L.; writing—original draft preparation, C.L. and G.P.; supervision, M.L.; project administration, S.B. and M.L. All authors have read and agreed to the published version of the manuscript.

Funding: This research received no external funding.

Institutional Review Board Statement: Not applicable.

Conflicts of Interest: The authors declare no conflict of interest.

References

1. Lu, R.; Yang, A.; Xue, Y.; Xu, L.; Zhu, C. Analysis of the key factors affecting the energy efficiency of batteries in electric vehicle. *World Electr. Veh. J.* **2010**, *4*, 9–13. [CrossRef]
2. EEA—European Environment Agency. Emissions of Air Pollutants from Transport. Available online: https://www.eea.europa.eu/data-and-maps/indicators/transport-emissions-of-air-pollutants-8/transport-emissions-of-air-pollutants-8 (accessed on 6 June 2020).
3. CO_2 Emissions from Heavy-Duty Vehicles—Preliminary CO_2 Baseline (Q3–Q4 2019). Available online: https://www.acea.auto/files/ACEA_preliminary_CO2_baseline_heavy-duty_vehicles.pdf (accessed on 7 July 2021).
4. EU. European Clean Bus Deployment Initiative. Mobility and Transport. 2020. Available online: https://ec.europa.eu/transport/themes/urban/cleanbus_en (accessed on 14 November 2020).
5. Majumder, S.; De, K.; Kumar, P.; Sengupta, B.; Biswas, P.K. Techno-commercial analysis of sustainable E-bus-based public transit systems: An Indian case study. *Renew. Sustain. Energy Rev.* **2021**, *144*, 111033. [CrossRef]
6. Guschinsky, N.; Kovalyov, M.Y.; Rozin, B.; Brauner, N. Fleet and charging infrastructure decisions for fast-charging city electric bus service. *Comput. Oper. Res.* **2021**, *135*, 105449. [CrossRef]
7. Vilppo, O.; Markkula, J. Feasibility of Electric Buses in Public Transport. *World Electr. Veh. J.* **2015**, *7*, 357–365. [CrossRef]
8. Carrilero, I.; González, M.; Anseán, D.; Viera, J.C.; Chacón, J.; Pereirinha, P.G. Redesigning European Public Transport: Impact of New Battery Technologies in the Design of Electric Bus Fleets. *Transp. Res. Procedia* **2018**, *33*, 195–202. [CrossRef]
9. Gao, Z.; Lin, Z.; LaClair, T.J.; Liu, C.; Li, J.-M.; Birky, A.; Ward, J. Battery capacity and recharging needs for electric buses in city transit service. *Energy* **2017**, *122*, 588–600. [CrossRef]
10. Roy, F.; Morency, C. Comparing Driving Cycle Development Methods Based on Markov Chains. *Transp. Res. Rec. J. Transp. Res. Board* **2021**, *2675*, 212–221. [CrossRef]
11. Wang, G.; Xie, X.; Zhang, F.; Liu, Y.; Zhang, D. BCharge: Data-Driven Real-Time Charging Scheduling for Large-Scale Electric Bus Fleets. In Proceedings of the 2018 IEEE Real-Time Systems Symposium (RTSS), Nashville, TN, USA, 11–14 December 2018; pp. 45–55.
12. Wenz, K.-P.; Serrano-Guerrero, X.; Barragán-Escandón, A.; González, L.; Clairand, J.-M. Route prioritization of urban public transportation from conventional to electric buses: A new methodology and a study of case in an intermediate city of Ecuador. *Renew. Sustain. Energy Rev.* **2021**, *148*, 111215. [CrossRef]
13. Global EV Outlook 2021—Analysis—IEA. Available online: https://www.iea.org/reports/global-ev-outlook-2021 (accessed on 7 July 2021).
14. Li, M.; He, J.; Demerdash, N.A. A flux-weakening control approach for interior permanent magnet synchronous motors based on Z-source inverters. In Proceedings of the 2014 IEEE Transportation Electrification Conference and Expo (ITEC) 2014, Detroit, MI, USA, 15–18 June 2014; pp. 1–6.
15. MATLAB and Simulink Racing Lounge: Vehicle Modeling—File Exchange—MATLAB Central. Available online: https://it.mathworks.com/matlabcentral/fileexchange/63823-matlab-and-simulink-racing-lounge-vehicle-modeling (accessed on 18 June 2021).
16. Phyphox—Physical Phone Experiments. Available online: https://phyphox.org/ (accessed on 18 June 2021).
17. Heydari, S.; Fajri, P.; Rasheduzzaman, M.; Sabzehgar, R. Maximizing Regenerative Braking Energy Recovery of Electric Vehicles Through Dynamic Low-Speed Cutoff Point Detection. *IEEE Trans. Transp. Electrif.* **2019**, *5*, 262–270. [CrossRef]
18. Leone, C.; Longo, M.; Foiadelli, F.; Bracco, S.; Piazza, G.; Delfino, F. Opportunity fast-charging of e-buses: A preliminary study for the city of Savona. In Proceedings of the 2020 AEIT International Conference of Electrical and Electronic Technologies for Automotive (AEIT AUTOMOTIVE), Turin, Italy, 18–20 November 2020. [CrossRef]
19. Green Car Congress. Available online: https://www.greencarcongress.com/ (accessed on 28 May 2021).

20. Basma, H.; Mansour, C.; Haddad, M.; Nemer, M.; Stabat, P. Comprehensive energy assessment of battery electric buses and diesel buses. In Proceedings of the 32nd International Conference on Efficiency, Cost, Optimization, Simulation and Environmental Impact of Energy Systems, Wroclaw, Poland, 23–28 June 2019.
21. Göhlich, D.; Ly, T.-A.; Kunith, A.; Jefferies, D. Economic Assessment of Different Air-conditioning and Heating Systems for Electric City Buses Based on Comprehensive Energetic Simulations. *World Electr. Veh. J.* **2015**, *7*, 398–406. [CrossRef]
22. Göhlich, D.; Fay, T.-A.; Jefferies, D.; Lauth, E.; Kunith, A.; Zhang, X. Design of urban electric bus systems. *Des. Sci.* **2018**, *4*. [CrossRef]
23. Al Haddad, R.; Basma, H.; Mansour, C. Analysis of heat pump performance in battery electric buses. In Proceedings of the 32nd International Conference on Efficiency, Cost, Optimization, Simulation and Environmental Impact of Energy Systems, Wroclaw, Poland, 23–28 June 2019.
24. Hasan, M.M.; Maas, J.; El Baghdadi, M.; de Groot, R.; Hegazy, O. Thermal Management Strategy of Electric Buses towards ECO Comfort. In Proceedings of the 8th Transport Research Arena Conference (TRA 2020), Helsinki, Finland, 27–30 April 2020.
25. IEEE Xplore Full-Text PDF. Available online: https://ieeexplore.ieee.org/stamp/stamp.jsp?arnumber=6389231 (accessed on 23 May 2021).
26. Guo, J.; Weihui, S.; Haoxuan, D.; Chao, T.U. Research of the Influence of Braking Conditions on Regenerative Braking Energy Recovery for Electric Vehicles. In Proceedings of the International conference on Energy, Ecology and Environment (ICEEE), Stockholm, Sweden, 26–29 July 2017.
27. Perrotta, D.; Ribeiro, B.; Rossetti, R.J.; Afonso, J.L. On the Potential of Regenerative Braking of Electric Buses as a Function of Their Itinerary. *Procedia Soc. Behav. Sci.* **2012**, *54*, 1156–1167. [CrossRef]
28. Zhang, J.; Lü, X.; Xue, J.; Li, B. Regenerative Braking System for Series Hybrid Electric City Bus. *World Electr. Veh. J.* **2008**, *2*, 363–369. [CrossRef]
29. Schwertner, M.; Weidmann, U. Comparison of Well-to-Wheel Efficiencies for Different Drivetrain Configurations of Transit Buses. *Transp. Res. Rec. J. Transp. Res. Board* **2016**, *2539*, 55–64. [CrossRef]
30. De Vroey, L.; Jahn, R.; El Baghdadi, M.; Van Mierlo, J. Plug-to-wheel energy balance—Results of a two years experience behind the wheel of electric vehicles. *World Electr. Veh. J.* **2013**, *6*, 130–134. [CrossRef]
31. Peter, M. Electric Buses for Mass Transit Seen as Cost Effective | American Public Power Association. 2019. Available online: https://www.publicpower.org/periodical/article/electric-buses-mass-transit-seen-cost-effective (accessed on 6 June 2021).
32. Chakraborty, D.; Nandi, A.K. Finding optimal deceleration with serial regenerative braking of electric vehicle using a multi-objective genetic algorithm. In Proceedings of the 2016 IEEE 1st International Conference on Power Electronics, Intelligent Control and Energy Systems (ICPEICES), Delhi, India, 4–6 July 2016. [CrossRef]
33. Edge, J.S.; O'Kane, S.; Prosser, R.; Kirkaldy, N.D.; Patel, A.N.; Hales, A.; Ghosh, A.; Ai, W.; Chen, J.; Yang, J.; et al. Lithium ion battery degradation: What you need to know. *Phys. Chem. Chem. Phys.* **2021**, *23*, 8200–8221. [CrossRef] [PubMed]
34. Transport Consortium of Andalucía. Available online: https://siu.ctmcg.es/horarios_lineas_tabla.php?from=1&linea=4 (accessed on 1 June 2021).

energies

Article

Robust Errorless-Control-Targeted Technique Based on MPC for Microgrid with Uncertain Electric Vehicle Energy Storage Systems

Yalin Liang, Yuyao He * and Yun Niu *

School of Marine Science and Technology, Northwestern Polytechnical University, Xi'an 710072, China; lyl_cassie@mail.nwpu.edu.cn
* Correspondence: heyyao@nwpu.edu.cn (Y.H.); niuyun010121@nwpu.edu.cn (Y.N.); Tel.: +86-13152151698 (Y.H.); +86-13991355690 (Y.N.)

Abstract: Regarding the microgrid with large-scale electric vehicle (EV) energy storage systems working at the vehicle-to-grid (V2G) mode, uncertain factors (e.g., the number of EVs feeding the microgrid shifts frequently) make the system unfixed, leading to the fact that it is difficult to precisely determine the real-time droop coefficients of the system, thereby degrading the performance of the traditional inverter control strategies that rely on the droop coefficients. To solve the problem, this paper proposes an errorless-control-targeted double control loop (DCL) technique based on robust MPC to control the microgrid with EV energy storage systems without using droop coefficients. Firstly, the structure of the DCL method is developed, with each component in the structure detailed. Compared to the traditional control strategies, the novel one regards the frequency, voltage, and currents as the control objectives instead of active/inactive power. It deserves to be mentioned that the frequency and voltage are regulated by proportional-integral controllers, while the currents are regulated by the finite control set model predictive control (FCS-MPC) method. Secondly, the impacts of system parameter uncertainties on the prediction accuracy of the FCS-MPC controller are analyzed clearly, illustrating that it is necessary to develop effective techniques to enhance the robustness of the controller. Thirdly, sliding mode observers (SMO) based on a novel hyperbolic function are constructed to detect the real-time disturbances, which can be used to generate voltage compensations by using automatic disturbance regulators. Then, the voltage compensations are adopted to establish a modified predicting plant model (PPM) used for the FCS-MPC controller. By using the proposed SMO-based disturbance detection and compensation techniques, the MPC controller gains a strong robustness against parameter uncertainties. Finally, a simulation is conducted on a microgrid system to verify the effectiveness of the proposed techniques, and the obtained results are compared with the traditional virtual synchronous machine (VSG) strategy relying on droop coefficients.

Keywords: microgrid; errorless control; model predictive control; robustness; sliding mode disturbance observer

Citation: Liang, Y.; He, Y.; Niu, Y. Robust Errorless-Control-Targeted Technique Based on MPC for Microgrid with Uncertain Electric Vehicle Energy Storage Systems. *Energies* **2022**, *15*, 1398. https://doi.org/10.3390/en15041398

Academic Editors: Mojtaba Ahmadieh Khanesar and Christian Ghiaus

Received: 13 December 2021
Accepted: 8 February 2022
Published: 15 February 2022

1. Introduction

Nowadays, the microgrid which contains distributed generation (e.g., photovoltaic, wind, and tidal generation, etc.) and energy storage (e.g., batteries and super-capacitors, etc.) is one of the most promising power generation and supply systems because it has the advantages of high flexibility, eco-friendliness and sustainability [1,2]. Apart from feeding and supporting the conventional grid (grid-connected mode), microgrids play an important role in forming the grid (island mode) [3]. When the microgrid operates at the island mode (see Figure 1), it needs to independently supply power for the local DC/AC loads, such as electric vehicles (EVs), household and industrial electric apparatuses, etc. [4]. Among those loads, EVs are special because they have rechargeable energy storage systems (RESS), and as a result, they can feed the DC-bus of the microgrid when the other types of loads are heavy. The technology is known as vehicle-to-grid (V2G) [5].

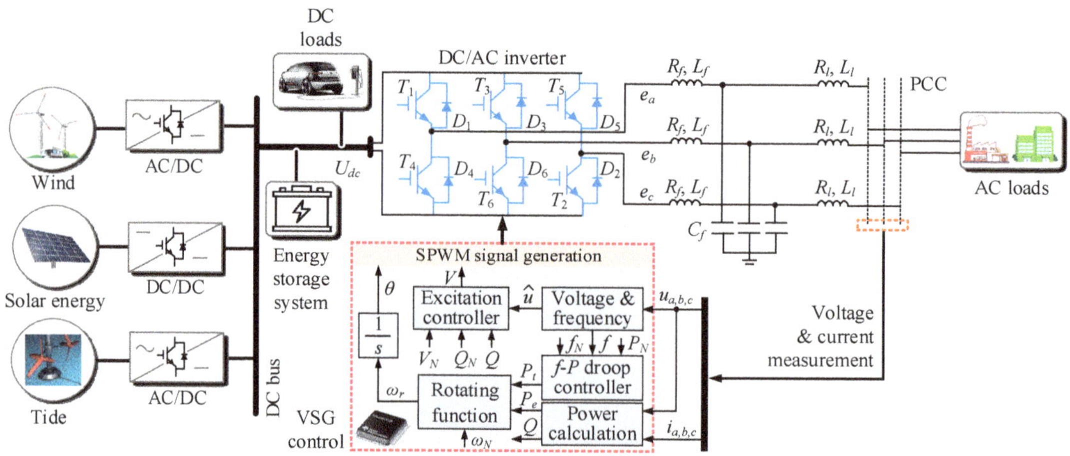

Figure 1. Structure of VSG control-based microgrid.

However, as for the EVs used in the applications of V2G, they are uncertain factors for the microgrid. The dynamic model of the EV energy storage system can be qualitatively represented as (1). It can be seen that the supporting power from the EVs is related to the number of charging piles n, the number of vehicles working at the V2G mode ($n \cdot c$), and the supporting period of the EVs (t_{su}). In detail, the uncertainties are reflected in the following four aspects. (1) More charging piles might be installed in one microgrid system. (2) The number of EVs connected with the microgrid varies in different periods. (3) Even though the charging piles installed within the microgrid are fully occupied, it does not mean that all of them are available for supporting the microgrid. Usually, whether a vehicle can provide energy to the grid will depend on the EV owners. (4) EVs that are supplying power to the grid may end at any time. These uncertainties lead to the fact that the power level (rated power) of the microgrid is not fixed, bringing about challenges for the DC/AC inverter control strategies [6]. Specifically, the commonly used inverter control techniques for the microgrid include droop control and virtual synchronous generator (VSG) control [7,8]. Droop control used for the microgrid originates from the conventional grid. It can be categorized into *P-f*/*Q-V* control and *f-P*/*V-Q* control, and the differences between them can be reflected by their control topology [9–11]. Considering that the droop control has the disadvantages of small inertia and insufficient damping, it will generate remarkable voltage and frequency fluctuations or deviations when the loads change [12]. To improve the inertia and damping of the system, VSG control technology that simulates the behaviors of the real synchronous generators has been studied since 2007 [13–19]. It deserves to be mentioned that as for the VSG control method (see Figure 1), *f-P*/*Q-V* droop control still needs to be incorporated. Overall, droop controllers cannot be eliminated for the above two traditional control methods, thereby requiring droop coefficients. However, when the uncertainties of the EVs are taken into account, the droop coefficients cannot maintain at a certain level because the power supply system (hardware and power level) is inclined to change [20]. In this case, the performance of the traditional control strategies based on droop control will decline unless the real-time droop coefficients cannot be provided.

$$P_{su} = f(n, c, t_{su}, t) \tag{1}$$

With reference to [21–25], to solve the uncertainty problem arising from the droop coefficients, two methods can be potentially adopted, that is, adaptive droop coefficients and coefficient elimination. In [21], the fuzzy logic-based and model reference-based adaptive droop coefficient design methods are presented to adjust the adaptability of the droop controllers, improving the transient response of the system. Reference [22] introduces a method that upgrades the droop coefficients based on a sliding mode strategy to optimize the power-sharing operations. In [23], for the sake of high performance, a stability-constrained adaptive droop approach is proposed for autonomous power-sharing of the grid. In [24], a VSG-based method with adaptive active power and DC voltage droop control is proposed to regulate the voltage and frequency of the grid. In this study, the droop coefficients are adaptively adjusted depending on the frequency margin of the system. Literature [25] illustrates that without adding a droop controlling unit, the power, frequency, and voltage can also be controlled by using the closed-loop feedback control technique, but no new control schemes are developed in this research. It deserves to be mentioned that compared to the research concerning adaptive droop coefficients, there are much fewer studies concerning the droop coefficient elimination strategies today. Hence, it is a valuable and timely measure to further investigate the control theories without using droop controllers and droop coefficients.

One main purpose of this paper is to develop a double-closed-loop (DCL) control strategy to directly regulate the voltage and frequency (the ultimate control objectives) of the microgrid with large-scale EVs connected. The outer loop will be the voltage and frequency control, while the inner loop will be the current control one. The goal of the proposed DCL control strategy is to ensure voltage and frequency are maintained at the desired level constantly without errors, regardless of the changes of load and EVs. Because the voltage, frequency, and currents are direct control targets, it is not necessary to employ droop controllers to generate voltage and frequency references. Therefore, the droop coefficients can be avoided. When achieving the errorless DCL control scheme, the most direct way is to adopt several (at least four) proportional-integral (PI) controllers to regulate the voltage, frequency, and currents. However, [26] and [27] clarify that the PI controllers have the disadvantages of complicated parameter tuning and low dynamics, leading to the fact that it is better to adopt the alternative controllers in the main structure.

Many optimization algorithms can be used to overcome the shortcomings of PI controllers. For example, [28] uses the model predictive control (MPC) method to manage energy resources efficiently. The author of [29] designs an optimal model predictive controller for the nonlinear multi-area power system, while [30] uses MPC to manage the power in a hydrogen-based microgrid. In [31], a new variable structure control method is proposed to regulate the frequency of the grid, while [32] proposes a gravitational search algorithm-based dual proportional-integral method to control the load frequency, which can be regarded as a typical application of the future search algorithms for optimization [33]. Among those strategies, MPC is characterized by online optimization and quick response, thereby being a promising alternative [34]. However, because MPC is a model-based control strategy, its performance highly relies on the accuracy of the system parameters, including line resistance and inductance of the microgrids. Unfortunately, considering the complexity of the wiring, connections and working environment of the microgrids, the system parameters might change continuously [35,36]. As for the traditional fixed-parameter-based MPC methods, once the parameter mismatch phenomenon occurs, the prediction accuracy of the MPC controllers will degrade significantly. On this premise, if MPC is used to achieve the DCL control scheme, it is vital to develop effective strategies to enhance the robustness of MPC controllers against parameter uncertainties.

In this paper, we propose a novel DCL strategy based on robust MPC theory to achieve errorless frequency and voltage regulation for the microgrids with uncertain EV energy storage systems. The novelties and contributions can be summarized as follows:

(1) The structure of the novel DCL control method is developed, with the details of each internal component presented. In comparison with the traditional droop control and VSG control strategies, droop controllers are not needed any longer, thereby avoiding the droop coefficients. As for the microgrids with uncertain EV energy storage systems, when the AC loads are heavy, the overall power provided by the EVs can be automatically determined by using the errorless-control-targeted scheme, and

the maximum power that can be provided by each EV can be restrained by setting the maximum allowable discharge current.

(2) After establishing a disturbance-based predicting plant model (PPM), the impacts of line resistance and inductance on the prediction accuracy of the FCS-MPC method are studied in the area of microgrid control, addressing the necessity of developing the robust MPC technique.

(3) Sliding mode observers (SMO) based on the novel hyperbolic function are developed to detect the real-time disturbances which can be further used to generate voltage compensations by controlling them to maintain at zero. This is new in the area of microgrid control. In order to make the SMOs stable, Lyapunov functions are constructed to design the parameters of the observers. It deserves to be mentioned that the voltage compensations need to be substituted into the PPM to select the optimal voltage vector applied to the inverter, improving the prediction accuracy. By using the proposed SMO-based disturbance detection and compensation techniques, the MPC controller is endowed with strong robustness against parameter uncertainties.

The structure of the rest of the paper is as follows: Section 2 introduces the structure of the proposed DCL control strategy with a traditional MPC controller in comparison to the VSG method. The mechanisms and implementations of each part in the structure are discussed. In Section 3, the impacts of the parameter mismatch on the prediction accuracy of the traditional MPC method are analyzed firstly. Then, the sliding mode disturbance observers (SMDO) are constructed with their stability analyzed. Additionally, the implementation procedures of the proposed DCL technique based on the robust MPC controller are presented in this part. Section 4 discusses the simulation results of the proposed algorithms, and Section 5 is the conclusion.

2. Proposed Errorless-Control-Targeted DCL Strategy with Traditional MPC Controller

This section introduces the structure of the proposed errorless-control-targeted DCL strategy in comparison with the traditional VSG method. by firstly revealing the novelty of the new strategy. Secondly, each part of the DCL structure is illustrated in detail. Lastly, the working mechanism of the hardware system is explained.

2.1. Structure of Proposed DCL Control Strategy

Figure 2 depicts the structure of the proposed DCL control strategy for the microgrid working in the island mode. Being similar to the distributed generation and energy storage systems, the EV energy storage systems are regarded as the power source of the microgrid in Figure 2. Based on the aforementioned analysis, the EVs are uncertain sources. The new control strategy comprises a phase-locked loop (PLL) used to detect the real-time frequency; phase angle and magnitude of the voltage; d, q-axis current calculation part; automatic voltage regulator (AVR); and an automatic frequency regulator (AFR), which is used to achieve errorless voltage and frequency control, respectively; an MPC controller used for current regulation and pulse width modulation (PWM) signal selection; and the hardware system, which is the controlled object. Their details will be presented in the latter parts of this section. The main rationale behind the control technique can be described as follows: The detected frequency and voltage are regulated by PI controllers to maintain at the desired level, generating d, q-axis current references. Then, the feedback control theory is adopted for current regulation by using the MPC controller. The advantage of this method is that no droop controllers are adopted, avoiding the influence of the power uncertainties caused by the EVs.

By comparing the proposed method with the traditional VSG control strategy shown in Figure 1, five important differences reflecting the novelties of the proposed method can be summarized as follows (see Table 1). Firstly, as the active power and inactive power are not calculated and regulated, so the f-P and Q-V droop controllers are not needed to calculate the power and voltage references. Considering that voltage and frequency are the ultimate control objectives, they are directly regulated by the PI controllers without considering the intermediate variables (power). Without using droop coefficients, the EV energy storage system uncertainties will have few impacts on the control strategy. Secondly, in Figure 1, the voltage control process is an open-loop regulation, while, as for the proposed method, voltage feedback regulation is adopted, thus contributing to the voltage-errorless control. Thirdly, in terms of the VSG strategy, the feedback frequency is used to generate the active power reference, and further, the power is regulated by using the rotating function of the VSG. Essentially, this is a power-targeted control strategy; however, the method in Figure 2 is a frequency-errorless control strategy. Fourthly, the inactive current and active current are regulated by an MPC controller (inner loop), which is not focused on in the traditional VSG strategy. Finally, the PWM signal generation procedures are different. As for the VSG method, after the magnitude and phase angle of the desired voltage are calculated, the control signals are generated relying on the sine PWM (SPWM) theories. However, in Figure 2, the cost function of the MPC controller can select the optimal control signal after substituting the candidate voltage vectors into the PPM. It deserves to be mentioned that the latter one is easier to implement because it is based on the intuitive optimal control theory.

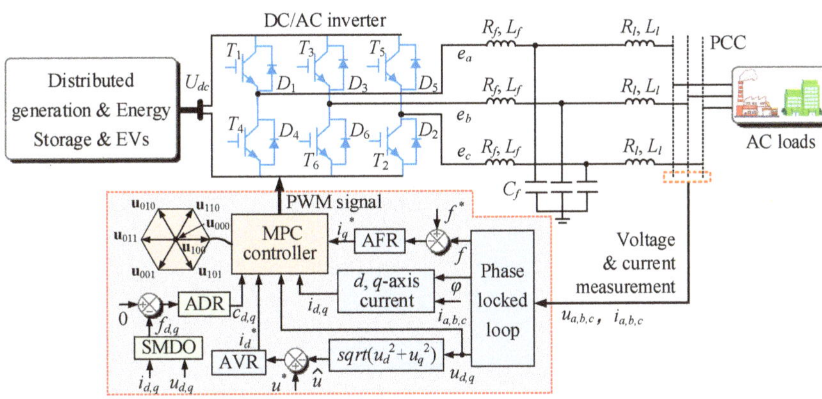

Figure 2. Structure of proposed DCL control strategy for microgrid working at island mode.

Table 1. Comparisons of traditional VSG strategy and proposed DCL strategy.

Methods / Differences	Traditional VSG Strategy	Proposed DCL Strategy
Regulated variables	active power and inactive power	frequency and voltage
Voltage regulation	open-loop voltage control	voltage-errorless feedback control
Frequency regulation	frequency-feedback-based power regulation	frequency-errorless feedback control
Current control	X	MPC-based current regulation
PWM signal generation	SPWM signal generation	optimal voltage vector selection-based PWM signal generation

2.2. Details of Each Part of Proposed DCL Control Strategy

2.2.1. Real-Time Frequency, Phase Angle, and *d*, *q*-axis Voltage Detection

To obtain the real-time frequency, phase angle, and voltage magnitude of the microgrid, a three-phase PLL, of which the structure is depicted in Figure 3, is adopted. It deserves to be mentioned that the coordinate transformation part (*abc*/*dq*) satisfies the following condition [37]:

$$\begin{bmatrix} u_d \\ u_q \end{bmatrix} = \frac{2}{3} T \cdot \begin{bmatrix} u_a \\ u_b \\ u_c \end{bmatrix} \tag{2}$$

where $\boldsymbol{T}$ is the transformation matrix, and it is

$$\boldsymbol{T} = \begin{bmatrix} \cos\varphi & \frac{\sqrt{3}\sin\varphi - \cos\varphi}{2} & \frac{-\sqrt{3}\sin\varphi - \cos\varphi}{2} \\ -\sin\varphi & \frac{\sin\varphi + \sqrt{3}\cos\varphi}{2} & \frac{\sin\varphi - \sqrt{3}\cos\varphi}{2} \end{bmatrix} \tag{3}$$

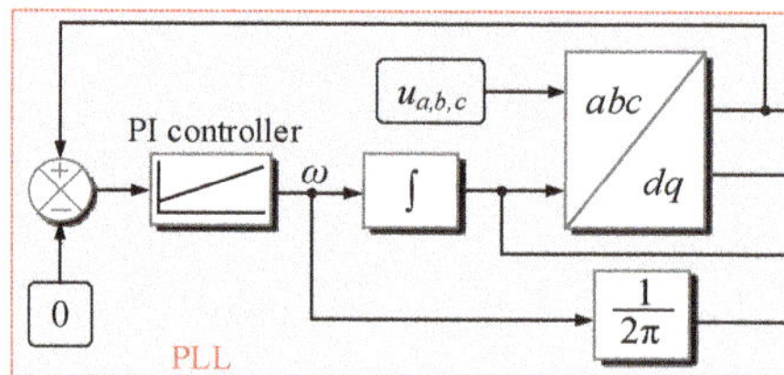

Figure 3. Three-phase PLL for frequency, phase angle, and voltage detection.

Moreover, in Figure 3, the PI controller is used to regulate the errors between the d-axis voltage and the reference (zero), and its output is the angular frequency (ω) of the microgrid. After imposing integral operation and proportional operation on the angular speed, the phase angle, and frequency (in Hz) of the microgrid can be obtained as:

$$\begin{cases} \varphi = \int_0^t \omega dt \\ f = \frac{\omega}{2\pi} \end{cases} \tag{4}$$

2.2.2. *d*, *q*-axis Current Calculation

After detecting the phase angle of the microgrid, the d, q-axis currents that are used for current regulation can be obtained by using the following equation:

$$\begin{bmatrix} i_d \\ i_q \end{bmatrix} = \frac{2}{3}T \cdot \begin{bmatrix} i_a \\ i_b \\ i_c \end{bmatrix} \tag{5}$$

2.2.3. Errorless Voltage and Frequency Control

In order to achieve errorless control, PI controllers are employed to regulate the voltage and frequency. For the AFR, input is the error between the frequency reference (rated frequency, e.g., 50 Hz) and the feedback frequency, while the output is set as the q-axis current reference $i_q{}^*$ (related to the active power). In terms of the AVR, its input is the error between the magnitude of the rated voltage and the magnitude of the feedback voltage that can be calculated by (6). As for the output of the AVR, it is the d-axis current reference $i_d{}^*$ which is related to the inactive power of the microgrid.

$$\hat{u} = \sqrt{u_d{}^2 + u_q{}^2} \tag{6}$$

The transfer functions of the AFR and AVR in the z-domain can be written as:

$$G_{AFR}(z) = k_{F\_p} + \frac{k_{F\_i}T_s z}{z-1},\ G_{AVR}(z) = k_{V\_p} + \frac{k_{V\_i}T_s z}{z-1} \tag{7}$$

In practice, the proportional and integral factors in (7) need to be tuned. However, as for the proposed DCL control strategy that contains both PI controllers and MPC controllers, there exist no mature theories for analysis and parameter tuning. Hence, the trial-and-error strategy is more suitable to design the parameters of the AFR and AVR [38].

2.2.4. Traditional MPC Control

In terms of the two kinds of commonly used MPC controllers in the field of power electronic systems, namely, the continuous control set model predictive control (CCS-MPC) controller and FCS-MPC controller [39], the latter one is simpler to implement, bringing about a remarkable computational complexity reduction. This benefits from the look-up table of the candidate voltages, which can be obtained offline. Comparatively speaking, the computations of the CCS-MPC methods are much more complicated when executing the modulation algorithms. Hence, the FCS-MPC controller is adopted in this research. A discrete PPM is the prerequisite for achieving an FCS-MPC algorithm. Based on Figure 2, the state-space model of the microgrid in the d, q-axis frame can be established as follows:

$$\begin{cases} \frac{di_d}{dt} = -\frac{R_f+R_l}{L_f+L_l} i_d + \frac{e_d-u_d}{L_f+L_l} \\ \frac{di_q}{dt} = -\frac{R_f+R_l}{L_f+L_l} i_q + \frac{e_q-u_q}{L_f+L_l} \end{cases} \tag{8}$$

Further, apply the Euler forward discretization algorithm to (8) in a time step of T_s (control period), and the future current states at t_{k+1} can be obtained as follows:

$$\begin{cases} i_d(k+1) = \frac{(L_f+L_l-R_fT_s-R_lT_s)}{L_f+L_l} i_d(k) - \frac{T_s}{L_f+L_l} u_d(k) + \frac{T_s}{L_f+L_l} e_d(k) \\ i_q(k+1) = \frac{(L_f+L_l-R_fT_s-R_lT_s)}{L_f+L_l} i_q(k) - \frac{T_s}{L_f+L_l} u_q(k) + \frac{T_s}{L_f+L_l} e_q(k) \end{cases} \tag{9}$$

In (9), $i_{d,q}(k)$ and $u_{d,q}(k)$ can be directly measured by using current and voltage sensors, while $e_{d,q}(k)$ should be calculated based on the candidate control voltages for an FCS-MPC controller. For instance, as far as a two-level inverter is concerned, when the FCS-MPC algorithm is implemented, seven candidate control voltages can be directly substituted into the model for prediction, which is denoted as $\mathbf{u}_{000}$, $\mathbf{u}_{100}$, $\mathbf{u}_{110}$, $\mathbf{u}_{010}$, $\mathbf{u}_{011}$, $\mathbf{u}_{001}$, and $\mathbf{u}_{101}$:

$$\boldsymbol{e}_{s_a s_b s_c} = \begin{bmatrix} e_a \\ e_b \\ e_c \end{bmatrix} = \frac{U_{dc}}{3} \begin{bmatrix} 2 & -1 & -1 \\ -1 & 2 & -1 \\ -1 & -1 & 2 \end{bmatrix} \begin{bmatrix} s_a \\ s_b \\ s_c \end{bmatrix} \tag{10}$$

where $[s_a, s_b, s_c]^{\mathrm{T}}$ includes $[0, 0, 0]^{\mathrm{T}}$, $[1, 0, 0]^{\mathrm{T}}$, $[1, 1, 0]^{\mathrm{T}}$, $[0, 1, 0]^{\mathrm{T}}$, $[0, 1, 1]^{\mathrm{T}}$, $[0, 0, 1]^{\mathrm{T}}$, and $[1, 0, 1]^{\mathrm{T}}$, and they are the switching states. With reference to (2), the control voltage sets used for prediction in the d, q-axis frame can be expressed as:

$$\begin{bmatrix} e_d(k) \\ e_q(k) \end{bmatrix} = \frac{2}{3} \boldsymbol{T} \cdot \boldsymbol{e}_{s_a s_b s_c} \tag{11}$$

After substituting the candidate voltages into (9) to calculate the predicted current values, a cost function should be used to select the optimal control voltage and the corresponding switching state to be applied to the inverter:

$$J = \left| i_d{}^* - i_d(k+1) \right| + \left| i_q{}^* - i_q(k+1) \right| \tag{12}$$

In terms of the FCS model predictive current control method, the following constraint needs to be incorporated:

$$\sqrt{i_d(k+1)^2 + i_q(k+1)^2} \le i_{\max} \tag{13}$$

As for the traditional FCS-MPC strategy, the resistances and inductances in (9) are fixed, which are usually measured (denoted as R_{f_mea}, L_{f_mea}, R_{l_mea}, and L_{l_mea}) by using the offline methods [40]. However, their real values are inclined to change as the working environment changes, degrading the prediction accuracy of the MPC controller. This can further bring about static voltage and frequency errors, failing the goal of errorless control. To solve the problem, a robust FCS-MPC control method against parameter uncertainties will be developed in Section 3.

2.2.5. Hardware System

As shown in Figure 2, the hardware directly controlled by the proposed DCL strategy is the DC/AC inverter, which will transmit the energy from the distributed generation system (DGS), inherent energy storage system, and EV storage system to the energy-consumption side. In the process of energy transmission, the goal is to automatically maintain the magnitude and frequency of the AC voltage at the rated levels through the proposed errorless control method. However, as for this system, there exist two crucial issues, (1) when the EV storage system feeds the microgrid, and (2) how to determine the maximum output power of one EV. In this paper, the working mechanisms of the EV storage system can be summarized as follows:

- Firstly, being similar to the traditional V2G technology, whether an EV begins to feed or stops feeding the microgrid, is determined by the vehicle owners. This can be achieved by using an app installed on their phones [41]. Secondly, the EV energy storage system works when it is needed. In detail, if the energy provided by the distributed generation and inherent storage systems is not enough to maintain the frequency and magnitude of the AC voltage at the rated level, the EV storage system will start to feed the grid. This phenomenon usually occurs when the AC loads are heavy, which reflect in the real-time voltage frequency and magnitude. Otherwise, the EV energy storage system is inactive.
- When the EVs provide energy to the microgrid, it is impossible to expect each vehicle to deliver unlimited power for the sake of safety. However, if no effective strategies are given to limit the output power of each vehicle, the horrible phenomenon might occur due to the proposed automatic control scheme when the microgrid voltage and frequency variations are extremely large. To solve the problem, considering that the output voltage of an EV energy storage system equals the DC-bus voltage U_{dc}, as long as the output current of each system can be confined within their safe limits (determined by each vehicle) the EVs can work safely in the V2G process. It deserves to be mentioned that the maximum allowable output current of each EV system needs to be provided by the vehicle designers or manufacturers.

3. Robust SMDO-Based MPC Method

This section introduces the impacts of the system parameter uncertainties on the MPC control firstly. Secondly, the SMDOs are constructed, with their stability discussed in an innovative way. Then, the proposed disturbance impact elimination method based on voltage compensation is presented. Finally, the implementation procedures of the DCL strategy with a modified MPC controller are detailed.

3.1. Impacts of Parameter Uncertainties on MPC Controller

Assume that the deviations of the filter resistance, filter inductance, line resistance, and line inductance are ΔR_f, ΔL_f, ΔR_l, and ΔL_l, respectively, and then, the real parameters of the microgrid can be described as:

$$R_f = R_{f\_mea} + \Delta R_f,\ L_f = L_{f\_mea} + \Delta L_f,\ R_l = R_{l\_mea} + \Delta R_l,\ L_l = L_{l\_mea} + \Delta L_l \tag{14}$$

Substitute (14) into (8), the accurate system model can be rewritten as:

$$\begin{cases} \frac{di_d}{dt} = -\frac{R_{f\_mea}+R_{l\_mea}}{L_{f\_mea}+L_{l\_mea}} i_d + \frac{e_d - u_d}{L_{f\_mea}+L_{l\_mea}} + f_d \\ \frac{di_q}{dt} = -\frac{R_{f\_mea}+R_{l\_mea}}{L_{f\_mea}+L_{l\_mea}} i_q + \frac{e_q - u_q}{L_{f\_mea}+L_{l\_mea}} + f_q \\ f_d = -\frac{\Delta R_l+\Delta R_f}{L_{f\_mea}+L_{l\_mea}} i_d - \frac{\Delta L_f+\Delta L_l}{L_{f\_mea}+L_{l\_mea}} \frac{di_d}{dt} \\ f_q = -\frac{\Delta R_l+\Delta R_f}{L_{f\_mea}+L_{l\_mea}} i_q - \frac{\Delta L_f+\Delta L_l}{L_{f\_mea}+L_{l\_mea}} \frac{di_q}{dt} \end{cases} \tag{15}$$

Further, by the use of the discretization algorithm, the discrete model can be obtained as:

$$\begin{cases} i_d(k+1) = \frac{(L_{f\_mea}+L_{l\_mea}-R_{f\_mea}T-R_{l\_mea}T_s)}{L_{f\_mea}+L_{l\_mea}} i_d(k) - \frac{T_s}{L_{f\_mea}+L_{l\_mea}} u_d(k) + \frac{T_s}{L_{f\_mea}+L_{l\_mea}} e_d(k) + f_d(k) \\ i_q(k+1) = \frac{(L_{f\_mea}+L_{l\_mea}-R_{f\_mea}T-R_{l\_mea}T_s)}{L_{f\_mea}+L_{l\_mea}} i_q(k) - \frac{T_s}{L_{f\_mea}+L_{l\_mea}} u_q(k) + \frac{T_s}{L_{f\_mea}+L_{l\_mea}} e_q(k) + f_q(k) \\ f_d(k) = -\frac{\Delta L_f+\Delta L_l}{L_{f\_mea}+L_{l\_mea}} i_d(k+1) + \frac{\Delta L_f+\Delta L_l-\Delta R_f T_s-\Delta R_l T_s}{L_{f\_mea}+L_{l\_mea}} i_d(k) \\ f_q(k) = -\frac{\Delta L_f+\Delta L_l}{L_{f\_mea}+L_{l\_mea}} i_q(k+1) + \frac{\Delta L_f+\Delta L_l-\Delta R_f T_s-\Delta R_l T_s}{L_{f\_mea}+L_{l\_mea}} i_q(k) \end{cases} \quad (16)$$

It should be noted that $f_d(k)$ and $f_q(k)$ represent the d, q-axis current estimation errors when subtracting the PPM (9) with the measured parameters R_{f_mea}, L_{f_mea}, R_{l_mea}, and L_{l_mea} incorporated from (16). Obviously, when the parameter mismatch issue arises, the prediction accuracy of the traditional FCS-MPC method will degrade. In order to intuitively illustrate the impacts of parameter mismatch on the prediction accuracy, a microgrid of which main parameters (without considering the EV energy storage system) are shown in Table 2 is employed for analysis. Assume that the d, q-axis currents are controlled to level off at 10 A ($i_d(k+1) = i_d(k) = 20$) and 20 A ($i_q(k+1) = i_q(k) = 20$), respectively, and Figure 4 shows the relationship between the current estimation errors and the parameter variations. Firstly, for the d, q-axis currents, the prediction errors are smaller than zero, and the larger the deviations, the larger the magnitudes of the prediction errors become. Secondly, the inductance variations see more severe impacts than the resistances. Thirdly, the d, q-axis current prediction errors reach around −45 A and −90 A, respectively, when ΔL_f approaches 1 H. Comparatively, the current prediction errors are small when the line inductance changes because the magnitude of the system inductance is small. However, for the system with large line inductance, this will change greatly. These represent that it is highly required to develop effective solutions to the parameter mismatch issue, otherwise the control performance of the FCS-MPC strategy will be poor.

Table 2. Main parameters of the microgrid.

Variable	Value	Unit
P_N	10	kVA
U_N	310	V
f_N	50	Hz
U_{DC}	800	V
L_f	0.18	H
R_f	0.16	Ω
C_f	10	μF
L_l	0.1	mH
R_l	1.2	Ω
T_s	0.001	s
k_ω	$-2\pi \times 10^{-4}$	-
k_v	0.0001	-

3.2. Disturbance Observation Based on Sliding Mode Theory

To eliminate the impacts of parameter uncertainties on the FCS-MPC control performance, as illustrated in Figure 2, the d, q-axis disturbances need to be detected at first, and then, they will be compensated by using a voltage compensation strategy. Hence, the prerequisite for this method is to construct disturbance observers. Considering that SM control has the advantages of fast response and high robustness [42], it is employed for disturbance detection in this paper.

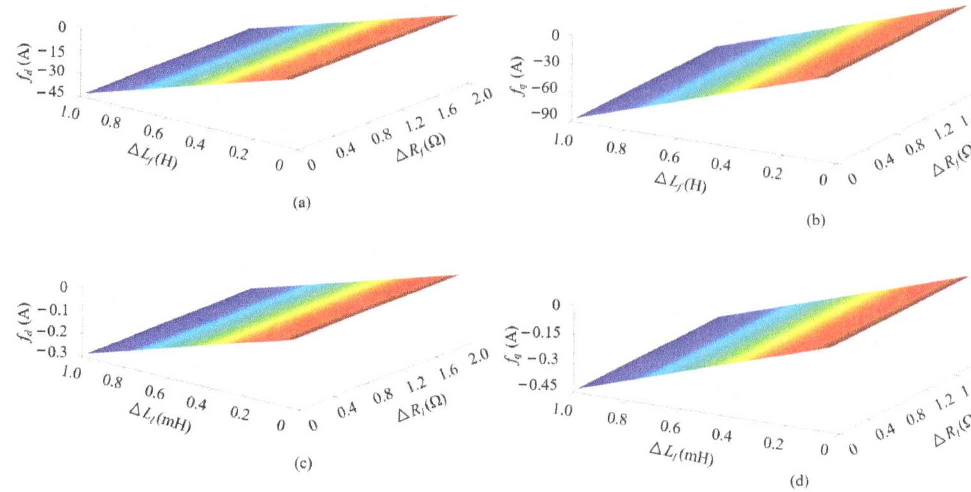

Figure 4. Relationship between the current estimation errors and the parameter variations. (**a**) Relationship between *d*-axis current estimation errors and ΔR_f, ΔL_f ($\Delta R_l = \Delta L_l = 0$). (**b**) Relationship between *d*-axis current estimation errors and ΔR_l, ΔL_l ($\Delta R_f = \Delta L_f = 0$). (**c**) Relationship between *q*-axis current estimation errors and ΔR_f, ΔL_f ($\Delta R_l = \Delta L_l = 0$). (**d**) Relationship between *q*-axis current estimation errors and ΔR_l, ΔL_l ($\Delta R_f = \Delta L_f = 0$).

3.2.1. SMDO

According to the sliding mode theory, the disturbance observer based on (15) can be constructed as:

$$\begin{cases} \frac{d\hat{i}_d}{dt} = -\frac{R_{f\_mea}+R_{l\_mea}}{L_{f\_mea}+L_{l\_mea}}\hat{i}_d + \frac{e_d-u_d}{L_{f\_mea}+L_{l\_mea}} + k_d F(\overline{i_d}) \\ \frac{d\hat{i}_q}{dt} = -\frac{R_{f\_mea}+R_{l\_mea}}{L_{f\_mea}+L_{l\_mea}}\hat{i}_q + \frac{e_q-u_q}{L_{f\_mea}+L_{l\_mea}} + k_q F(\overline{i_q}) \end{cases} \tag{17}$$

In (17), $F(\overline{i_d})$ and $F(\overline{i_q})$ are the switching functions that can be represented as:

$$\begin{bmatrix} F(\overline{i_d}) \\ F(\overline{i_q}) \end{bmatrix} = \begin{bmatrix} \frac{e^{m\cdot\overline{i_d}}-e^{-m\cdot\overline{i_d}}}{e^{m\cdot\overline{i_d}}+e^{-m\cdot\overline{i_d}}} \\ \frac{e^{m\cdot\overline{i_q}}-e^{-m\cdot\overline{i_q}}}{e^{m\cdot\overline{i_q}}+e^{-m\cdot\overline{i_q}}} \end{bmatrix} \tag{18}$$

It deserves to be mentioned that the switching function is a novel hyperbolic function rather than the traditional signum function [42], of which properties are mainly determined by the boundary-layer constant m. The reason why the hyperbolic function is designed in this paper is that it is more inclined to suppress the chattering of the SMOs. When the system becomes stable, the d, q-axis disturbances can be calculated by:

$$\begin{bmatrix} f_d \\ f_q \end{bmatrix} = \begin{bmatrix} k_d F(\overline{i_d}) \\ k_q F(\overline{i_q}) \end{bmatrix} \tag{19}$$

3.2.2. Stability Analysis

To construct the Lyapunov function to analyze the stability of the proposed SMOs, sliding mode surfaces of d, q-axis currents are defined as:

$$\mathbf{S} = \begin{bmatrix} S_d \\ S_q \end{bmatrix} = \begin{bmatrix} \overline{i_d} \\ \overline{i_q} \end{bmatrix} = 0 \tag{20}$$

Then, the Lyapunov function can be constructed as:

$$L_y = \frac{1}{2}\mathbf{S}\cdot\mathbf{S}^T = \frac{1}{2}\overline{i_d}^2 + \frac{1}{2}\overline{i_q}^2 \tag{21}$$

Undoubtedly, $L_y > 0$. Based on the Lyapunov theorem of stability, as long as the derivative of L_y is smaller than zero, the proposed SMDOs are stable. The derivative of (20) is:

$$\frac{dL_y}{dt} = \overline{i_d}\cdot\frac{d\overline{i_d}}{dt} + \overline{i_q}\cdot\frac{d\overline{i_q}}{dt} \tag{22}$$

Substitute (15) and (17) into (22), and then it can be obtained that:

$$\frac{dL_y}{dt} = \underbrace{-\frac{R_{f\_mea} + R_{l\_mea}}{L_{f\_mea} + L_{l\_mea}}(\overline{i_d}^2 + \overline{i_q}^2)}_{term\ 1<0} + \underbrace{(k_d F(\overline{i_d}) - f_d)\overline{i_d} + (k_q F(\overline{i_q}) - f_q)\overline{i_q}}_{term\ 2} \tag{23}$$

In (23), *term* 1 is less than zero. To make the system stable, it can be deduced from *term* 2 that the following condition should be satisfied:

$$\begin{cases} (k_d F(\overline{i_d}) - f_d)\overline{i_d} < 0 \\ (k_q F(\overline{i_q}) - f_q)\overline{i_q} < 0 \end{cases} \tag{24}$$

Considering the signs (positive or negative) of $\overline{i_d}$ and $\overline{i_q}$, it can be derived that:

$$\begin{cases} k_d < \frac{f_d}{F(\overline{i_d})},\ if\ \overline{i_d} > 0 \\ k_d < -\frac{f_d}{F(\overline{i_d})},\ if\ \overline{i_d} < 0 \end{cases} \rightarrow k_d < -|\frac{f_d}{F(\overline{i_d})}| \tag{25}$$

$$\begin{cases} k_q < \frac{f_T}{F(\overline{i_q})},\ if\ \overline{i_q} > 0 \\ k_q < -\frac{f_q}{F(\overline{i_q})},\ if\ \overline{i_q} < 0 \end{cases} \rightarrow k_q < -|\frac{f_q}{F(\overline{i_q})}| \tag{26}$$

Overall, the stability condition of the system can be summarized as:

$$k_d, k_q < \min(-|\frac{f_d}{F(\overline{i_d})}|, -|\frac{f_q}{F(\overline{i_q})}|) \tag{27}$$

Unlike the signum function of which the output is either −1 or 1, the output of the hyperbolic function ranges from −1 and 1. When the $\overline{i_d}$ and $\overline{i_q}$ approaches zero, the magnitudes of k_d and k_q should be extremely large according to (27). In theory, there is not one constant for k_d and k_q satisfying (27) constantly. To solve this problem, considering that the current estimation errors $\overline{i_d}$ and $\overline{i_q}$ physically represent how close the system statuses are to the sliding surfaces, we can manually design their tolerance values. In this paper, the smallest allowable estimation error is denoted as λ, that is:

$$\lambda = \min(|\overline{i_d}|, |\overline{i_q}|) \tag{28}$$

Further, the minimum value of the switching function can be calculated as:

$$\min F = \frac{e^{m\cdot\lambda} - e^{-m\cdot\lambda}}{e^{m\cdot\lambda} + e^{-m\cdot\lambda}} \tag{29}$$

So far, the observer gains k_d and k_q can be designed as:

$$k_d = k_q = \min(-|\frac{f_d}{\min F}|, -|\frac{f_q}{\min F}|) \tag{30}$$

In practice, as long as the system is able to become stable, the disturbances are finite. Hence, there must exist a constant (for k_d and k_q) that makes the observers stable. During the control process, although $\overline{i_d}$ and $\overline{i_q}$ are possible to be less than the pre-set value of λ, the proposed SMDOs can re-converge once their values increase [42].

3.3. Disturbance Impact Elimination-Based Voltage Compensation for MPC

After observing the real-time disturbances by using the SMDOs, the most direct way to eliminate them is to control them by maintaining them at zero. In this paper, automatic disturbance regulators (ADR) are adopted to achieve this goal. The inputs of the ADRs are the errors between zero and the detected disturbances, while the outputs are the voltage compensations c_d and c_q. As for the proposed disturbance impact elimination method, two aspects need to be addressed. Firstly, the ADRs are PI controllers and their transfer functions are similar to those in (7), and the parameters of the controllers can be tuned by using the trial-and-error strategy. Secondly, the voltage compensations generated by the ADRs are used to compensate the control voltages e_d and e_q of the MPC controller. In detail, after c_d and c_q are taken into account, the PPM used for prediction should be modified as:

$$\begin{cases} i_d(k+1) = \frac{(L_f+L_l-R_fT_s-R_lT_s)}{L_f+L_l} i_d(k) - \frac{T_s}{L_f+L_l} u_d(k) + \frac{T_s}{L_f+L_l}(e_d(k)+c_d(k)) \\ i_q(k+1) = \frac{(L_f+L_l-R_fT_s-R_lT_s)}{L_f+L_l} i_q(k) - \frac{T_s}{L_f+L_l} u_q(k) + \frac{T_s}{L_f+L_l}(e_q(k)+c_q(k)) \end{cases} \tag{31}$$

In (31), the resistances and inductances are also designed as the measured ones (R_{f_mea}, L_{f_mea}, R_{l_mea}, and L_{l_mea}). However, being different from the traditional MPC strategy, the proposed method is able to eliminate the impacts of the parameter mismatch.

Based on the above analysis and Figure 2, which clearly shows how the SMDO and MPC controller are integrated into the control system, at the kth period, the implementations of DCL strategy with a robust MPC control incorporated for the microgrid can be summarized as follows:

(a) Measurement: The three-phase currents $i_{a,b,c}$ and voltages $u_{a,b,c}$ are detected by using current and voltage sensors.
(b) Frequency, phase angle, and voltage extraction: The measured phase voltages are used to extract the frequency f, phase angle φ, and d, q-axis voltages $u_{d,q}$ by using the PLL.
(c) For the abc/dq transformation of current: The measured phase currents are transformed to the d, q-axis currents $i_{d,q}$ relying on the detected phase angle φ.
(d) Calculation of magnitude of voltage: Use (6) to calculate the magnitude of voltage $\hat{u}$
(e) The d, q-axis disturbance observation: The proposed SMDOs are employed to detect the real-time disturbances $f_{d,q}$.
(f) Calculation of voltage compensations: Use the ADRs to calculate the voltage compensations $c_{d,q}$ according to $f_{d,q}$.
(g) Prediction: Use the modified PPM (31) to estimate the future current states for each candidate voltage vector $\mathbf{u}_{000}$, $\mathbf{u}_{100}$, $\mathbf{u}_{110}$, $\mathbf{u}_{010}$, $\mathbf{u}_{011}$, $\mathbf{u}_{001}$, and $\mathbf{u}_{101}$, and select the voltage vectors that can make the future current satisfy the constraint condition (13).
(h) Evaluation: Substitute the seven predicted values satisfying (13) into the cost function, selecting the optimal control voltage and the corresponding switching states.
(i) Actuation: Apply the optimum switching states to the DC/AC inverter.

Overall, the proposed SMDO is able to detect the disturbances caused by the parameter mismatch, so theoretically there are no certain bounds of uncertainties in resistances and inductances. However, considering that when the proposed SMDO-based MPC algorithms are executed, the normal current, voltage, and frequency need to be measured in advance; the proposed method is available as long as the uncertainties in the resistances and inductances do not influence the system stability.

4. Simulation Results

The simulation was conducted on a microgrid system of which parameters are given in Table 2 to verify the effectiveness of the proposed errorless-control-targeted technique based on the robust MPC controller. The simulation model was established in MATLAB/Simulink 2018b (The Mathworks inc., Natick, MA, USA), which is shown in Figure 5. In terms of the simulation setups, four aspects needed to be addressed: (1) The control step was set as $7e^{-6}$ s (sampling period); (2) Resistances were used to simulate the AC loads, and they could be changed in simulation; (3) Instead of establishing one model for every EV in the microgrid discretely, an integrated DC source was adopted as the EV energy storage system, of which maximum allowable output DC current was 10 A. Thus, the maximum power provided by EVs was 8 kVA. Concerning the four uncertainties of the EV energy storage systems mentioned in Section I, as for one fixed microgrid system, the number of EVs that could feed the microgrid was the most crucial factor influencing the power level of the system. Hence, in the simulation, the uncertainties of EV energy storage systems were simulated by controlling the maximum allowable current (e.g., when the allowable current is 0, it represents that there are no vehicles that can feed the microgrid); (4) The DGS was equivalent to a DC source as well; (5) To better illustrate the effectiveness of the proposed strategy, the simulation results of the traditional VSG control method mentioned in [43] (see Figure 1) and the improved VSG with adaptive droop coefficients in [24] are presented for comparison. It deserves to be mentioned that in addition to the parameters in Table 2, the inertia and damping factors used for the VSG strategies were J = 0.5 and D = 0.5, respectively, but in terms of the improved VSG, the values of the initial droop coefficients double those in Table 2. Besides, the main parameters of the proposed DCL control strategy are as follows: $k_{F\_p} = 10$, $k_{F\_i} = 100$, $k_{V\_p} = 20$, $k_{V\_i} = 250$, $k_d = k_q = -3000$.

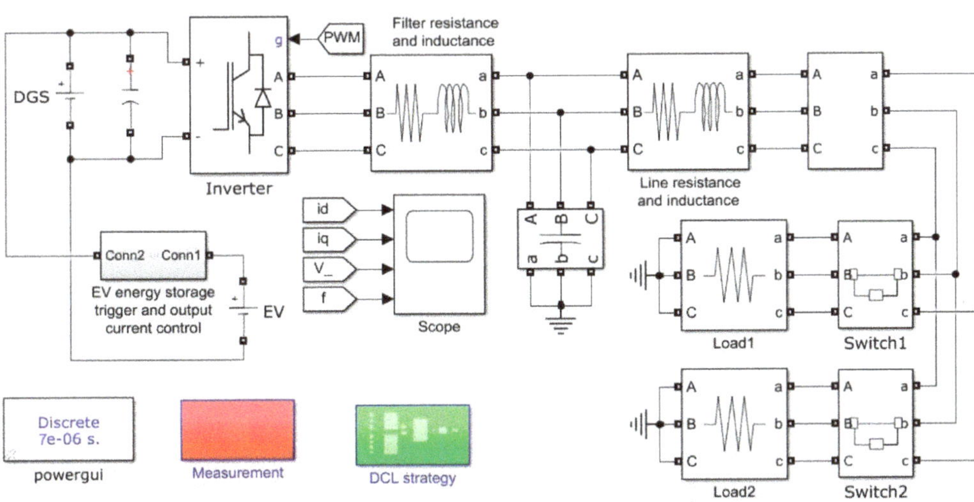

Figure 5. Simulation setup for the proposed errorless-control-targeted technique based on the robust MPC.

For the sake of comprehensiveness, the simulation can be divided into three parts. Firstly, assuming that the EV energy storage system can provide the maximum power and the parameter mismatch phenomenon does not occur, the control performance characteristics of the proposed DCL strategy, the traditional VSG method, and the improved VSG method under different loads were compared, proving that the proposed method is effective. Secondly, to verify that the proposed method is able to maintain the voltage and frequency at the desired levels with EV uncertainties, assuming that the microgrid operates under the heavy load, the control performance of the proposed DCL strategy (no parameter mismatch) when the maximum allowable output DC current changes, is presented. Thirdly, to illustrate that it is necessary to develop the robust MPC controller against parameter mismatch and the proposed SMDO-based control method is effective, the comparative results of the traditional fixed-parameter based FCS-MPC controller and the proposed disturbance compensation method (the EV energy storage system can provide the maximum power) are given.

4.1. Validation of Proposed DCL Method without Parameter Mismatch and EV Uncertainties Considered

In this part, between 0 and 0.5 s, Load 1 in Figure 5 is set as 14.44 Ω (rated load) while Load 2 (72.2 Ω) does not need to be connected with the microgrid. At 0.5 s, Load 2, together with Load 1, is suddenly imposed on the system. Figures 6–8 illustrates the control performance of the traditional VSG (frequency, active power, and reactive power references are set as the rated values), improved VSG (frequency, active power, and reactive power references are set as the rated values), and proposed DCL strategies (frequency and voltage references are set as the rated values), respectively. As for the traditional VSG strategy, under the rated load, the frequency and voltage can maintain the rated levels. When the load becomes heavy at 0.5 s, the frequency decreases to 49.5 Hz, and after 0.3 s (settling time t_{se} = 0.3 s), the frequency becomes 49.8 Hz, which is not the rated level. Under the heavy load condition, the voltage magnitude sees no change after short-term fluctuations. In terms of the d-axis currents, they are 21.5 A and 26 A under the rated load and heavy load, respectively, while the q-axis current keeps around 0.2 A during the test. In terms of the improved VSG method, the static performance characteristics of the frequency, voltage, and current under the rated load condition are very similar to those in Figure 6, though the THD of the phase current is slightly higher. However, the dynamic performance (at around 0.5 s) showed more obvious differences. In detail, the settling time of the improved VSG becomes 0.09 s shorter, and the lowest frequency is 49.4 s. Moreover, under the heavy load condition, the frequency stabilizes at 49.9 Hz, which is slightly larger than that in Figure 6; these happen because the droop coefficients for the two methods are different. In Figure 8, the performance characteristics of the proposed DCL strategy are totally different from those in Figures 6 and 7. Specifically, the frequency levels off at 50 Hz even when the load is suddenly imposed on the system, however, the voltage is a little smaller than the reference value. Interestingly, the d-axis currents are smaller than those of the traditional VSG strategy in both the rated load and heavy load conditions, which are 20.5 A and 24 A, respectively. As for the q-axis current, it measures around zero, which is also smaller than that in Figures 6 and 7. Interestingly, in Figure 8, there are small oscillations in the currents, and the total harmonic distortion (THD) is much larger than those in Figures 6 and 7, regardless of the rated and heavy load conditions. One main reason why this happens is the modulation method of the VSG strategies (SPWM) is inclined to reduce the harmonics. Another interesting phenomenon is that in Figure 8, under the heavy load, the current oscillations are inclined to decline, which deserves more attention in a future study.

4.2. Validation of Proposed DCL Method with EV Uncertainties Considered

To verify that the proposed DCL method is effective when the EVs that can provide energy to the microgrid change, at 0.5 s, the maximum output current of the EV energy storage system is set as 0.5 A (0.4 kVA) from 10 A (8 kVA). This can simulate two working conditions, that is, when the maximum output current is 10 A, all EVs and charging piles can feed the microgrid, while only a small number of EVs can feed the microgrid when the maximum output current is set as 0.5 A. During the test, Load 1 and Load 2 are simultaneously connected with the microgrid (heavy load). Figure 9 depicts the control performance of the proposed DCL strategy with the EV uncertainties considered. It can be seen that firstly, before 0.5 s, the output DC current of the EV energy storage system is 1.5 A, but after 0.5 s, it becomes 0.5 A as designed. Secondly, the voltage and frequency do not see obvious change before and after the number of EVs is simulated to decline. Thirdly, the d-axis current decreases to 22.5 A from 24 A in Figure 8 as the total power of the microgrid drops, which is reasonable. These illustrate that the proposed DCL method can maintain the frequency and voltage at the original values, even when the output power of the EVs lowers. It needs to be mentioned that the harmonics (THD) in the currents are similar to those (heavy load condition) in Figure 8. These prove that the proposed method is effective.

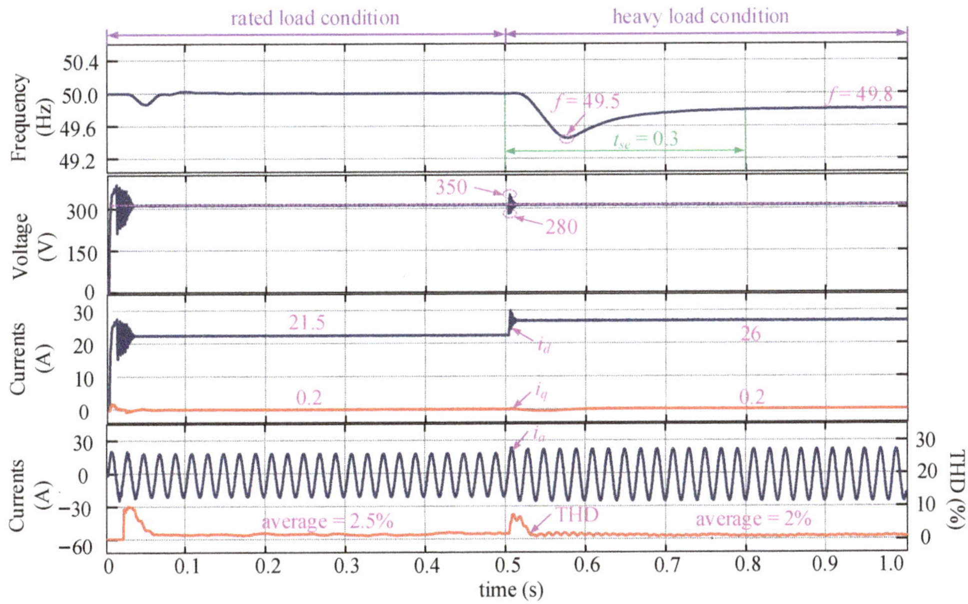

Figure 6. Control performance of the traditional VSG method without parameter mismatch and EV uncertainties considered.

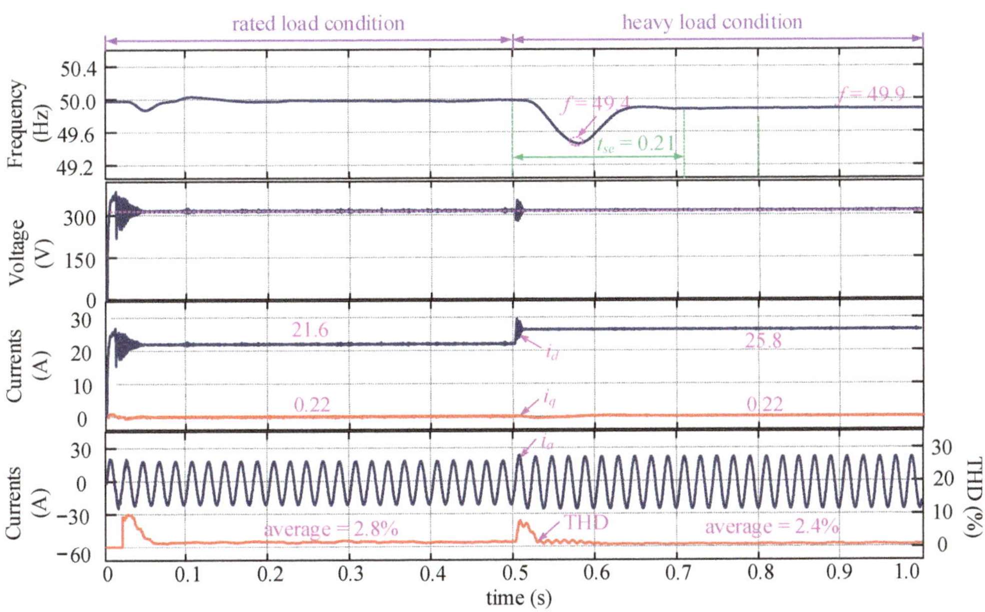

Figure 7. Control performance of the improved VSG method without parameter mismatch and EV uncertainties considered.

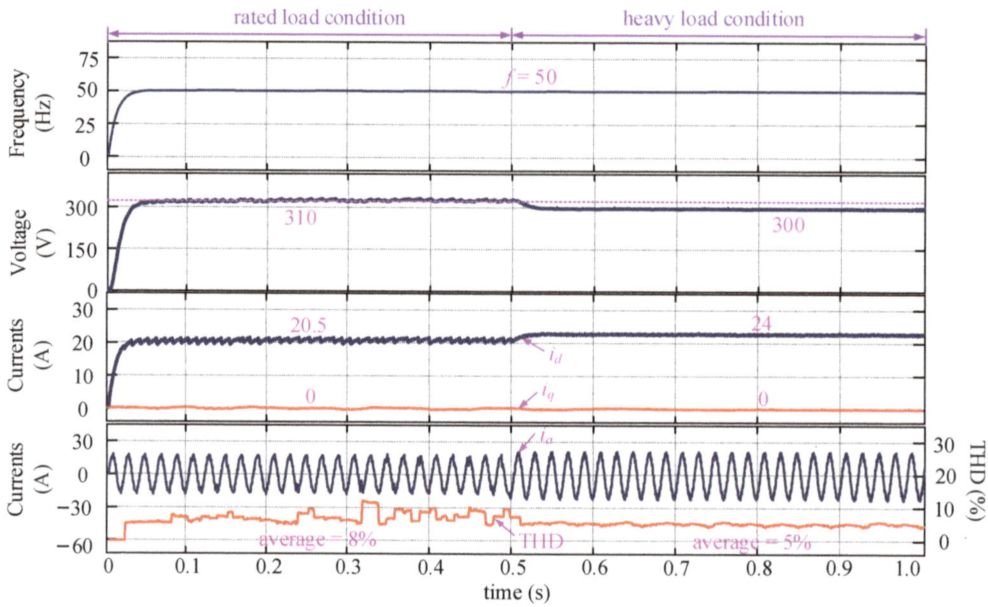

Figure 8. Control performance of the proposed DCL method without parameter mismatch and EV uncertainties considered.

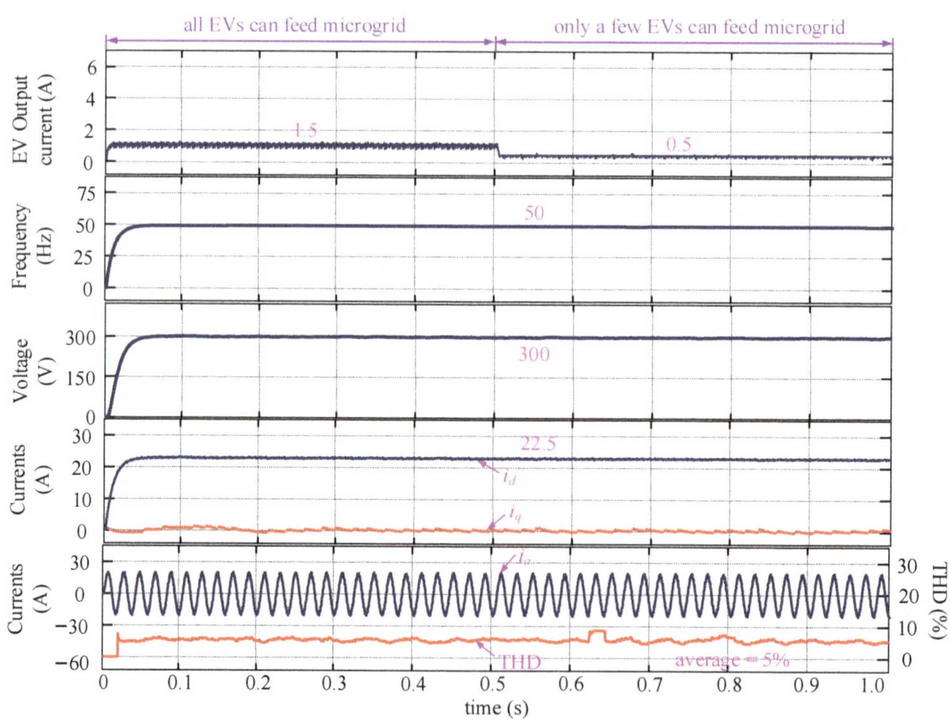

Figure 9. Control performance of the proposed DCL method with EV uncertainties considered.

4.3. Validation of Proposed DCL Method with Parameter Mismatch Considered

4.3.1. $L_{f\_mea} = 0.2L_f$, $L_{l\_mea} = 0.2L_l$, $R_{f\_mea} = 2R_f$, $R_{l\_mea} = 2R_l$

The simulation setups (load) are the same as those in Section 4.1, and besides, assume that the measured inductances used for the FCS-MPC controllers are 20% of the real values while the measured resistances used for control are twice as large as the real values. That is, $L_{f\_mea} = 0.2L_f = 0.036$ H, $L_{l\_mea} = 0.2L_l = 0.02$ mH, $R_{f\_mea} = 2R_f = 0.32\ \Omega$, and $R_{l\_mea} = 2R_l = 2.4\ \Omega$. Figure 10 shows the control performance of the proposed DCL method based on the traditional FCS-MPC controller. It can be seen that, when parameter mismatch occurs, the control performance degrades significantly in comparison with that in Figure 8. In detail, the frequency of the system cannot level off at 50 Hz, leading to the THD of the AC current continually increasing. Additionally, the currents become larger when the parameters witness variations. These represent that the system cannot work under the optimal states, and thus, it is necessary to develop an effective method to deal with the parameter uncertainty problem. Figure 11 presents the system performance when the proposed SMDO-based MPC controller is used for achieving the DCL strategy. Compared to the results in Figure 10, the frequency can stabilize at the desired position, and the AC current (THD) becomes normal. Additionally, the currents in the system are more similar to those in Figure 8, proving that the proposed SMDO-based disturbance detection and compensation strategy are effective. Moreover, Figure 12 compares the current prediction errors before and after the proposed SMDO-based MPC method is employed for control, which is calculated using (16). Between 0 and 0.5 s, the SMDO does not work, while after 0.5 s, the SMDO works. It can be seen that the prediction errors are slightly changed. This happens because the purpose of the proposed control strategy is to eliminate the impacts of the disturbances rather than directly reject the disturbances, which can also be explicitly derived from Section 3.

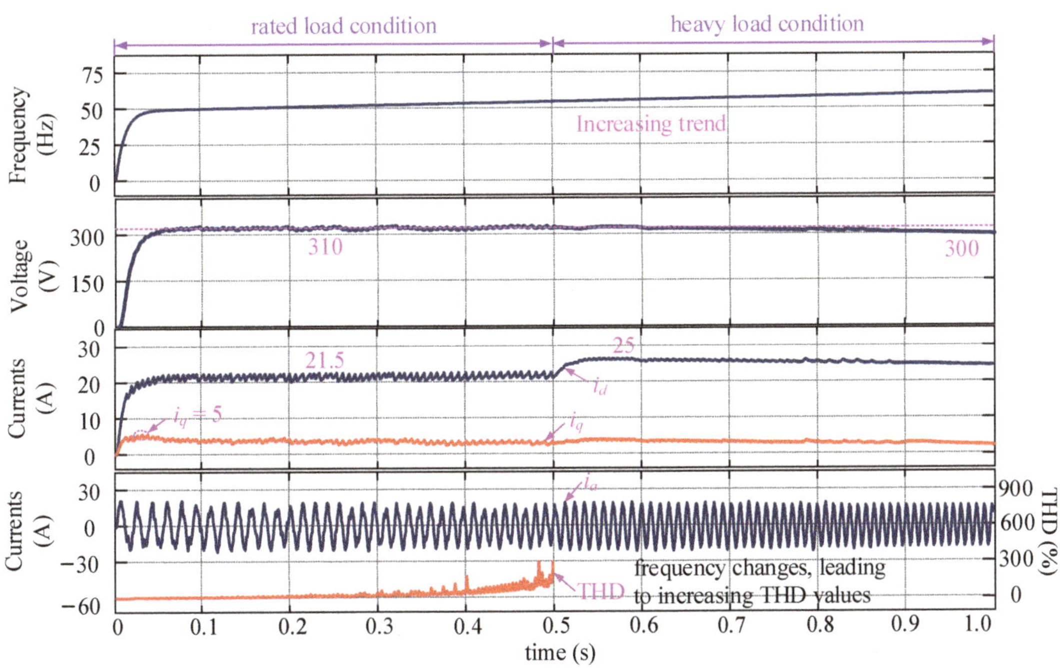

Figure 10. Control performance of the proposed DCL method with parameter mismatch considered (the SMDO does not work, $L_{f\_mea} = 0.2L_f$, $L_{l\_mea} = 0.2L_l$, $R_{f\_mea} = 2R_f$, $R_{l\_mea} = 2R_l$).

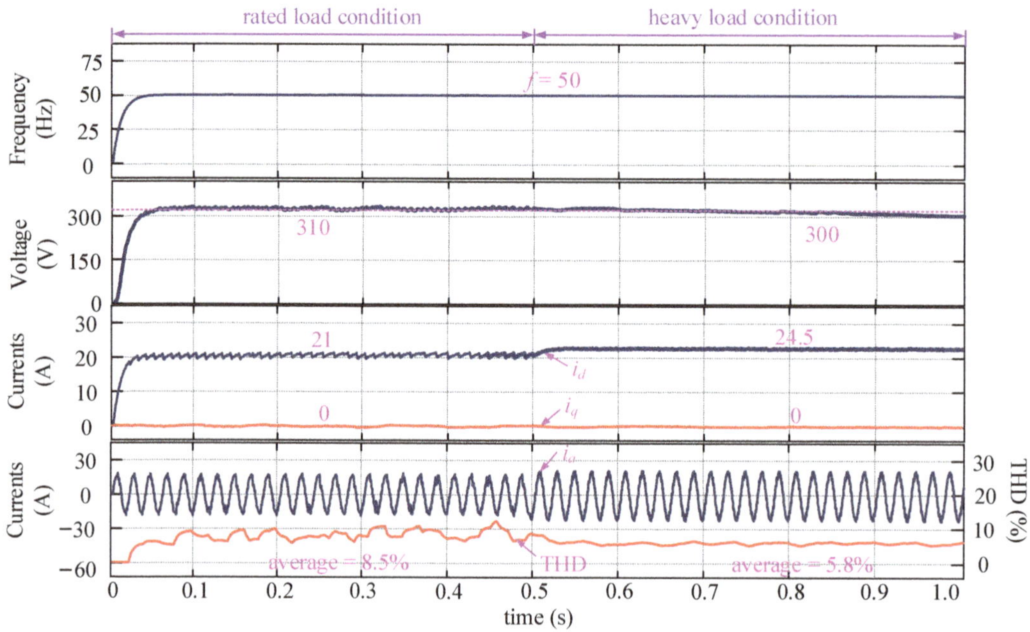

Figure 11. Control performance of the proposed DCL method with parameter mismatch considered (the SMDO works, $L_{f\_mea} = 0.2L_f$, $L_{l\_mea} = 0.2L_l$, $R_{f\_mea} = 2R_f$, $R_{l\_mea} = 2R_l$).

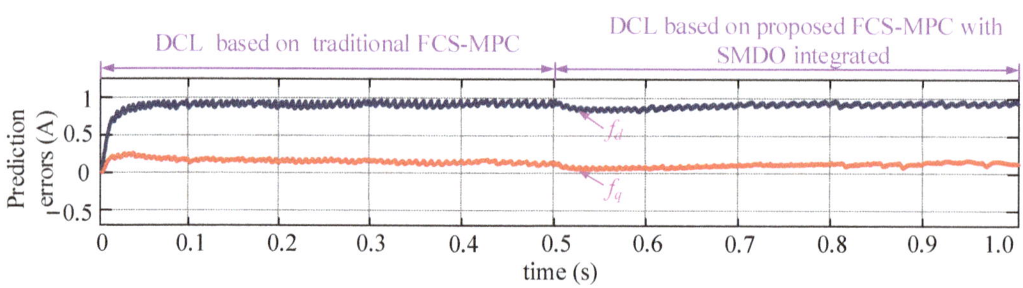

Figure 12. Comparative results of the current prediction errors before and after the proposed SMDO-based MPC method is employed ($L_{f\_mea} = 0.2L_f$, $L_{l\_mea} = 0.2L_l$, $R_{f\_mea} = 2R_f$, $R_{l\_mea} = 2R_l$).

4.3.2. $L_{f\_mea} = 2L_f$, $L_{l\_mea} = 1.5L_l$, $R_{f\_mea} = 0.5R_f$, $R_{l\_mea} = 0.3R_1$

Assume that the measured filter inductance, line inductance, filter resistance, and line resistance used for the FCS-MPC controller are 4, 3, 0.5, and 0.3 times as large as the real values, respectively. Namely, $L_{f\_mea} = 2L_f = 0.36$ H, $L_{l\_mea} = 3L_l = 0.3$ mH, $R_{f\_mea} = 0.5R_f = 0.08\ \Omega$, and $R_{l\_mea} = 0.3R_l = 0.36\ \Omega$. Figures 13 and 14 show the performance of the DCL strategy based on the traditional FCS-MPC and the DCL strategy based on the proposed FCS-MPC with SMDO integrated. Figure 15 shows the current prediction errors. Being different from condition a), the control performance in Figures 13 and 14 is very similar. This happens because, when in this parameter mismatch condition, the prediction errors are pretty small (see Figure 15). It deserves to be mentioned that even the prediction errors are nearly zero, and the proposed technique is able to achieve voltage and frequency-errorless control without sacrificing the control performance.

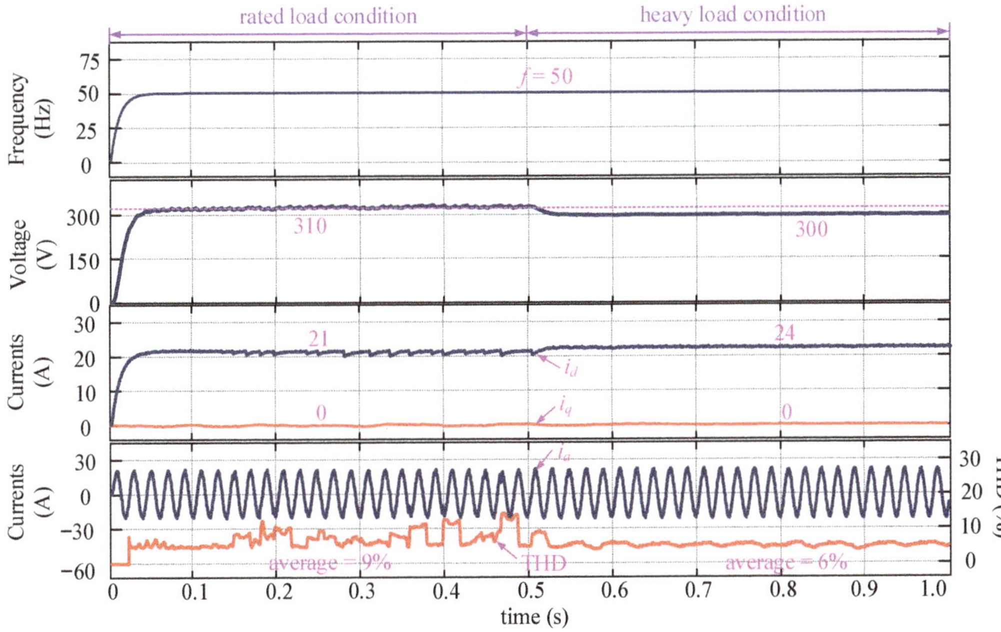

Figure 13. Control performance of the proposed DCL method with parameter mismatch considered (the SMDO does not work, $L_{f\_mea} = 4L_f$, $L_{l\_mea} = 3L_l$, $R_{f\_mea} = 0.5R_f$, $R_{l\_mea} = 0.3R_l$).

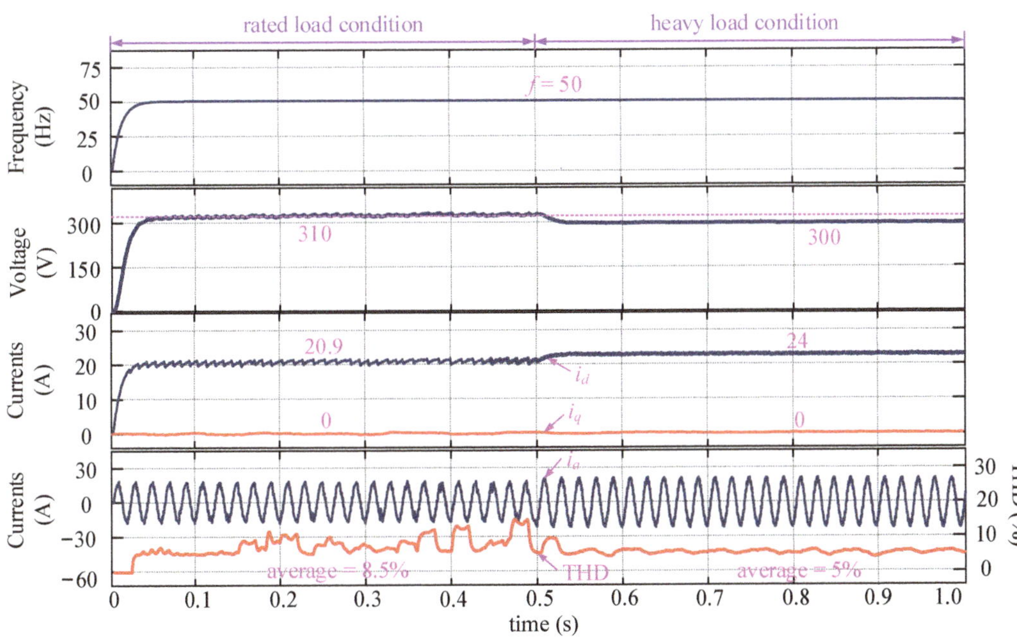

Figure 14. Control performance of the proposed DCL method with parameter mismatch considered (the SMDO works, $L_{f\_mea} = 4L_f$, $L_{l\_mea} = 3L_l$, $R_{f\_mea} = 0.5R_f$, $R_{l\_mea} = 0.3R_l$).

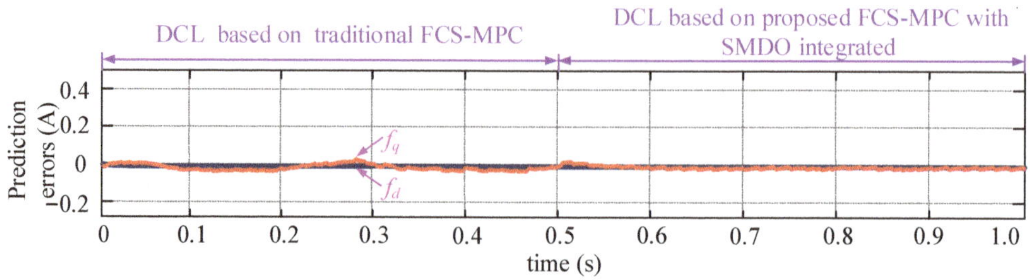

Figure 15. Comparative results of the current prediction errors before and after the proposed SMDO-based MPC method are employed ($L_{f\_mea} = 4L_f$, $L_{l\_mea} = 3L_l$, $R_{f\_mea} = 0.5R_f$, $R_{l\_mea} = 0.3R_l$).

5. Conclusions

This paper proposes a robust errorless-control-targeted technique based on FCS-MPC for a microgrid with uncertain electric vehicle energy storage systems, which works under the island mode. The main contributions and novelties are as follows:

(1) A novel DCL control method is proposed, with each part of it detailed. In comparison with the traditional control strategies, such as droop control and VSG control, the proposed one aims to directly regulate the frequency, voltage, and currents rather than the power of the system. As for this strategy, droop coefficients are no longer needed, solving the problem that it is difficult to determine the droop coefficients of the microgrid with uncertain EV energy storage systems.

(2) The impacts of the system parameter mismatch on the prediction accuracy of the FCS-MPC are analyzed explicitly, posing the necessity of developing the robust MPC controller.

(3) SMDOs based on the hyperbolic function are developed to detect the d, q-axis disturbances, with their stability discussed innovatively. The detected disturbances are controlled to maintain at zero by using ADRs, generating voltage compensations used for modifying the PPM of the MPC controller, thus achieving the goal of disturbance impact elimination. Simulation results prove that the proposed strategies are effective.

As far as the proposed strategies in this research are concerned, there are several interesting problems that deserve future study. Firstly, in addition to the FCS-MPC controller, many advanced controllers, such as the variable structure control and future search algorithm, etc., can be adopted to achieve the DCL method, which might also be efficient. Secondly, although the stability of the proposed SMDO is analyzed, the stability of the whole closed-loop control strategy with the FCS-MPC and SMDO integrated is not addressed in this paper. Thirdly, a trial-and-error strategy is used to tune the parameters of the PI controllers used in the DCL control topology, which needs to be replaced by a theoretical analysis strategy. Fourthly, the simulation results show that there are small oscillations that are related to the load conditions in the currents of the proposed method. This is an interesting phenomenon, and it might be caused by more different reasons that need to be urgently explored.

Author Contributions: Supervision, Y.H., and Y.N.; writing—original draft, Y.L. All authors have read and agreed to the published version of the manuscript.

Funding: This research received no external funding.

Institutional Review Board Statement: Not applicable.

Informed Consent Statement: Not applicable.

Conflicts of Interest: The authors declare no conflict of interest.

Nomenclature

n	Number of charging piles in one microgrid
c	Ratio of EVs working at V2G mode to the maximum number of EVs
t_{su}, t	Period of EVs working at V2G mode, time
P_{su}	Supporting power from EVs
$u_{a,b,c}$, $i_{a,b,c}$	Three-phase voltages and currents
U_{dc}	Bus voltage
$T_1, T_2, \ldots, T_6$	Transistors
$D_1, D_2, \ldots, D_6$	Diodes
R_f, L_f, C_f, R_l, L_l	Filter and line resistances, inductances and capacitance
R_{f_mea}, L_{f_mea}, R_{l_mea}, L_{l_mea}	Measured filter and line resistances and inductances
ΔR_f, ΔL_f, ΔR_l, ΔL_l	Deviations of filter and line resistances and inductances
e_a, e_b, e_c	Output voltage of the inverter
e_d, e_q	*d*, *q*-axis control voltages
u_d, u_q	*d*, *q*-axis voltages
i_d, i_q	*d*, *q*-axis currents
$i_d{}^*$, $i_q{}^*$	*d*, *q*-axis current references
u^*	Voltage reference
φ	Phase angle
f, $\hat{u}$	Frequency and voltage magnitude
P_t, P_e, Q	Mechanical power and active power, reactive power in VSG control
ω, ω_N, P_N, Q_N, V_N, f_N	Angular frequency, nominal speed, power, voltage and frequency
k_w, k_v	Droop coefficients for frequency and voltage in VSG control
θ, V	Angle and voltage amplitude for control
J	Virtual inertia for VSG control
D	Virtual damping for VSG control
f_d, f_q	*d*, *q*-axis disturbances
c_d, c_q	*d*, *q*-axis voltage compensations
k_{F_p}, k_{F_i}	Proportional and integral factors for AFR
k_{V_p}, k_{V_i}	Proportional and integral factors for AVR
k_d, k_q	Gain factors of sliding mode observers
$\hat{i}_d$, $\hat{i}_q$	Estimated currents of SMDOs
$\bar{i}_d$, $\bar{i}_q$	Errors between estimated currents and real currents in SMDOs
T_s	Control period
m	Boundary-layer constant
S	Sliding mode surfaces
L_y	Lyapunov function
λ	The smallest allowable estimation error
$i_{\max}$	The maximum allowable current of the system

References

1. Wang, T.; O'Neill, D.; Kamath, H. Dynamic Control and Optimization of Distributed Energy Resources in a Microgrid. *IEEE Trans. Smart Grid* **2015**, *6*, 2884–2894. [CrossRef]
2. Benhalima, S.; Miloud, R.; Chandra, A. Real-Time Implementation of Robust Control Strategies Based on Sliding Mode Control for Standalone Microgrids Supplying Non-Linear Loads. *Energies* **2018**, *11*, 2590. [CrossRef]
3. Mueller, J.A.; Kimball, J.W. An Efficient Method of Determining Operating Points of Droop-Controlled Microgrids. *IEEE Trans. Energy Convers.* **2017**, *32*, 1432–1446. [CrossRef]
4. Camacho, A.; Castilla, M.; Canziani, F.; Moreira, C.; Coelho, P.; Gomes, M.; Mercado, P.E. Performance Comparison of Grid-Faulty Control Schemes for Inverter-Based Industrial Microgrids. *Energies* **2017**, *10*, 2096. [CrossRef]
5. Liu, C.; Chau, K.T.; Wu, D.; Gao, S. Opportunities and Challenges of Vehicle-to-Home, Vehicle-to-Vehicle, and Vehicle-to-Grid Technologies. *Proc. IEEE* **2013**, *101*, 2409–2427. [CrossRef]
6. Ouramdane, O.; Elbouchikhi, E.; Amirat, Y.; Sedgh Gooya, E. Optimal Sizing and Energy Management of Microgrids with Vehicle-to-Grid Technology: A Critical Review and Future Trends. *Energies* **2021**, *14*, 4166. [CrossRef]
7. Shuai, Z.; Mo, S.; Wang, J.; Shen, Z.J.; Tian, W.; Feng, Y. Droop control method for load share and voltage regulation in high-voltage microgrids. *J. Mod. Power Syst. Clean Energy* **2016**, *4*, 76–86. [CrossRef]
8. Beck, H.P.; Hesse, R. Virtual Synchronous Machine. In Proceedings of the International Conference on Electrical Power Quality & Utilisation, Barcelona, Spain, 9–11 October 2007; pp. 1–6.

9. Zhao, Y.; Guo, L. Dynamical Simulation of Laboratory MicroGrid. In Proceedings of the IEEE 2009 Asia-Pacific Power and Energy Engineering Conference (APPEEC), Wuhan, China, 28–30 March 2009; pp. 1–5.
10. Deng, Y.; Tao, Y.; Chen, G.; Li, G.; He, X. Enhanced Power Flow Control for Grid-Connected Droop-Controlled Inverters with Improved Stability. *IEEE Trans. Ind. Electron.* **2017**, *64*, 5919–5929. [CrossRef]
11. Ashabani, M.; Ibrahim-Mohamed, Y.; Mirsalim, M.; Aghashabani, M. Multivariable Droop Control of Synchronous Current Converters in Weak Grids/Microgrids with Decoupled dq-Axes Currents. *IEEE Trans. Smart Grid* **2015**, *6*, 1610–1620. [CrossRef]
12. Gorijeevaram Reddy, P.K.; Dasarathan, S.; Krishnasamy, V. Investigation of Adaptive Droop Control Applied to Low-Voltage DC Microgrid. *Energies* **2021**, *14*, 5356. [CrossRef]
13. Li, G.; Ma, F.; Luo, A.; He, Z.; Wu, W.; Wei, X.; Zhu, Z.; Guo, J. Virtual impedance-based virtual synchronous generator control for grid-connected inverter under the weak grid situations. *IET Power Electron.* **2018**, *11*, 2125–2132. [CrossRef]
14. Wu, W.; Chen, Y.; Zhou, L.; Luo, A.; Zhou, X.; He, Z.; Yang, L.; Xie, Z.; Liu, J.; Zhang, M. Sequence Impedance Modeling and Stability Comparative Analysis of Voltage-Controlled VSGs and Current-Controlled VSGs. *IEEE Trans. Ind. Electron.* **2018**, *66*, 6460–6472. [CrossRef]
15. Zhong, Q.; Weiss, G. Synchronverters: Inverters That Mimic Synchronous Generators. *IEEE Trans. Ind. Electron.* **2011**, *58*, 1259–1267. [CrossRef]
16. Jin, N.; Pan, C.; Li, Y.; Hu, S.; Fang, J. Model Predictive Control for Virtual Synchronous Generator with Improved Vector Selection and Reconstructed Current. *Energies* **2020**, *13*, 5435. [CrossRef]
17. Magdy, G.; Bakeer, A.; Nour, M.; Petlenkov, E. A New Virtual Synchronous Generator Design Based on the SMES System for Frequency Stability of Low-Inertia Power Grids. *Energies* **2020**, *13*, 5641. [CrossRef]
18. Mo, O.; D'Arco, S.; Suul, J.A. Evaluation of Virtual Synchronous Machines with Dynamic or Quasi-Stationary Machine Models. *IEEE Tran. Ind. Electron.* **2017**, *64*, 5952–5962. [CrossRef]
19. D'Arco, S.; Suul, J.A. Equivalence of Virtual Synchronous Machines and Frequency-Droops for Converter-Based MicroGrids. *IEEE Trans. Smart Grid* **2014**, *5*, 394–395. [CrossRef]
20. Li, B.; Li, Q.; Wang, Y.; Wen, W.; Li, B.; Xu, L. A Novel Method to Determine Droop Coefficients of DC Voltage Control for VSC-MTDC System. *IEEE Trans. Power Deliv.* **2020**, *35*, 2196–2211. [CrossRef]
21. Pavan Kumar, Y.V.; Bhimasingu, R. Modern Control Methods for Adaptive Droop Coefficients' Design. In *Microgrid: Operation, Control, Monitoring and Protection*; Springer: Singapore, 2020; pp. 111–148.
22. Saleh-Ahmadi, A.; Moattari, M.; Gahedi, A.; Pouresmaeil, E. Droop Method Development for Microgrids Control Considering Higher Order Sliding Mode Control Approach and Feeder Impedance Variation. *Appl. Sci.* **2021**, *11*, 967. [CrossRef]
23. Yogarathinam, A.; Chaudhuri, N.R. Stability-Constrained Adaptive Droop for Power Sharing in AC-MTDC Grids. *IEEE Trans. Power Syst.* **2019**, *34*, 1955–1965. [CrossRef]
24. Wang, R.; Chen, L.; Zheng, T.; Mei, S. VSG-based adaptive droop control for frequency and active power regulation in the MTDC system. *CSEE J. Power Energy Syst.* **2017**, *3*, 260–268. [CrossRef]
25. Baudoin, S.; Vechiu, I.; Camblong, H. A review of voltage and frequency control strategies for islanded microgrid. In Proceedings of the 2012 16th International Conference on System Theory, Control and Computing (ICSTCC), Sinaia, Romania, 12–14 October 2012; pp. 1–5.
26. Kos, T.; Huba, M.; Vrančić, D. Parametric and Nonparametric PI Controller Tuning Method for Integrating Processes Based on Magnitude Optimum. *Appl. Sci.* **2020**, *10*, 1443. [CrossRef]
27. Leal, I.S.; Abeykoon, C.; Perera, Y.S. Design, Simulation, Analysis and Optimization of PID and Fuzzy Based Control Systems for a Quadcopter. *Electronics* **2021**, *10*, 2218. [CrossRef]
28. Simmini, F.; Caldognetto, T.; Bruschetta, M.; Mion, E.; Carli, R. Model Predictive Control for Efficient Management of Energy Resources in Smart Buildings. *Energies* **2021**, *14*, 5592. [CrossRef]
29. Elsisi, M.; Abeoelela, M.; Soliman, M.; Mansor, W. Design of Optimal Model Predictive Controller for LFC of Nonlinear Multi-area Power System with Energy Storage Devices. *Electr. Power Compon. Syst.* **2018**, *46*, 1300–1311. [CrossRef]
30. Valverde, L.; Bordons, C.; Rosa, F. Power Management using Model Predictive Control in a Hydrogen-based Microgrid. In Proceedings of the IECON 2012—38th Annual Conference on IEEE Industrial Electronics Society, Montreal, QC, Canada, 25–28 October 2012; pp. 5669–5676.
31. Elsisi, M. New Variable Structure Control based on Different Meta-Heuristics Algorithms for Frequency Regulation Considering Nonlinearities Effects. *Intern. Trans. Electr. Energy Syst.* **2020**, *30*, 12428. [CrossRef]
32. Elsisi, M.; Soliman, M.; Aboelela, M.A.S.; Mansour, W. GSA-Based Design of Dual Proportional Integral Load Frequency Controllers for Nonlinear Hydrothermal Power System. *World Acad. Sci. Eng. Technol.* **2015**, *9*, 1–7.
33. Elsisi, M. Future Search Algorithm for Optimization. *Evol. Intell.* **2019**, *12*, 21–31. [CrossRef]
34. Han, Y.; Gong, C.; Yan, L.; Wen, H.; Wang, Y.; Shen, K. Multiobjective Finite Control Set Model Predictive Control Using Novel Delay Compensation Technique for PMSM. *IEEE Trans. Power Electron.* **2020**, *35*, 11193–11204. [CrossRef]
35. Valibeygi, A.; Habib, A.H.; de Callafon, R.A. Robust Power Scheduling for Microgrids with Uncertainty in Renewable Energy Generation. In Proceedings of the 2019 IEEE Power & Energy Society Innovative Smart Grid Technologies Conference (ISGT), Gramado, Brazil, 18–21 February 2019; pp. 1–5.
36. Davari, M.; Mohamed, Y.A.I. Robust Multi-Objective Control of VSC-Based DC-Voltage Power Port in Hybrid AC/DC Multi-Terminal Micro-Grids. *IEEE Trans. Smart Grid* **2013**, *4*, 1597–1612. [CrossRef]

37. Aryani, D.R.; Song, H. Coordination Control Strategy for AC/DC Hybrid Microgrids in Stand-Alone Mode. *Energies* **2016**, *9*, 469. [CrossRef]
38. Shi, X.; Lan, T.; Hu, J.P. PID controller tuning using optimization based on gradient-immune algorithm. In Proceedings of the 2012 International Symposium on Instrumentation & Measurement, Sensor Network and Automation (IMSNA), Sanya, China, 25–28 August 2012; pp. 173–175.
39. Gao, J.; Gong, C.; Li, W.; Liu, J. Novel Compensation Strategy for Calculation Delay of Finite Control Set Model Predictive Current Control in PMSM. *IEEE Trans. Ind. Electron.* **2019**, *67*, 5816–5819. [CrossRef]
40. Hennane, Y.; Berdai, A.; Martin, J.-P.; Pierfederici, S.; Meibody-Tabar, F. New Decentralized Control of Mesh AC Microgrids: Study, Stability, and Robustness Analysis. *Sustainability* **2021**, *13*, 2243. [CrossRef]
41. Krueger, H.; Cruden, A. Integration of electric vehicle user charging preferences into Vehicle-to-Grid aggregator controls. *Energy Rep.* **2020**, *6*, 86–95. [CrossRef]
42. Gong, C.; Hu, Y.; Gao, J.; Wang, Y.; Yan, L. An Improved Delay-Suppressed Sliding-Mode Observer for Sensorless Vector-Controlled PMSM. *IEEE Trans. Ind. Electron.* **2020**, *67*, 5913–5923. [CrossRef]
43. Liang, Y.; He, Y.; Niu, Y. Microgrid Frequency Fluctuation Attenuation Using Improved Fuzzy Adaptive Damping-Based VSG Considering Dynamics and Allowable Deviation. *Energies* **2020**, *13*, 4885. [CrossRef]

energies

Article

Analysis of Acoustic Characteristics under Battery External Short Circuit Based on Acoustic Emission

Nan Zhou [1,2,3,*], Xiulong Cui [4], Changhao Han [5] and Zhou Yang [1]

1 School of Mechanical Engineering and Automation, Northeastern University, Shenyang 110819, China; yangzhou@mail.neu.edu.cn
2 Key Laboratory of Vibration and Control of Aero-Propulsion System, Northeastern University, Shenyang 110819, China
3 Vehicle Measurement, Control and Safety Key Laboratory of Sichuan Province, Chengdu 610039, China
4 School of Mechatronics Engineering, Harbin Institute of Technology, Harbin 150001, China; cuixiulong8985@163.com
5 School of Mechanical Engineering, Beijing Institute of Technology, Beijing 100081, China; chwyt516080282@163.com
* Correspondence: zhounan@me.neu.edu.cn; Tel.: +86-(24)-8369-1095

Abstract: The safety of power batteries has received more and more attention in promoting electric vehicles. The external short circuit is particularly prominent as an abnormal and harmful event of a battery, and the exploration of in-situ low-cost detection technology for such an event is the starting point of this paper. By building an experimental bench that could detect the external short circuit of the battery and obtain the acoustic, electrode, and temperature responses, the resulting acoustic analysis would establish an internal connection with the electrode and temperature measurement when the external short circuit occurs. The respective acoustic response characteristics of different initial battery states of charge were analyzed by selecting appropriate acoustic characteristic parameters in the time and frequency domains. The acoustic measurement could represent the battery abnormality synchronously like the electrode measurement, and the results of the damage and rearrangement of the internal of the battery are easy to characterize through a moderate amplification of the acoustic response. The different initial state of charge (SOC) state reflects noticeable differences in the acoustic characteristics. Therefore, it is considered that the acoustic emission technology might have potential battery condition assessment capabilities and be a tool for in-situ battery fault diagnosis.

Keywords: lithium-ion battery; acoustic emission; state of charge; external short circuit; acoustic characteristic

Citation: Zhou, N.; Cui, X.; Han, C.; Yang, Z. Analysis of Acoustic Characteristics under Battery External Short Circuit Based on Acoustic Emission. *Energies* **2022**, *15*, 1775. https://doi.org/10.3390/en15051775

Academic Editor: Daniel-Ioan Stroe

Received: 24 January 2022
Accepted: 25 February 2022
Published: 28 February 2022

1. Introduction

With the vigorous development of lithium battery chemistry technology, pure electric vehicles (PEVs) have gradually begun to replace fuel vehicles. Lithium batteries are widely used in electric vehicles due to a good balance of many aspects, such as cost, capacity, safety, and other excellent characteristics. People have anticipated higher requirements for the overall performance indicators of lithium batteries, especially in terms of volume/mass-energy density, power density, life cycle, cost, safety performance, etc.

Battery-embedded sensors are used to collect valuable data to optimize battery operation strategies [1,2]. The data will help the battery management system to acquire the correct status of the battery, optimize the battery performance, and detect abnormal conditions of the battery in time. At the same time, from the perspective of vehicle battery driving range and power requirements, the demand for the number of lithium-ion batteries is increasing, and the performance of these batteries, such as fast charge and discharge capabilities and safety issues [3–5], are also constantly improving. To meet the needs of the market, enhance the performance and safety of batteries, and shorten the development cycle of new battery systems, it is necessary to carry out failure analysis and fault diagnosis of lithium batteries in the course of battery operation. The diagnostic techniques for Li-ion batteries are increasingly emphasizing a balance of economy and reliability due to their need to target larger battery sizes and the cost of batteries deployed in increasingly expensive applications, including electric vehicles.

1.1. Literature Review

Lithium-ion battery cells usually have variable electrical characteristics and are very sensitive to operating and environmental temperatures. With diagnostic tools and sophisticated battery management strategies, battery performance has been improved by confining battery operation within specified limits [6–8]. Those limits balance the battery's performance and alleviate the problem of rapid degradation. The causes of rapid degradation include high-current charging and discharging, low-efficiency temperature environments, and low or high battery charge timing. The state of the lithium battery would affect the life of the battery itself and be directly linked to the performance of the vehicle. The characteristic response of the battery electrode as a real-time diagnostic tool usually involves complex algorithm processing, and the high-precision hardware sensors for the battery management systems (BMS) are expensive. In this case, although the detection of electrode characteristics could identify abnormal data, it is challenging to locate the specific fault location, especially when the batteries are arranged in groups or packages. It can be used as a supplement to the judgment of battery failure by combining visualizing the inside of the battery. In recent years, acoustic-related testing methods have gradually gained more and more attention in lithium battery fault diagnosis due to their versatility, ease of operation, and acceptable accuracy of results.

In situ and ex situ diagnostic techniques have been broadly used in the fabrication, planning, performance assessment, and tracking of electrochemical devices [9,10]. There are four main methods for measuring lithium-ion battery (LIB) failure [11], including destructive testing, imaging methods, methods based on battery swelling, and methods based on acoustic measurements. The experimental study of lithium-ion battery failure is primarily destructive testing, which means that the battery is damaged or severely affected in its operational behavior and safety; it is crucial for the study of the mechanical and electrochemical relationships of the battery, but it rarely applies to troubleshooting.

Imaging methods such as X-ray or computed tomography are used to gain insight into the physical structure of a battery or material and can measure layer thickness through changes in the state of charge (SOC) or state of health (SOH) and identify structural defects in materials and battery components. Historically, X-ray tomography has been commonly used to obtain 3D images of batteries [12–14]. The introduction of lock-in thermal imaging technology also enables internal visible battery inspections. Combined with the thermal imaging technology developed in recent years, a clearer view of the internal failure of the battery would be achieved [15]. However, this usually requires a lot of laboratory equipment, space, and funds, so there is a high threshold for promotion and application. Another limitation is that the components to be tested must absorb X-rays to ensure the feasibility of detection [16,17].

The research conducted by the swelling method is mainly based on the contact measurement for the pouch battery. It is an economical way to reveal most of the phenomena in LIBs and is widely used in laboratory-stage research. The research includes the correlation of SOC and charge-discharge C rate, considering the battery structure, battery SOH estimation, and the effect of temperature on swelling, etc. [18]. There is no report on the measurement of swelling under abuse conditions. Experimental modal analysis (EMA) is a relatively new topic in LIB measurements. No firm conclusions could yet be drawn about SOC, SOH, and other effects on EMA behavior, but can be used as an integrated diagnostic development tool.

A class of acoustic-based methods transmits mechanical waves into the battery cell through an actuator, and the receiver sensor measures the response of the device to an external stimulus, called ultrasonic testing (US) [19]. During scanning, the distance between the transducer and the object's surface to be measured remains constant. The energy of sound waves decays faster in gas than in liquids and solids, which will lead to low accuracy of measurement data. To enhance the transfer energy of sound waves in the sample and avoid external interference of the measured object, both the transducer and the battery are usually invaded into the coupling medium (gel or water) to reduce the effect of reflections at the outer interface on the measurement. The US can detect not only operating values, such as SOC and SOH, but also abusive behaviors, such as high temperature, electroplating, and overcharging, and applies to a wide range of frequencies and various form factor studies [20,21]. As a non-destructive battery diagnostic tool, some literature has explored acoustic methods to characterize Li-ion pouch cells, with sound as an indicator of battery structural health and state-of-charge-related parameters. Time of fly (TOF) offset and total signal amplitude as index parameters for ultrasonic analysis are used to predict the battery state and state of health. Combined with traditional electrode detection data, TOF could present direct structural information of actual batteries in real-time [22,23]. At the same time, many macroscopic-level static strains of batteries have been associated with the operating state in the battery, which shows a strong correlation with the volume change in some pouch batteries. Acoustic measurements have also been used to detect possible macroscopic crack detection, and microscopic measurements have been used to further clarify the cause of the shock response in the electrochemical reaction of the battery [24,25]. However, critical explanations for battery fault diagnosis and acoustic detection feature extraction are still lacking.

Acoustic Emission (AE) is another acoustic-based measurement method that differs from ultrasonic testing in that the object to be measured is not subjected to external mechanical excitation, but the material itself is measured by a connected (surface) sensor collecting fast releasing mechanical stress. The resulting sound waves are widely used to monitor fatigue, crack formation, and mechanical damage in building materials. Also included under external stimuli (e.g., changes in pressure, strain, temperature, or load), local sources within the material trigger energy release in the form of stress waves, which can be detected and converted on the surface of the measured object as electrical signals for subsequent studies [26,27]. Therefore, the use of AE to detect the electrochemical process of the battery is also a further application of its technical characteristics. For LIB batteries, the particle fracture will inevitably fluctuate in acoustic signals, which can be obtained theoretically through the selection of appropriate equipment and the tuning of filtering algorithms. However, it is accompanied by an increase in acquisition cost and higher requirements for robustness. Therefore, from the perspective of balancing the cost of signal acquisition and the ability to predict abnormal events, the recognition object of the acoustic signal may also need to be carefully selected. Potential sources of AE events in LIBs include electrode cracking, transition metal dissolution, cathode electrolyte interface (CEI) formation, solid electrolyte interface (SEI) formation, etc. On a cylindrical 18,650 cell, some researchers measured AE during full cell discharge [28], detected SOH by AE [29], and observed the trend of the AE intensity for each window as a function of a cycle. Other researchers used the AE to probe the generation, change process of solid electrolyte interphase (SEI) in batteries [30]. The acquired AE signals could also be divided into various types. Targeting for different analysis purposes, appropriate characteristic indicators should be selected carefully to explore the damage mechanism of lithium-ion batteries [30,31]. Generally speaking, with the occurrence of electrochemical reactions inside the battery, the transformation of almost any energy level will be associated with the rearrangement of the local structure. An acoustic signal will inevitably be emitted in this rearrangement process, which also means that AE can be regarded as a relatively simple principle diagnostic tool on the one hand; on the other hand, their broad applicability is accompanied by high sensitivity issues. At the same time, according to the different anode and cathode material selection and rational design processes mentioned in the literature [32], it is necessary to fully understand the acoustic characteristics of a single structure or component of the battery. In the future, in the face of acoustic events and characteristics with low discrimination, it is necessary to introduce more acoustic sensors for simultaneous observation of anodes and cathodes, and the introduction of half-cell will also be an ideal research object.

In general, a variety of battery fault diagnosis methods have been widely used in the research of LIB batteries, not only in the analysis of the regular operation of the battery but also in the real-time monitoring of various failure precursors. Destructive testing means that the battery is damaged or seriously affected in its operating behavior and safety. It is primarily used in laboratory research and unsuitable for in situ detection needs. It is of great help to study the mechanical and electrochemical relationship of batteries in the lab, while it is not applicable for troubleshooting in reality. Generally speaking, non-destructive testing (NDT) is of great significance for verifying battery quality and safety standards [33]. The detection of the swelling method has a good effect on a specific pouch battery. When the measured object is a prismatic battery or a cylindrical battery, the swelling application effect will be significantly reduced. The other three methods are classic non-destructive methods, which would provide potentially practical information and has the possibility of application in in-situ and ex-situ operations, helping to overcome the difficulties and the problems to be solved in traditional battery state estimation and fault diagnosis operations based on electrode measurements. Acoustic-based sensing technology is mainly used for SOC and SOH measurement of batteries, but the associativity of temperature, C rate, and abuse conditions are less considered. In terms of monitoring time, in comparison, imaging methods require less measurement time, but it is not easy to form a long-term continuous analysis process. Acoustic measurement has excellent advantages in long-term monitoring data accumulation. On that basis, deep learning and mathematical statistics have great potential as data-driven estimation methods, and they work best when combined with traditional electrode measurements such as voltage and current observations. So, the acoustic measurement is of great significance for battery state estimation and fault diagnosis based on battery electrode measurements.

1.2. Motivation and Original Contribution

From the perspective of battery safety, the short circuit caused by external circuit failure is a relatively severe electrical abuse behavior. The subsequent performance failure of the battery is easy to generate, which is reflected in capacity attenuation, short cycle life, poor consistency, easy self-discharge, high- and low-temperature performance degradation, etc.; on the other hand, the battery is prone to safety risks such as thermal runaway, gas leakage, and expansion deformation. From an acoustic point of view, the microstructure of the battery must be accompanied by relatively drastic changes in the short-circuit process of high current discharge, which is reflected in the transformation of the microstructure of the cathode and anode materials, breakdown of the separator accompanied by continuous heat release of the battery, temperature rise, and other behaviors. Changes in the micromechanical structure of the battery will negatively affect the current state and subsequent performance of the battery. Although the data measured by the electrode can detect the abnormality of the current data, the ability to capture some short-term abnormal signals is weak, and the subsequent impact cannot be continuously monitored and identified. For the short-circuit behavior of different initial SOC, the evaluation method should also require diversified measurements. In terms of the acquisition of acoustic signals and analysis of parametric variables, few studies have conducted an in-depth analysis of the high current caused by short circuits. Therefore, this paper analyzes the acoustic response characteristics of the battery cells with external short circuits (ESC) under different SOC initial conditions, obtains the acoustic discrimination criteria for the short-circuit fault diagnosis of 18,650 batteries, and conducts a possible correlation analysis. The research would provide a method for the failure analysis of lithium batteries, also a new promising tool.

1.3. Configuration of This Paper

The remainder of this paper is organized as follows: Section 2 illustrates the experiment's setup. Preliminary Analysis of Acoustic Signals introduces the primary methods for this study. Results and discussion present the results of the analysis for the acoustic response of the external short-circuit battery and develop the necessary discussions. Finally, the conclusion of our approach is given in Conclusion. All term abbreviations used in the study could be found in Nomenclature.

2. Experimental Section

Acoustic measurements were performed on commercial 18,650-type cylindrical nickel manganese cobalt oxide (NCM) lithium cells. According to the test conditions and safety considerations, two cases of initial battery SOC of 80% and 50% were selected. When the battery is placed in the temperature chamber, the ambient temperature is set to 20 °C to facilitate later observation of temperature changes and establish the connection between temperature, electrode measurement, and acoustic response. The battery-cell characteristics are shown in Table 1. Taking Lithium NCM111as an example, the electrochemical reaction during the charging and discharging process is shown in (1). Figure 1 shows the installed lithium-ion battery external short-circuit the experimental system. The computer is connected to the Motohawk via controller area network (CAN) and controls the relays in the circuit. Hall current sensors are used to measure short-circuit current. The ultrasonic needle sensor is from ndtXducer, and the miniature needle-type sensor is for industrial and laboratory use in ultrasonic and acoustic measurements with characteristics such as low cost, flat sensitivity, and good directivity, and its maximum operating temperature is around 100 °C. The sensor is used to collect the vibration signal generated by the change of the internal structure of the battery, and the output signal is a high-frequency voltage signal. The voltage, current, and vibration signals were all recorded by the Nioki MR6000 data acquisition device, and the short-circuit duration was 150 s. To reduce the external interference to the acoustic signal of the battery, a vibration isolation material is arranged around the battery and the acoustic sensor; to enhance the sensor's ability to receive signals, the Vaseline gel is applied at the contact position between the battery body and the sensor.

$$LiNi_{\frac{1}{3}}Co_{\frac{1}{3}}Mn_{\frac{1}{3}}O_2 \underset{discharge}{\overset{charge}{\rightleftarrows}} Li_{1-x}\left(Ni^{2+}_{\frac{1}{3}-x}Ni^{3+}_{x}\right)Co_{\frac{1}{3}}Mn_{\frac{1}{3}}O_2 + xLi^{+} + xe^{-}\left(0 \le x \le \frac{1}{3}, 3.8\text{–}4.1\right) \quad (1)$$

Table 1. List of main parameters of the experiment.

Type of the Battery Cell	18,650-Type Cylindrical NCM Lithium Cells
Nominal cell capacity (0.3 C)	2.0 Ah
Average battery cell voltage	3.6 V
End of discharge voltage	2.5 V
High voltage protection	4.2 V
Operation temperature range	−20–55 °C
Cathode materials	$LiNi_{1-y-z}Co_yMn_zO_2$
Anode materials	Graphite

The collected raw voltage, current, acoustic, and temperature signals are shown in Figure 2. The time range of the short-circuit occurrence can be clearly defined by the fluctuating region of the voltage change. The voltage has the remarkable characteristics of initial sag, fluctuation retention, and secondary sag. The corresponding current has a similar change process of the initial jump, mid-wave, and terminal drop, which is a phenomenon that can be clearly distinguished from the short-circuit condition at the electrode measurement. Corresponding to the temperature signal, the initial occurrence of the short circuit could obtain the rise of the temperature for the positive and negative electrodes and the slight decrease of the temperature after the short circuit is completed. Corresponding to the acoustic signal, the initial signal observation would roughly distinguish between the beginning and the end of the short circuit (compared to the range determined by the voltage or current); the signal has an initial apparent change (at the beginning of the ESC), and a relatively weak mutation (at the end of the ESC); other acoustic features are not very obvious; it is difficult to distinguish from the data on the figure.

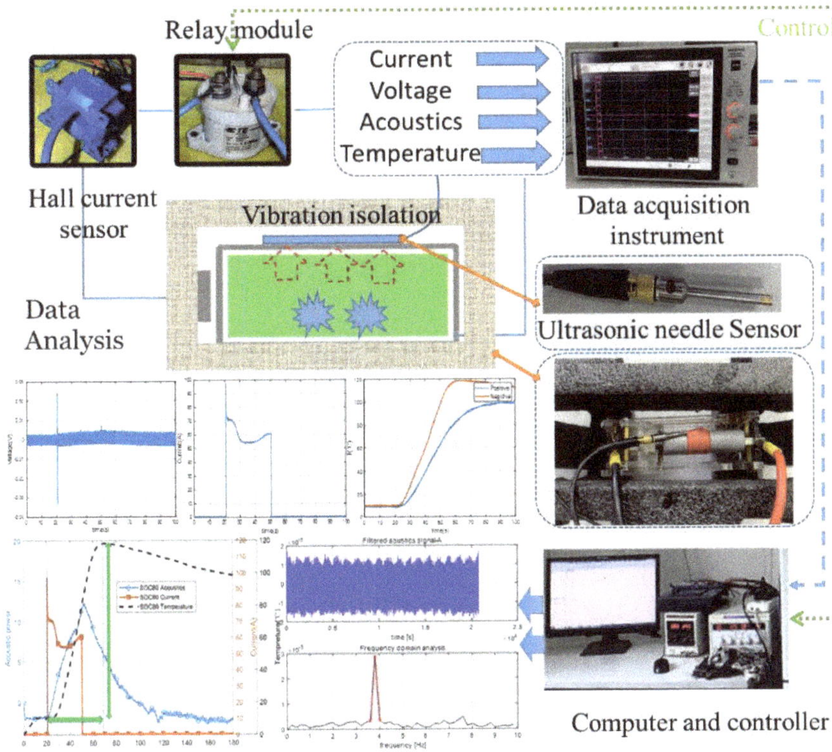

Figure 1. Schematic diagram of the ESC experimental setup based on AE.

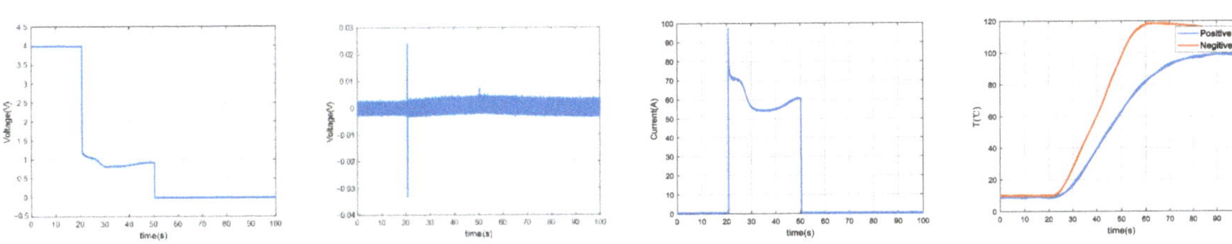

Figure 2. Voltage, current, acoustic, and temperature data curve across the entire ESC.

Therefore, in terms of acoustic signal data analysis, phase spectrum analysis reduces various influencing factors to obtain signal amplitude and phase values for further research. The frequency-domain characteristics of the AE signal were analyzed from the time-domain acoustic signal by Fast Fourier Transform (FFT).

3. Preliminary Analysis of Acoustic Signals

According to the change of the corresponding current, the acoustic signal is preliminarily divided into four time slots in Figure 3, where A is before the ESC, B is the continuous process of the ESC, C is the near-end after the ESC, and D is the far end after the ESC, which are used to observe the possible acoustic responses of the battery after a short circuit occurs. The following results are obtained after filtering the PPT analysis for the acoustic characteristic data segment of the above time slot.

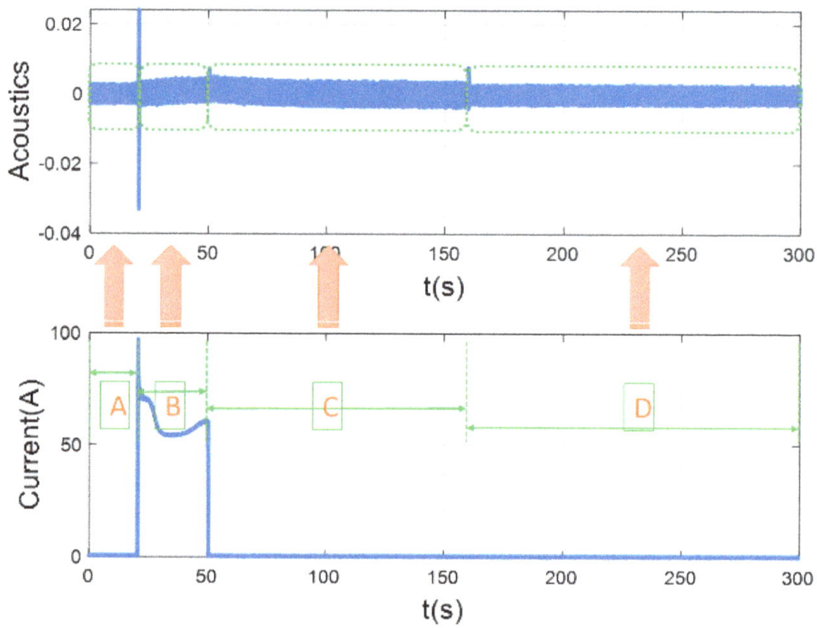

Figure 3. Segmentation diagram of an acoustic signal.

In Figure 4, there is an apparent difference between the amplitude change of short-circuit occurrence and the short-circuit occurrence near-end, which reflects a difference value of more than ten times. From the perspective of frequency, the frequency of the inherent acoustic characteristics of the battery cell is relatively concentrated before the short circuit occurs, which is between 3–4 Hz. The frequency characteristics of the ESC at the near-end are shifted to the left, indicating that the internal structure of the battery is changed due to the breakdown of the short-circuit current and the acoustic features are changed, and the material fracture continues after the ESC in the proximal short-circuit area C. It may be due to residual stress, which is reflected in the continuous increase in the amplitude of the center frequency. The continuous temperature rise also proves that the structure continues to change after the short circuit, and there is continuous heat release. The characteristics of the acoustic response of region D become blurred due to the rearrangement of the structure and lose its original characteristics, which is reflected in the chaos of the microstructure and change. These indicate a dramatic change in the mechanical properties of the electrode, which is consistent with the static mechanical analysis of the electrode [24,26,30].

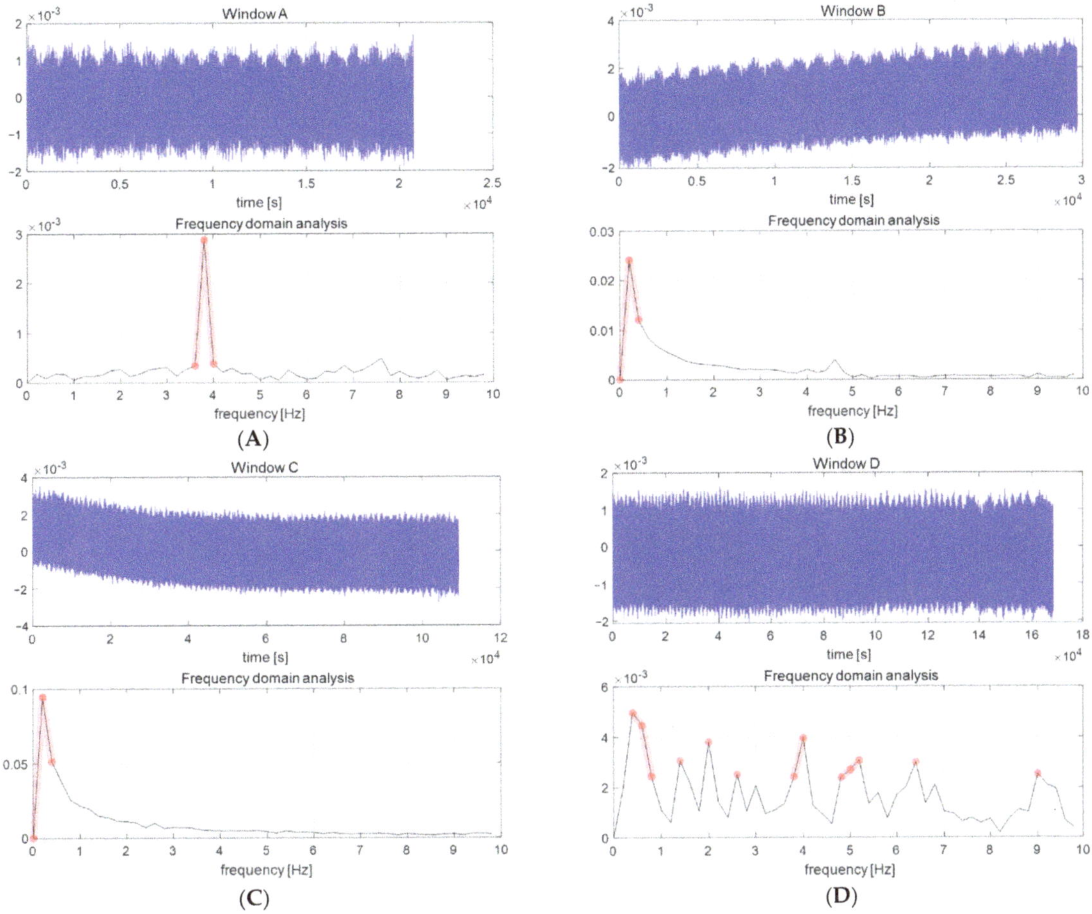

Figure 4. Original acoustic signal and frequency domain characteristics of window **A–D**. (**A**), Filtered acoustic signal before the ESC; (**B**), Filtered acoustic signal during the ESC; (**C**), Filtered acoustic signal near-end after the ESC; (**D**), Filtered acoustic signal far-end after the ESC.

From the perspective of the acoustic characteristics of a single short-circuit battery, in terms of the near-end frequency domain characteristics before, during, and after the ESC (window A–C), it is preliminarily shown that the value of the center frequency and its amplitude, which are representative, would characterize the change of the internal structure and the change of the mechanical properties when the short circuit occurs to a certain extent. The signal at the far (window D) end becomes more ambiguous due to the internal change of its acoustic characteristics after the short circuit occurs so that it can be temporarily ignored in selecting the ESC data.

Therefore, in this paper, the correlation analysis of the measured values of electrodes, acoustics, and temperature is carried out in both time and frequency domains. According to the relationship between voltage, current, and temperature, the implication between the response of the battery in different time slots during the ESC process and the characteristic acoustic parameters would be obtained; the representation to the initial SOC at the occurrence of the ESC might be clear; while the link between cell structural damage and rearrangement and the acoustic response could be discussed. The overall framework of the presented ideas is shown in Figure 5.

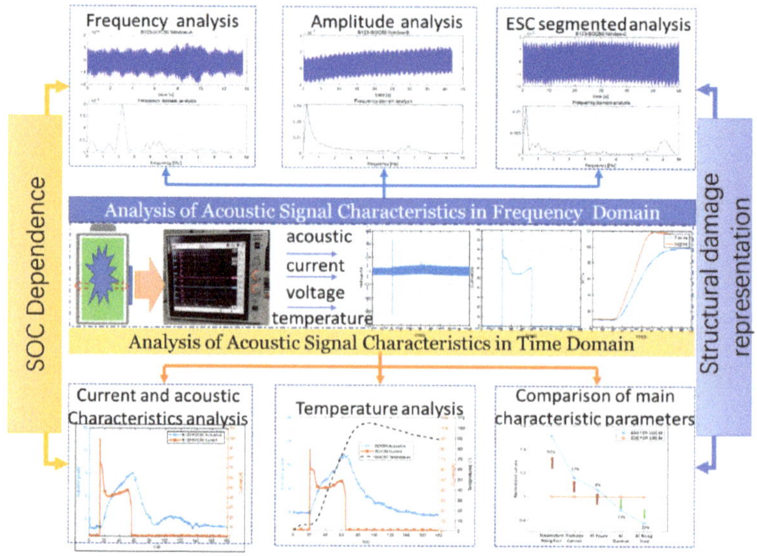

Figure 5. Framework for the acoustic analysis in ESC.

4. Results and Discussion for Acoustic Characteristics of Battery Short Circuit under Different SOC Conditions

4.1. Analysis of Acoustic Signal Characteristics in Time Domain

Characteristic parameters that are important for battery acoustics include rise time, the time between the first threshold crossing and the peak amplitude; "duration" or the time interval between the first and last threshold crossing in a burst signal; and the amplitude of the response acoustic intensity, etc. The preferred peak value (amplitude) reflects the response of the whole battery cell to the high-discharge current process in the short-circuit process, which is closely related to the damage to the battery's internal structure when the short circuit occurs. What the acoustic sensor would receive is damage to the internal structure of the battery. Although the specific position of the AE (sound radiation intensity) cannot be determined under passive conditions, its amplitude reflects the characteristics of the special battery discharge conditions. It would be used for comprehensive judgment of the battery state.

The original acoustic signal obtained by the sensor has a low degree of recognition with poor signal legibility. To better analyze the acoustic signal, the coaxial variation curve of the current and the acoustic signal is obtained after processing through a rational transfer function. The acoustic signal strength is normalized after filtering to ensure good readability of the values. Overall, in the time range of interest, regardless of the initial SOC (SOC80 OR SOC50) value, the acoustic signal immediately responds when a short circuit current occurs. Still, the trend of the acoustic signal is a step-by-step process, and its peak value shows at the end of the short-circuit process. In Figure 6, it appears later; that is, it could be found from the figure that the acoustic curve reaches its peak value in about 5 s after the ESC current drops to 0. The current change is generally in a relatively short time, but the acoustic change starts at the same time as the current; while it is slightly achieved, a delayed peak point reflects that the rearrangement of the internal structure of the battery is the consequence of the short-circuit current shock. The subsequent decline process demonstrates the process of the destruction and rearrangement of the battery content structure. More importantly, from the perspective of detection, the initially short-term electrode behavior would be transformed into a more obvious acoustic change process, reflecting the vital characterization significance of the battery acoustic characteristics for the abnormal state of the battery.

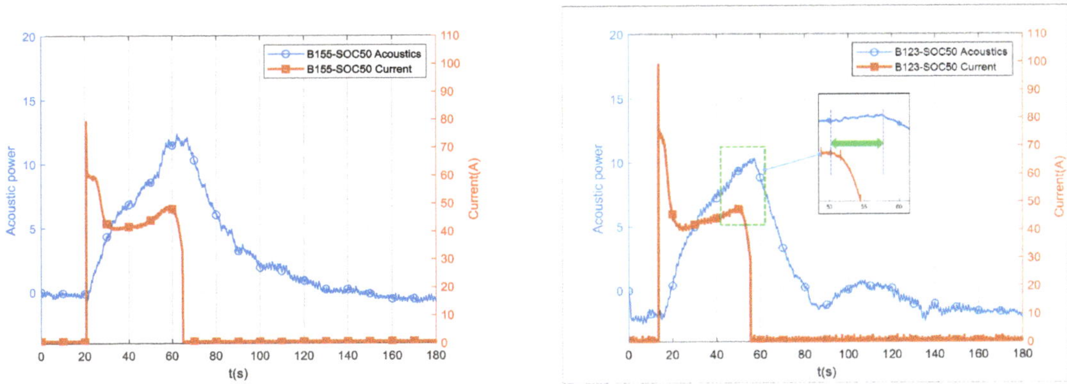

Figure 6. Current and acoustic characteristics of ESC under initial SOC50 condition.

Through observation for Figure 7, there is a specific SOC correlation between the acoustic intensities detected in SOC80 and SOC50. The acoustic radiation intensity of SOC80 is higher than that of SOC50 by more than 6%, which corresponds to the case of high current discharge. Compared with SOC50, the external short-circuit discharge current of SOC80 also increases by 17%. Of course, this is in line with our expectations for the initial state of the battery. For the analysis of battery AE duration, the structural characteristics of the battery itself reflect the response to the short-circuit process. Due to the difference in the battery's state, the degree of damage to the battery resulting from high-current discharge is also different. In Figure 8, SOC80 and SOC50, from the perspective of the two initial states, the acoustic response process of SOC80 is shorter, reflecting the severity of the structural damage process. The 22% reduction for the AE rise time in SOC80 also demonstrates the severity of structural damage inside of the battery, which does not optimistically speculate the possibility of structural rearrangement and repair. In the acoustic response of B123 battery SOC50 Figure 6, the fluctuation of the acoustic signal could be observed in the later stage, which may indicate the possibility of repair after structural damage. In the case of SOC80, no noticeable acoustic signal fluctuations were found subsequently.

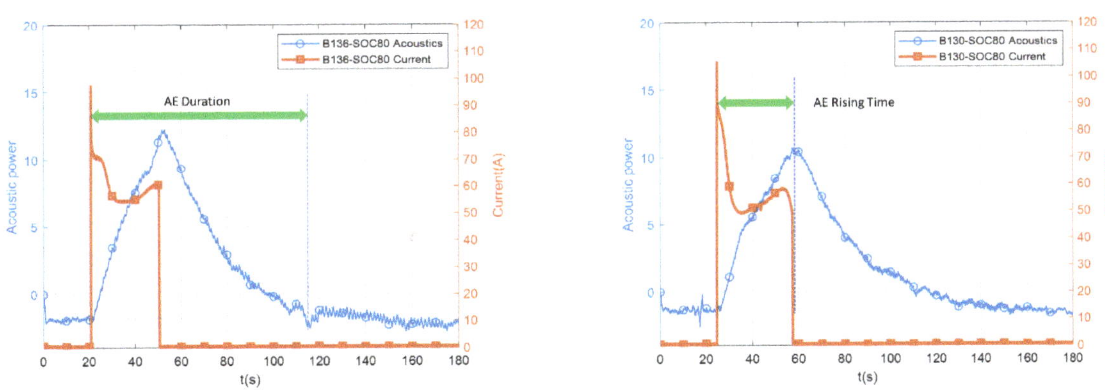

Figure 7. Current and acoustic characteristics of ESC with initial SOC80 condition.

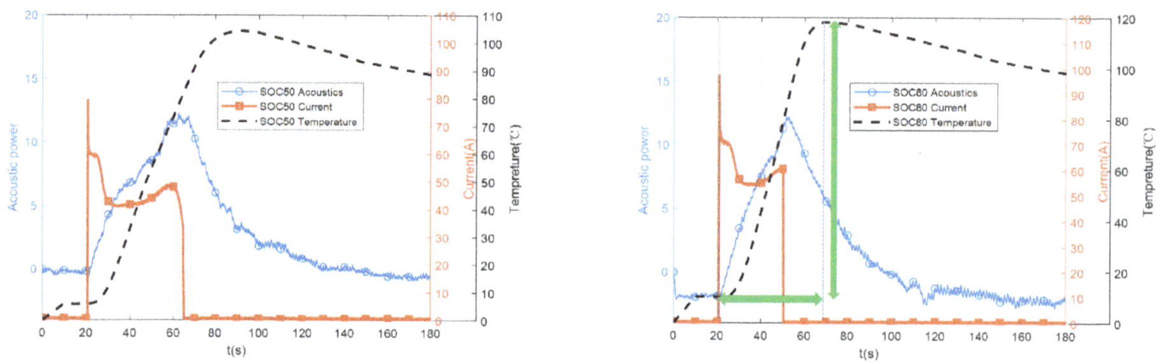

Figure 8. Current, acoustic, and temperature average characteristics of ESC under initial SOC50 and SOC80 conditions.

To observe the relationship between electrode measurement (current), temperature, and acoustic signal more intuitively, the data of SOC50 and SOC80 are averaged to obtain a comparative analysis of coaxial curves under different SOC conditions, as shown in Figure 8. Finally, from the perspective of the temperature change during the short-circuit process, multi-position high-strength structural damage probably occurs inside the battery parallelly. So, the temperature rise rate for the case of SOC80 is 52% higher than that of SOC50, while the time of its peak arrival is later than the acoustic peak. The reason for the subsequent temperature descent is more complicated. It is presumed to be the result of the combined effect of two factors: on the one hand, with the end of the primary short-circuit reaction, the large-scale structural damage stops; on the other hand, with some small-scale structural repairs, while some continuous destruction of the structural rearrangement results in the constant local release of heat, the observed result is a slower decline rate after the peak. Therefore, the judgment of the abnormal state of the battery from the perspective of temperature seems to have a relatively evident hysteresis in terms of timeliness, as the major damage to the internal structure of the battery has been completed. In this case, the analysis of the acoustic properties would present the abnormal behavior of the battery ahead of the temperature. Considering that the battery temperature will also be undulate due to heat accumulation during the regular operation of the battery, the direct response of the acoustic characteristics to the changes in the battery structure has certain advantages. The acoustic state can be used as the basis for judging the abnormal state of the battery. In Figure 9, all the values are normalized, and the SOC50 is used as the comparison benchmark, reflecting the comparison results of the above-selected parameters.

4.2. Analysis of Acoustic Signal Characteristics in Frequency Domain

The correlation analysis of the external short-circuit acoustic characteristics of the battery with two initial conditions of battery SOC of 80% and SOC of 50% was carried out, and three time windows (before, during, and after the ESC) were selected as the main observation time slot. Here, the magnitude of the amplitude in the frequency domain is used to represent the strength of the acoustic characteristics.

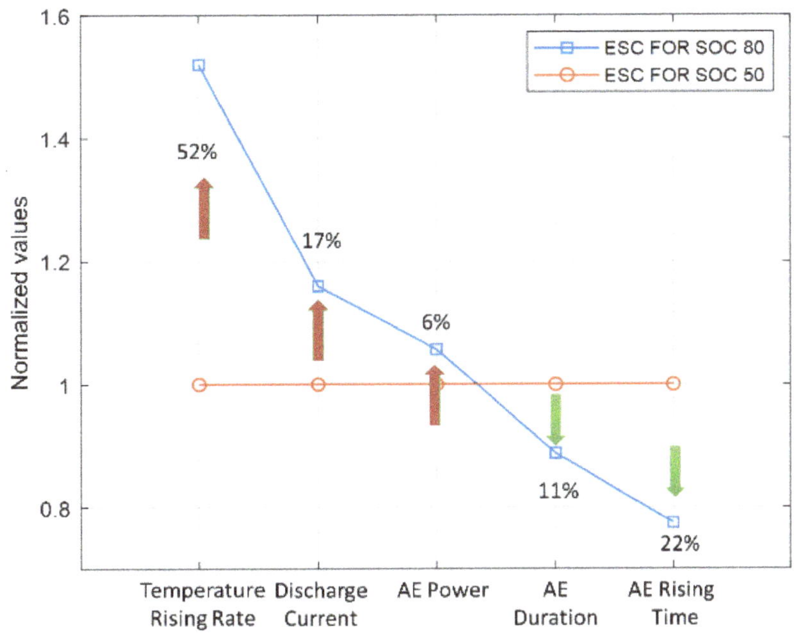

Figure 9. Comparison of main characteristic parameters of ESC occurrence.

In Figure 10, based on the acoustic signal characteristics in the three time windows concerned, the short-circuit acoustic features window-A of SOC80 Figure 10(a1, b1) generally shows relatively stable acoustic characteristics. The center frequency is between 3–6 Hz, which refers to the background acoustic aspects of the operation of the experimental device. Depending on the arrangement of the device, it would be used as an initial calibration reference in the future considering improving the accuracy of the identification of the acoustic characteristics for different types of batteries or setups. In the acoustic characteristics of the window-B range Figure 10(a2, b2), the center frequency moves to the low-frequency area, and the amplitude is ten times higher than that of the window-A field. The recognition is with the remarkable feature, representing that the system structure responds strongly and concentratedly to ESCs, which may be due to the transformation of the internal structure when the high current, inducing the breakdown of the separator, or the structural rearrangement. The system is damaged, but the subsequent structure might continue to be damaged or maintained. It is necessary to move the data analysis of area C. The acoustic characteristics of the window C interval do not change significantly from the perspective of the center frequency, but there are specific differences in the amplitude changes. A set of data shows that the C interval in B130 Figure 10(b3) amplitude is unchanged, but the C interval in B136 Figure 10(a3) data shows an increase of about five times the acoustic signal intensity, which indicates that even if there is no subsequent operation of the electrode terminal after the short circuit occurs, the trend of structural rearrangement and change continues after the ESC. The fluctuation may be different for individuals, but the trend should be the change of battery structure. As the current device cannot guarantee the safety of subsequent battery electrode tests after the ESC, further observation and analysis of the acoustic characteristics of the battery electrode operation are still to be confirmed, and more experimental data are needed.

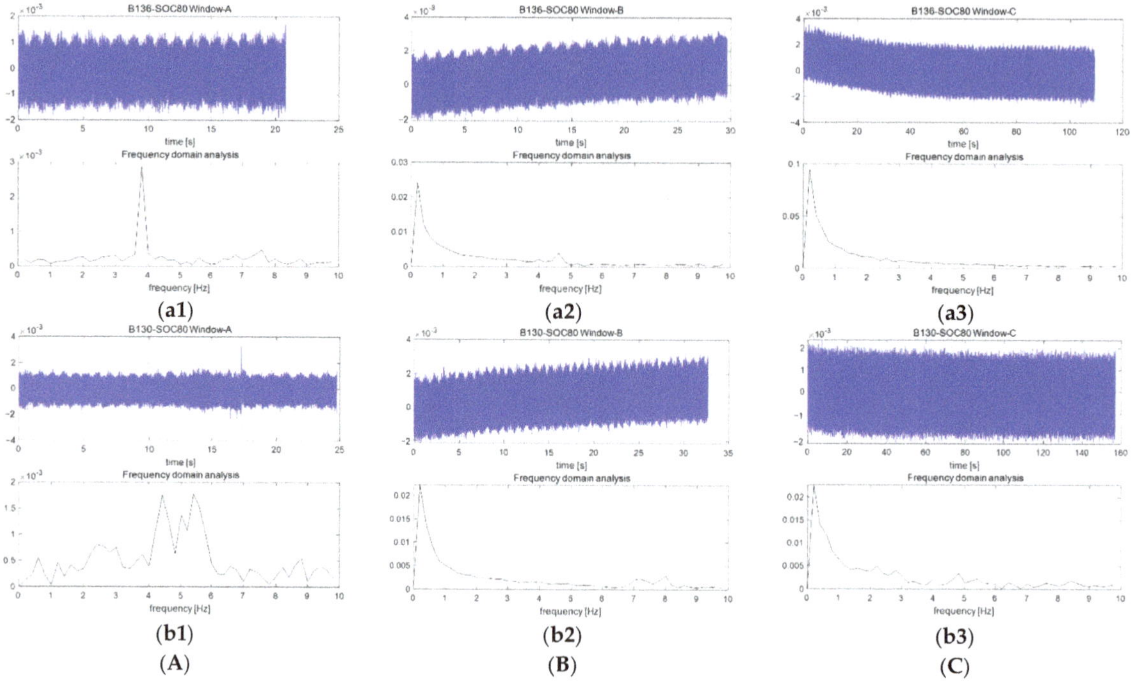

Figure 10. Two cases frequency domain characteristics of window A–C with initial SOC80 condition. (**A**), acoustic signal before the ESC in (**a1**,**b1**); (**B**), acoustic signal during the ESC in (**a2**,**b2**); (**C**), acoustic signal near-end after the ESC in (**a3**,**b3**).

From the analysis of the initial battery short-circuit test results of SOC50, the overall trend is similar to that of SOC80. Still, the specific characteristics also demonstrate strong SOC dependence, in Figure 11. For example, the value of the center frequency moves along the low-frequency direction, the approximate frequency range is between 2–4 Hz Figure 11(a1, b1), and the amplitude of the center frequency is less than the case of SOC80, which reflects a relatively obvious SOC correlation. In the process of the ESC, that is, in window B Figure 11(a2, b2), the amplitude of its center frequency also appears to move to the low-frequency end, and the amplitude is lower than that of SOC80. Still, it has an increase of about 15 times compared with the amplitude of SOC50 window A. The short-circuited near-end acoustic signal in window C shows that the structure remains unchanged Figure 11(b3) or further aggravates the damage Figure 11(a3). Based on the observation of acoustic data, there is no observation that the acoustic characteristics represented by the structure would be restored to the initial state (window A) after a short circuit occurs. Even if it is left standing for a long enough time, no similar supporting data are observed. It is explained that the microstructure damage of the battery cells after the high current short circuit is irreversible, and it should be isolated or replaced in time from the perspective of battery safety.

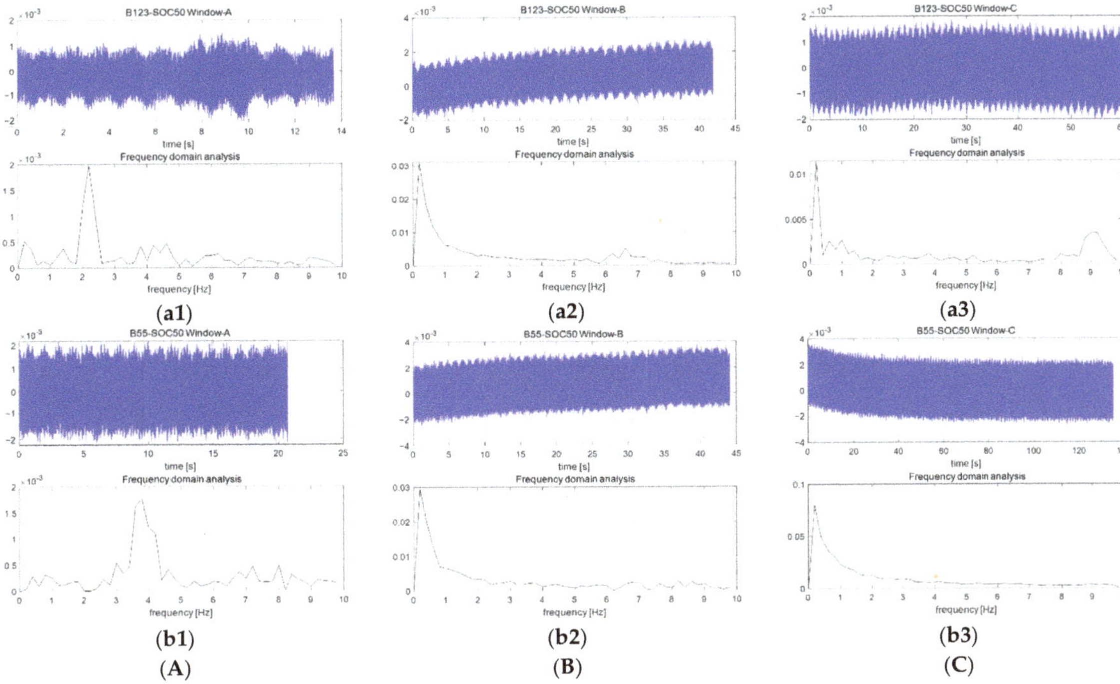

Figure 11. Two cases frequency domain characteristics of windows A–C with initial SOC50 condition. (**A**), acoustic signal before the ESC in (**a1**,**b1**); (**B**), acoustic signal during the ESC in (**a2**,**b2**); (**C**), acoustic signal near-end after the ESC in (**a3**,**b3**).

5. Conclusions

This study investigates the AE characteristics of cylindrical lithium batteries when an external short circuit occurs. Firstly, a test bench capable of measuring the acoustic characteristics of the external short-circuit of the battery is built to obtain the electrode, temperature, and acoustic data of the battery when the ESC occurs. Then, the analysis is carried out from two aspects of the time domain and frequency domain. In the time domain, the acoustic signal of the battery appears synchronously with the voltage/current change during the ESC. With the destruction and rearrangement of the internal structure during the short circuit, the acoustic response gradually increases and reaches a peak after the end of the ESC. The critical acoustic characteristic values of rising time, amplitude, and duration were selected, the SOC correlation analysis was carried out for batteries with different initial SOC, and the mapping relationship with battery electrode and temperature characteristics was further considered. In terms of frequency domain analysis, the acoustic response of the battery is divided into three observation windows before, during, and after the ESC, and the center frequency and amplitude of the acoustic signal in the three stages are discussed. The correlation between the initial SOC of the short-circuit occurrence and the acoustic response is verified. The expectation of the SOC dependence in the time domain is confirmed. Therefore, by selecting appropriate acoustic characteristic parameters, AE technology has the ability to diagnose abnormal battery status together with the electrode and temperature measurement.

From the observation results of the ESC in analysis, the discharge current hopping from 0 to 20–30 C-rate, the change of the acoustic intensity could be observed, and there is still a lack of acoustic-related active constant discharge current test data at different levels to make further judgments. According to the results observed so far, the correlation between C-rate and acoustic energy is not clear, or the acoustic energy is relatively independent of the ESC in our research. Compared with electrode measurement, it would extend the duration of abnormal events and improve the detection rate of suspicious events. Compared with temperature measurement, AE can reflect the damage of abnormal circumstances to the structure earlier, to gain time for subsequent operations. In the future, combined with the analysis of the charge and discharge acoustic characteristics of different currents of the different SOC states, and supplemented by the correction of temperature effects, a more comprehensive battery acoustic and electrode characteristics would be established. The mapping relationship of characteristics will provide more basis for battery state estimation, life span estimation, and battery energy management.

Author Contributions: Conceptualization, N.Z.; methodology, N.Z.; software, N.Z.; validation, Z.Y., X.C. and C.H.; formal analysis, N.Z.; investigation, X.C. and C.H.; resources, X.C. and C.H.; data curation, X.C.; writing—original draft preparation, N.Z.; writing—review and editing, X.C. and C.H.; visualization, Z.Y.; supervision, N.Z.; project administration, N.Z.; funding acquisition, N.Z. and Z.Y. All authors have read and agreed to the published version of the manuscript.

Funding: This research was funded by the National Natural Science Foundation of China (52172401), Vehicle Measurement, Control and Safety Key Laboratory of Sichuan Province (QCCK2021-005).

Institutional Review Board Statement: Not applicable.

Informed Consent Statement: Not applicable.

Data Availability Statement: Please contact the corresponding author for the data from this study.

Acknowledgments: Acknowledge affiliation 2 to provide the site and some equipment for experiments.

Conflicts of Interest: The authors declare no conflict of interest.

Nomenclature

PEVs	Pure electric vehicles
SOC	State of Charge
BMS	Battery management system
LIB	Lithium-ion battery
TOF	Time of fly
ESC	External short-circuits
SOH	State of health
EMA	Experimental modal analysis
US	Ultrasonic testing
AE	Acoustic emission
CEI	Cathode electrolyte interface
SEI	Solid electrolyte interface
EVs	Electric vehicles
NDT	Non-destructive testing
FFT	Fast Fourier transform
CAN	Controller area network
NCM	Nickel Manganese Cobalt Oxide

References

1. Wei, Z.; Zhao, J.; He, H.; Ding, G.; Cui, H.; Liu, L. Future smart battery and management: Advanced sensing from external to embedded multi-dimensional measurement. *J. Power Sources* **2021**, *489*, 229462. [CrossRef]
2. Wei, Z.; Zhao, J.; Ji, D.; Tseng, K.J. A multi-timescale estimator for battery state of charge and capacity dual estimation based on an online identified model. *Appl. Energy* **2017**, *204*, 1264–1274. [CrossRef]
3. Zhu, G.; Zhao, C.; Huang, J.; He, C.; Zhang, J.; Chen, S.; Xu, L.; Yuan, H.; Zhang, Q. Fast Charging Lithium Batteries: Recent Progress and Future Prospects. *Small* **2019**, *15*, 1805389. [CrossRef] [PubMed]
4. Tomaszewska, A.; Chu, Z.; Feng, X.; O'Kane, S.; Liu, X.; Chen, J.; Ji, C.; Endler, E.; Li, R.; Liu, L.; et al. Lithium-ion battery fast charging: A review. *ETransportation* **2019**, *1*, 100011. [CrossRef]
5. Liu, Y.; Zhu, Y.; Cui, Y. Challenges and opportunities towards fast-charging battery materials. *Nat. Energy* **2019**, *4*, 540–550. [CrossRef]
6. Birkl, C.R.; Roberts, M.R.; McTurk, E.; Bruce, P.G.; Howey, D.A. Degradation diagnostics for lithium ion cells. *J. Power Sources* **2017**, *341*, 373–386. [CrossRef]

7. Chen, Z.; Zhou, N.; Liang, H.; Cui, J.; Fang, Z. A fusion-based method of state-of-charge online estimation for lithium-ion batteries under low capacity conditions. *Front. Energy Res.* **2021**, *9*, 797.
8. Divakaran, A.M.; Hamilton, D.; Manjunatha, K.N.; Minakshi, M. Design, development and thermal analysis of reusable Li-ion battery module for future mobile and stationary applications. *Energies* **2020**, *13*, 1477. [CrossRef]
9. Liu, D.; Shadike, Z.; Lin, R.; Qian, K.; Li, H.; Li, K.; Wang, S.; Yu, Q.; Liu, M.; Ganapathy, S.; et al. Review of recent development of in situ/operando characterization techniques for lithium battery research. *Adv. Mater.* **2019**, *31*, 1806620. [CrossRef]
10. Lu, J.; Wu, T.; Amine, K. State-of-the-art characterization techniques for advanced lithium-ion batteries. *Nat. Energy* **2017**, *2*, 17011. [CrossRef]
11. Wang, L.; Yin, S.; Zhang, C.; Huan, Y.; Xu, J. Mechanical characterization and modeling for anodes and cathodes in lithium-ion batteries. *J. Power Sources* **2018**, *392*, 265–273. [CrossRef]
12. Finegan, D.P.; Scheel, M.; Robinson, J.B.; Tjaden, B.; Di Michiel, M.; Hinds, G.; Brett, D.J.; Shearing, P.R. Investigating lithium-ion battery materials during overcharge-induced thermal runaway: An operando and multi-scale X-ray CT study. *Phys. Chem. Chem. Phys.* **2016**, *18*, 30912–30919. [CrossRef]
13. Heenan, T.M.; Tan, C.; Hack, J.; Brett, D.J.; Shearing, P.R. Developments in X-ray tomography characterization for electrochemical devices. *Mater. Today* **2019**, *31*, 69–85. [CrossRef]
14. Pfrang, A.; Kersys, A.; Kriston, A.; Sauer, D.U.; Rahe, C.; Käbitz, S.; Figgemeier, E. Geometrical in homogeneities as cause of mechanical failure in commercial 18,650 lithium ion cells. *J. Electrochem. Soc.* **2019**, *166*, A3745. [CrossRef]
15. Wu, Y.; Saxena, S.; Xing, Y.; Wang, Y.; Li, C.; Yung, W.K.; Pecht, M. Analysis of manufacturing-induced defects and structural deformations in lithium-ion batteries using computed tomography. *Energies* **2018**, *11*, 925. [CrossRef]
16. Robinson, J.B.; Engebretsen, E.; Finegan, D.P.; Darr, J.; Hinds, G.; Shearing, P.R.; Brett, D.J. Detection of internal defects in lithium-ion batteries using lock-in thermography. *ECS Electrochem. Lett.* **2015**, *4*, A106. [CrossRef]
17. Friesen, A.; Schultz, C.; Brunklaus, G.; Rodehorst, U.; Wilken, A.; Haetge, J.; Winter, M.; Schappacher, F. Long term aging of automotive type lithium-ion cells. *ECS Trans.* **2015**, *69*, 89. [CrossRef]
18. Tsutsui, W. The Mechanical Response of Multifunctional Battery Systems. Ph.D. Thesis, Purdue University, West Lafayette, IN, USA, 2017.
19. Gao, X.; He, P.; Ren, J.; Xu, J. Modeling of contact stress among compound particles in high energy lithium-ion battery. *Energy Storage Matter.* **2019**, *18*, 23–33. [CrossRef]
20. Zhu, J.E.; Li, W.; Xia, Y.; Sahraei, E. Testing and modeling the mechanical properties of the granular materials of graphite anode. *J. Electrochem. Soc.* **2018**, *165*, A1160–A1168. [CrossRef]
21. Higa, K.; Srinivasan, V. Stress and strain in silicon electrode models. *J. Electrochem. Soc.* **2015**, *162*, A1111–A1122. [CrossRef]
22. Cannarella, J.; Arnold, C.B. The effects of defects on localized plating in lithium-ion batteries. *J. Electrochem. Soc.* **2015**, *162*, A1365. [CrossRef]
23. Etiemble, A.; Idrissi, H.; Meille, S.; Roué, L. In situ investigation of the volume change and pulverization of hydride materials for Ni-MH batteries by concomitant generated force and acoustic emission measurements. *J. Power Sources* **2012**, *205*, 500–505. [CrossRef]
24. Majasan, J.; Robinson, J.; Owen, R.; Maier, M.; Radhakrishnan, A.N.P.; Pham, M.; Tranter, T.G.; Zhang, Y.; Shearing, P.; Brett, D. Recent advances in acoustic diagnostics for electrochemical power systems. *J. Phys. Energy* **2021**, *3*, 032011. [CrossRef]
25. Woodford, W.H.; Chiang, Y.; Carter, W.C. Quantifying reliability statistics for electrochemical shock of brittle materials. *J. Mech. Phys. Solids* **2014**, *70*, 71–83. [CrossRef]
26. Nakamura, H.; Ohtsu, M.; Enoki, M.; Mizutani, Y.; Shigeishi, M.; Inaba, H.; Nakano, M.; Shiotani, T.; Yuyama, S.; Sugimoto, S. *Practical Acoustic Emission Testing*; Springer: Tokyo, Japan, 2016.
27. Kircheva, N.; Genies, S.; Brun-Buisson, D.; Thivel, P.X. Study of solid electrolyte interface formation and lithium intercalation in Li-Ion batteries by acoustic emission. *J. Electrochem. Soc.* **2011**, *159*, A18. [CrossRef]
28. Popp, H.; Koller, M.; Jahn, M.; Bergmann, A. Mechanical methods for state determination of Lithium-Ion secondary batteries: A review. *J. Energy Storage* **2020**, *32*, 101859. [CrossRef]
29. Kim, C.S.; Yoo, J.S.; Jeong, K.M.; Kim, K.; Yi, C.W. Investigation on internal short circuits of lithium polymer batteries with a ceramic-coated separator during nail penetration. *J. Power Sources* **2015**, *289*, 41–49. [CrossRef]
30. Choe, C.Y.; Jung, W.S.; Byeon, J.W. Damage evaluation in lithium cobalt oxide/carbon electrodes of secondary battery by acoustic emission monitoring. *Mater. Trans.* **2015**, *56*, 269–273. [CrossRef]
31. Villevieille, C.; Boinet, M.; Monconduit, L. Direct evidence of morphological changes in conversion type electrodes in Li-ion battery by acoustic emission. *Electrochem. Commun.* **2010**, *12*, 1336–1339. [CrossRef]
32. Divakaran, A.M.; Minakshi, M.; Bahri, P.A.; Paul, S.; Kumari, P.; Divakaran, A.M.; Manjunatha, K.N. Rational design on materials for developing next generation lithium-ion secondary battery. *Prog. Solid State Chem.* **2021**, *62*, 100298. [CrossRef]
33. Bauermann, L.P.; Mesquita, L.V.; Bischoff, C.; Drews, M.; Fitz, O.; Heuer, A.; Biro, D. Scanning acoustic microscopy as a non-destructive imaging tool to localize defects inside battery cells. *J. Power Sources Adv.* **2020**, *6*, 100035. [CrossRef]

energies

MDPI

Article

Parametric Investigation on the Performance of a Battery Thermal Management System with Immersion Cooling

Yuxin Zhou [1,2,3], Zhengkun Wang [1,2,3], Zongfa Xie [1,2,3,*] and Yanan Wang [1,2,3,*]

1 School of Mechanical Engineering, Shandong University, Jinan 250061, China; zhouyuxin@mail.sdu.edu.cn (Y.Z.); wangzhengkun@mail.sdu.edu.cn (Z.W.)
2 Key Laboratory of High-Efficiency and Clean Mechanical Manufacture, Ministry of Education, Shandong University, Jinan 250061, China
3 National Demonstration Center for Experimental Mechanical Engineering Education, Shandong University, Jinan 250061, China
* Correspondence: zongfax@sdu.edu.cn (Z.X.); wyn@sdu.edu.cn (Y.W.)

Abstract: Lithium-ion batteries will generate a large amount of heat during high-rate charging and discharging. By transferring the heat to the environment in time, the batteries can be kept in a suitable temperature range. This allows them to work normally, prolongs their cycle life, and reduces the risk of thermal runaway. Immersion cooling is a simple and efficient thermal management method. In this paper, a battery thermal management system (BTMS) with immersion cooling was designed by immersing the lithium-ion cells in the non-conductive coolant—dimethyl silicone oil. The electric–thermal coupled model was adopted to obtain the heat production and temperature distribution of the cell during discharging, and the performance of the system was obtained by numerical calculation. It was found that, compared with natural cooling, immersion cooling could significantly reduce both the maximum temperature (MAT) of the cell and the temperature of the tabs during the 3C discharging process. However, the maximum temperature difference (MATD) of the cell was significantly increased. To solve this problem, the effects of the flow rate, viscosity, specific heat capacity, and thermal conductivity of the coolant on the performance of immersion cooling were further investigated and discussed, including the MAT and MATD of the cell, and the pressure drop of the coolant. The method and results could provide references for the design and application of the BTMS with immersion cooling in the future.

Keywords: immersion cooling; lithium-ion battery; thermal management; temperature; pressure drop

Citation: Zhou, Y.; Wang, Z.; Xie, Z.; Wang, Y. Parametric Investigation on the Performance of a Battery Thermal Management System with Immersion Cooling. *Energies* **2022**, *15*, 2554. https://doi.org/10.3390/en15072554

Academic Editor: Christian Veje

Received: 17 February 2022
Accepted: 30 March 2022
Published: 31 March 2022

1. Introduction

Lithium-ion batteries are widely used in electric vehicles due to their great energy density, high voltage, and small self-discharge rate [1,2]. However, lithium-ion batteries should be working under certain temperature conditions. The best temperature range is usually between 20 °C and 40 °C, and the maximum temperature difference should be controlled within 5 °C [3,4]. Excessively higher or lower temperatures could affect the normal working of the battery, reduce its capacity, and shorten its life [5]. When the temperature is extremely high, thermal runaway may even occur, resulting in security incidents [6]. In addition, lithium-ion batteries will release a large amount of heat during long-term cycles and rapid charging and discharging processes. Insufficient and uneven heat dissipation can easily lead to local heat accumulations, which cause uneven temperature distributions in the batteries and in the battery pack, and affect their consistency and safety [7]. Therefore, it is essential to employ an effective battery thermal management system (BTMS) to regulate the temperatures and equalize the heat in the batteries.

According to the kinds of heat transfer mediums, the BTMSs commonly used in electric vehicles mainly include air cooling, liquid cooling, heat pipe cooling, PCM cooling, and hybrid cooling. The air-cooling type relies on the air as the heat transfer medium [8].

However, due to the smaller specific heat capacity and lower thermal conductivity of the air, its temperature regulation and thermal equalization capabilities are usually not as good as those of the liquid cooling type. The liquid cooling type can be further classified into indirect cooling and direct cooling (also known as immersion cooling). Indirect cooling is to attach one or a few cooling plate(s) or pipe(s) on the cells, and the heat is taken away by the flowing coolant (such as a water–glycol solution) in the channel(s) of the cooling plate(s) or pipe(s) [9–12]. Direct cooling is usually to immerse the cells in the coolant, so it can directly contact the cells, and absorb the heat from them. According to the working state of the coolant, the BTMS with immersion cooling can be further classified into liquid-filled BTMS and liquid-circulated BTMS [13]. In the former, the coolant is stationary, and in the latter, the coolant is circulating. In addition, phase change materials (PCMs) and heat pipes are also commonly used in the research on BTMSs [14,15]. Moreover, hybrid cooling is used to combine two or more cooling methods to improve the cooling efficiency of the whole system. The characteristics of these BTMS technologies are summarized in Table 1, including their application levels, advantages, and limitations [16–20].

Table 1. Existing BTMS technologies.

Cooling Method	Application Level	Advantages	Limitations
Air cooling	Middle	The structure is relatively simple, the design is easy to implement, and the adaptability is good.	1. The thermal conductivity of air is low, the specific heat capacity is small, and the temperature control ability under high-rate charge and discharge conditions is relatively weak. The temperature uniformity in the battery pack is relatively poor; 2. When the active cooling method meets the thermal management requirements, the power consumption is greater, the space required for the system is larger, and the energy density of the battery pack is smaller; 3. The potential for improvement is relatively small, the achievable temperature control limit is relatively low. It is mainly suitable for the battery pack with small energy density and low charge and discharge rates.
Liquid cooling (indirect)	High	1. Compared with those of air, the specific heat capacity and thermal conductivity of liquid are usually greater, and the cooling effect is usually better under the same power consumption; 2. The cooling effect can be effectively improved through the flow flux, channel design and material properties of the coolant. The potential for improvement is relatively great; 3. The temperature uniformity is usually better when the pipes or cooling plates are in contact with the side of the cells.	1. The structure of the system is more complex, the overall weight is greater, and the cost is higher; 2. In order to prevent leakage and short circuits, the coolant should be in indirect contact with the cells, which increases the thermal resistance and suppresses the cooling effect; 3. A pump is required to drive the coolant to circulate, and an additional cooling system is usually implemented to chill the coolant; 4. The pipes or cooling plates usually have great thermal conductivity, which is not conducive to inhibiting the thermal runaway propagation in the battery pack.

Table 1. *Cont.*

Cooling Method	Application Level	Advantages	Limitations
Liquid cooling (direct)	Low	1. Simple and compact structure, light weight and low cost; 2. The coolant is in direct contact with the cells, so the convective heat transfer is stronger, and the cooling effect is further improved; 3. The coolant is dielectric, which could avoid short circuiting and inhibit the thermal runaway propagation; 4. The pressure drop of the coolant is usually lower, and the power consumption is usually smaller.	1. The sealing requirements for the battery pack are higher, and conductive media are not allowed to enter the system; 2. A pump and a cooling system are usually required to drive the coolant and reduce its temperature.
Phase Change Material (PCM) cooling	Low	1. Simple structure, light weight, low cost, and easy to implement; 2. The PCM absorbs heat to reduce the temperature, and does not require additional cooling system. So, the power consumption is small; 3. The shape of the PCM is easy to change, the layout of the system is simple, and the temperature uniformity is relatively good; 4. PCMs usually have good insulation resistivity and can act as insulating materials to reduce the risk of short circuits.	1. After phase transition, the volume of the PCM usually changes obviously, increasing the possibility of leakage; 2. Most PCMs have low thermal conductivity and poor sensitivity to temperature changes. Adding materials that can improve the thermal conductivity may reduce the insulation resistivity of PCMs; 3. In the case of continuous cycles, the cooling effect declines. An additional cooling system is required to take away the heat absorbed by the PCM; 4. The heat absorbed by a certain amount of PCM is limited. Increasing the PCM volume could improve the cooling efficiency, but it will also increase the mass and power consumption.
Heat pipe cooling	Low	1. Excellent thermal conductivity and wide application range. Sensitive to temperature changes, which can effectively control the temperature in real time; 2. The working of the heat pipe alone does not require additional power consumption.	1. The structure of the system is complex, and is difficult to manufacture; 2. Higher cost and risk of leakage. Small heat capacity; 3. An additional cooling system is usually required to chill the condenser section of the heat pipe; 4. The contact area with the cells is small, and additional cooling plates are usually needed to improve the temperature uniformity.
Hybrid cooling	Middle	1. Different kinds of thermal management methods are combined with complementary advantages and limitations. It is conducive to improving the overall performance of the system; 2. The power consumption of the system can be reduced to a certain extent.	1. Compared with the single thermal management method, the volume and structure complexity of the hybrid cooling system increases, and the manufacture and maintenance costs increase accordingly; 2. The control of the hybrid cooling systems is more difficult. The coupling between different kinds of cooling methods and the adaptability of various technologies need to be further studied.

In immersion cooling, the coolant is in direct contact with the cells, eliminating the thermal resistance caused by the air gaps and thermal conductive materials. So, theoretically its heat transfer efficiency will be high. Chen et al. [21] compared the thermal management performance of several different cooling methods by experiment and calculation, including air cooling, plate cooling, and immersion cooling. It was found that when obtaining the same cooling effects, the air cooling consumed more energy, while the two liquid cooling methods consumed less. However, under small mass flow rates, the immersion cooling worked better in controlling the maximum temperature and maximum temperature difference of the cells than the plate cooling. Dubey et al. [22] compared the performance of immersion cooling and plate cooling applied on a 21,700 lithium-ion battery module by the CFD method. It was found that the performance of the two methods was basically the same at a lower discharge rate. However, at a higher discharge rate, the maximum temperature and maximum temperature difference in the battery module with immersion cooling were both remarkably lower than those in the battery module with plate cooling. Besides, with the increase in the flow rate of the coolant, the pressure drop in the cooling plate rose significantly, up to 15–25 times that of the immersion cooling.

However, since the cells are immersed in the coolant directly, the coolant must be a fluid with insulation resistivity. So, conductive coolants are no longer suitable for the BTMS with immersion cooling. For example, Sundin et al. [23] investigated the application of a non-conductive AmpCool AC-100 coolant developed by Engineered Fluids in the BTMS with immersion cooling. It was found that the average temperature of the cells was maintained at about 22.5 °C during rapid charging and discharging cycles, and the temperature fluctuation of each monitor was very small. In contrast, when forced air cooling was used, the average temperature was 28.7 °C, and the temperature fluctuation was much greater. Bhattacharjee et al. [24] designed an immersion cooling system for a lithium-ion battery stack at the scale of kWs. The system also used the AmpCool AC-100 coolant developed by Engineered Fluids. Simulation results showed that, compared with air cooling and plate cooling, the immersion cooling could significantly reduce the maximum temperature of the battery stack during discharging. Jithin et al. [25] applied deionized water, mineral oil, and AmpCool AC-100 to the BTMS with immersion cooling. It was found that the deionized water could control the average temperature rise of the cells within 5 °C under low mass flow rate (0.003 kg/s). Its cooling effect was the best, and its power consumption was the lowest. The cooling effects of mineral oil and AmpCool AC-100 were similar, but the power consumption of AmpCool AC-100 was less.

Tian et al. [26] applied hydrofluoroether (HFE-6120) to a BTMS with immersion cooling, and found that compared with those of air cooling, the maximum temperature, maximum temperature difference, and temperature standard deviation of the battery module with immersion cooling were significantly reduced under the same power consumption. Wang et al. [27] used a transformer oil with high insulation resistivity as the coolant, and designed an immersion cooling thermal management system for pouch-type lithium-ion cells. The effects of the immersion depth and the flow rate on the maximum temperature and the maximum temperature difference were investigated. It was found that when the cells were completely immersed in the coolant, the cooling performance was the best. By increasing the flow rate of the coolant, the cooling performance could be further improved. Patil et al. [28] immersed the pouch-type lithium-ion cells in a non-conductive coolant, and designed a BTMS with immersion cooling. It was found that when the coolant was flowing and tab cooling was assisted, the cells could obtain the lowest temperature rise and the smallest temperature difference. Meanwhile, the effects of the discharge rate, spacing between the cells, type and flow rate of the coolant, and inlet and outlet positions on the thermal management performance were investigated.

Typically, the coolant of the BTMS with immersion is in direct contact with the cells. However, Luo et al. [29] designed a special sealing structure for the battery module. Water was injected into the cooling channels as the coolant, which could directly contact the cells without causing a short circuit. It was found that the maximum temperature of the cells discharged at 2C could be controlled below 50 °C at the flow rate of 200 mL/min, and the maximum temperature difference could be controlled within 5 °C at the flow rate of 500 mL/min.

Thermal safety design for cells is also essential in BTMSs [30]. The immersion cooling with the application of the dielectric coolant can effectively inhibit the propagation if the thermal runaway of a single cell occurs [20]. For example, Wu et al. [31] used silicone oil as the coolant to design a BTMS based on immersion cooling. The temperatures of the battery modules with immersion cooling and with tubular liquid cooling were compared through numerical simulations. It was found that the maximum temperature rise and maximum temperature difference of the battery module with immersion cooling were significantly lower than those with tubular liquid cooling, and the immersion cooling could effectively suppress the spread of thermal runaway.

In summary, immersion cooling has a simpler design structure and lower pressure drop, and could achieve better performance in BTMSs. With the continuous increase in consumers' demands for the power of electric vehicles, as well as the rapid application of fast charging technology, higher requirements have been surfaced for BTMSs under high-rate charge and discharge conditions. It is necessary to investigate the performance and design method of immersion cooling in the BTMS more deeply. However, the current studies on BTMSs with immersion cooling did not investigate the effects of the viscosity, specific heat capacity, thermal conductivity, and other material parameters of the coolant on the thermal management performance of the system, nor did they discuss the pressure drop of the coolant. To fill the above knowledge gaps, this research designed a BTMS with immersion cooling based on the pouch-type lithium-ion cells and the coolant of dimethyl silicone oil. The coolant was in direct contact with the cells and absorbed the heat. The performance of the BTMS was obtained by a numerical method and compared with that of the natural cooling method. The effects of the flow rate, viscosity, specific heat capacity, and thermal conductivity of the coolant on the maximum temperature (MAT) and the maximum temperature difference (MATD) of the cell, as well as the pressure drop of the coolant, were further investigated, which could provide references for the design and application of BTMSs with immersion cooling in the future.

2. Models and Governing Equations

2.1. Battery

The battery used in this research was a pouch-type lithium-ion cell of 24 Ah. The basic parameters of the cell are shown in Table 2. The shape and dimensions of the cell are shown in Figure 1.

Table 2. Basic parameters of the cell [32].

Parameters	Value
Capacity (Ah)	25
Rated voltage (V)	3.8
Cathode material	$LiMn_xCo_yNi_zO_2$, $LiMn_2O_4$
Anode material	graphite
Charging cut-off voltage (V)	4.2
Discharging cut-off voltage (V)	3.0
Width (mm)	166
Height (mm)	205
Thickness (mm)	7.2

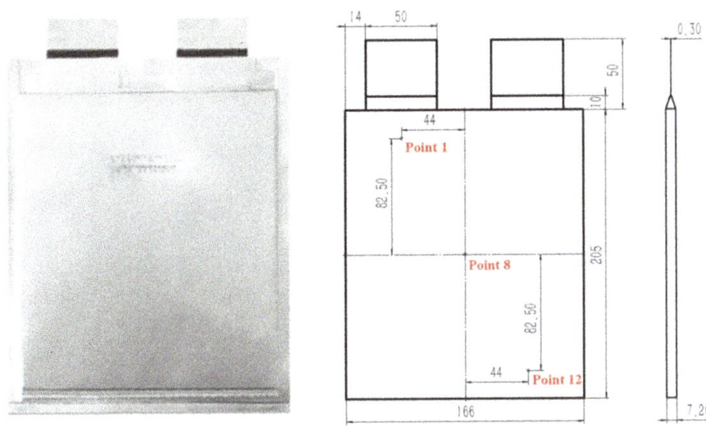

(a) **(b)**

Figure 1. (**a**) Shape of the cell; (**b**) dimensions of the cell (mm).

The electric and thermal behaviors of the cell were obtained by the electric–thermal coupled model. The electric model was the NTGK model in ANSYS/Fluent. This model has great advantages in the parameter setting and building process, and is widely used in the performance simulation of lithium-ion batteries [33]. Assuming that the current in the current collector is distributed on the spanwise plane of the cell, the potential in the current collector is:

$$\sigma_i \frac{\partial^2 \varnothing_i}{\partial x^2} + \sigma_i \frac{\partial^2 \varnothing_i}{\partial y^2} + \frac{J \cdot \vec{n}}{\delta_i} = 0 \; (i = p, n) \tag{1}$$

where σ_i is the conductivity of current collector i (S·m^{-1}), $\varnothing_i$ is the potential in current collector i (V), δ_i is the thickness of the current collector i (m), and p or n in the subscript indicates the positive or negative electrode. J is the current density across the separator (A·m^{-2}), and $\vec{n}$ is the normal unit vector perpendicular to the current collector.

The boundary condition at the junction of the current collector and the tab in the positive electrode is:

$$\vec{n} \cdot (-\sigma_p \nabla \varnothing_p) = \frac{I_{cell}}{L_{tab} \delta_p} \tag{2}$$

where I_{cell} is the current of a single pair of electrodes (A), L_{tab} is the width of the tab (m), and δ_p is the thickness of the current collector in the positive electrode (m).

The boundary condition at the junction of the current collector and the tab in the negative electrode is:

$$\varnothing_n = 0 \tag{3}$$

According to the actual situation, other boundary conditions of the electric model are insulation conditions.

The current density passing through the separator can be calculated by the equivalent resistance:

$$J = \frac{U_{oc} - (\varnothing_p - \varnothing_n)}{R} \tag{4}$$

where R is the internal resistance per unit area in the cell (Ω·m^{-2}), and U_{oc} is the open-circuit voltage of the cell (V).

The temperature distribution in the cell is:

$$\rho C_p \frac{\partial T}{\partial t} = \lambda_{in} \frac{\partial^2 T}{\partial X^2} + \lambda_{in} \frac{\partial^2 T}{\partial Y^2} + \lambda_{th} \frac{\partial^2 T}{\partial Z^2} + q \tag{5}$$

where ρ is the density (kg·m^{-3}), C_p is the specific heat capacity at constant pressure (J·kg^{-1}·K^{-1}), and T is the thermodynamic temperature (K). λ_{in} is the thermal conductivity parallel with the spanwise plane of the cell (W·m^{-1}·K^{-1}), λ_{th} is the thermal conductivity perpendicular to the spanwise plane of the cell (W·m^{-1}·K^{-1}), and q is the heat production rate of the cell (W·m^{-3}).

The heat production rate in the Bernardi's heat production model is expressed as:

$$q = q_{rev} + q_{irev} + q_{CC,p} + q_{CC,n} \tag{6}$$

$$q_{rev} = -\frac{J/2}{(d_{PE} + d_S + d_{NE})} T \frac{\partial U_{oc}}{\partial T} \tag{7}$$

$$q_{irev} = \frac{J/2}{(d_{PE} + d_S + d_{NE})} (U_{oc} - \varnothing_p + \varnothing_n) = \frac{J^2 R/2}{(d_{PE} + d_S + d_{NE})} \tag{8}$$

$$q_{CC,i} = \frac{\sigma_i (\nabla \varnothing_i)^2}{d_{CC,i}/2} \tag{9}$$

where q_{rev} is the reversible heat production rate (W·m^{-3}), q_{irev} is the irreversible heat production rate (W·m^{-3}), and $q_{CC,\,i}$ is the ohmic heat production rate of the current collector (W·m^{-3}). d_{PE}, d_S, and d_{NE} are the thicknesses of the positive electrode, separator, and negative electrode (m).

The heat production at the tab area is originated from the impedance of the tab and the contact impedance between the tab and the wire:

$$q_{tab,i} = I_{tab}^2 \cdot R_{tab,i} \tag{10}$$

$$R_{tab,i} = H_{tab} \left(\frac{1}{\sigma_{tab,i}} + \frac{1}{\sigma_{c,i}} \right) \tag{11}$$

where $q_{tab,i}$ is the heat production rate of the tab (W), I_{tab} is the current passing through the tab (A), $R_{tab,i}$ is the impedance of the tab (Ω), and H_{tab} is the height of the tab (m). $\sigma_{tab,i}$ is the conductivity of the tab (S·m^{-1}), $\sigma_{c,i}$ is the converted conductivity of the contact impedance between the wire and the tab (S·m^{-1}).

The boundary condition of the thermal model is:

$$Q_c = hA_c \left(T_w - T_f \right) \tag{12}$$

where Q_c is the convective heat exchange (W), h is the convective heat transfer coefficient (W·m^{-2}·K^{-1}), A_c is the convective heat transfer area (m^2), T_w is the temperature on the surface of the cell (K), and T_f is the temperature of the fluid (K).

We imported the three-dimensional cell model in Figure 1b into ANSYS and performed meshing. The result is shown in Figure 2a. The number of grids was 44,536 and the number of nodes was 57,704. Then, the electric–thermal coupled model of the cell was built in ANSYS/Fluent. The relevant material parameters are shown in Tables 3–5. The relationships of the open-circuit potential and internal resistance with the state of charge (SOC) are referenced from the experiment results in [32].

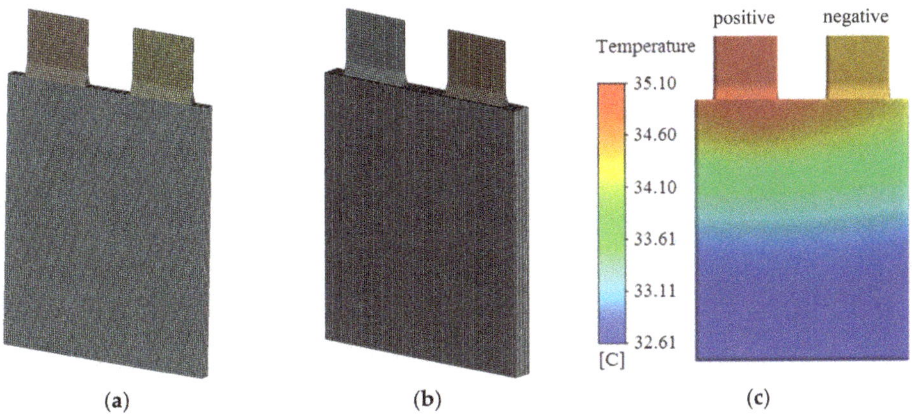

Figure 2. (**a**) Grids of the cell; (**b**) grids of the model used for calculation; (**c**) temperature distribution on the cell at the end of 1.5 C discharge.

Table 3. Material parameters of the cell [32].

Cell	Value
Density ρ_{cell} (kg·m^{-3})	2300
Specific heat capacity $C_{p,cell}$ (J·kg^{-1}·K^{-1})	1243
Convective heat transfer coefficient h_{cell} (W·m^{-2}·K^{-1})	3.5
Thermal conductivity λ_{in} (W·m^{-1}·K^{-1})	21
Thermal conductivity λ_{th} (W·m^{-1}·K^{-1})	0.48

Table 4. Material parameters of the current collectors [32].

Current Collector	Positive	Negative
Thickness δ_i (μm)	20	14
Electric conductivity σ_i (S·m^{-1})	37.8×10^6	59.6×10^6

Table 5. Material parameters of the tabs [32].

Tab	Positive	Negative
Width L_{tab} (mm)	50	50
Height H_{tab} (mm)	50	50
Thickness δ_{tab} (mm)	0.3	0.3
Density ρ_{tab} (kg·m^{-3})	8700	2700
Specific heat capacity $C_{p,tab}$ (J·kg^{-1}·K^{-1})	385	900
Electric conductivity σ_{tab}(S·m^{-1})	59.6×10^6	37.8×10^6
Thermal conductivity λ_{tab} (W·m^{-1}·K^{-1})	401	237
Convective heat transfer coefficient h_{tab} (W·m^{-2}·K^{-1})	4	4
Converted conductivity $\sigma_{c,tab}$(S·m^{-1})	80×10^6	70×10^6

Based on the experiment conditions given in [32], the voltages of the cell at three discharge rates (0.5 C, 1 C, 1.5 C) were calculated and compared with the experiment results given in [32], as shown in Figure 3a. It could be seen that the calculated results were basically consistent with the experiment results. Meanwhile, the temperature distribution on the cell at the end of 1.5 C discharge was calculated, as shown in Figure 2c. The temperature gradually decreased from the top to the bottom, and the temperature at the junction of the positive tab and the cell was the highest. The maximum temperature, minimum temperature, and maximum temperature difference were essentially consistent with the calculation results given in [32]. Due to the different software used (Comsol was used in [32]), the color distributions on the temperature scales are not completely identical, so the display styles of the temperature distributions on the cell are slightly different. As shown in Figure 1b, three temperature monitor points were set at the corresponding positions of the model. The temperatures of the three points during 1.5 C discharge were compared with the experiment results provided in [32], as shown in Figure 3b. It can be seen that the calculated results were in good consistency with the experiment results.

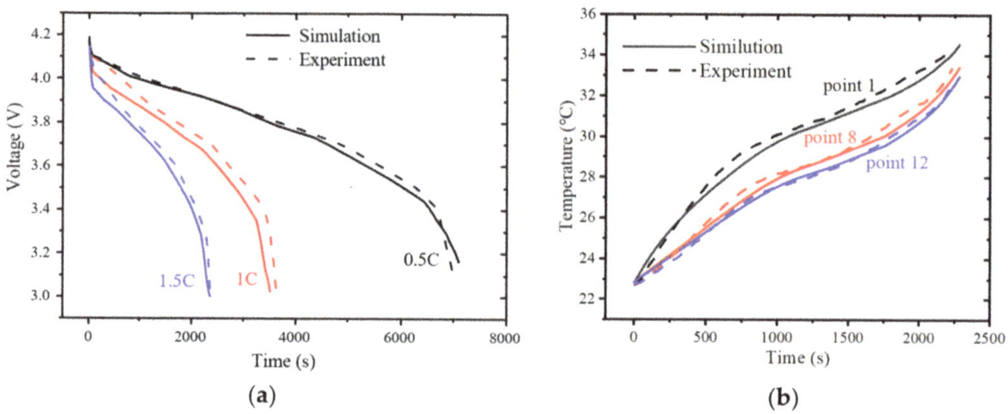

Figure 3. (**a**) Voltages of the cell during discharge at 0.5 C, 1 C and 1.5 C; (**b**) temperatures at point 1, point 8 and point 12 during discharge at 1.5 C.

When an object is placed in static air for natural convection, the variation range of the convective heat transfer coefficient is usually between 5 and 25. So, we set the convective heat transfer coefficient h_{cell} to 5, 10, 15, 20 and 25, respectively. The sensitivity analysis was performed with the maximum temperature and minimum temperature of the cell as the indexes. The results are shown in Figure 4. With the increase in h_{cell}, the maximum temperature and minimum temperature of the cells both demonstrated significant downward trends. This indicates that the convective heat transfer between the cell and the air can significantly affect the temperature distribution in the cell. As shown in Figure 1, the thickness of the pouch cell was small, but its surface area was large.

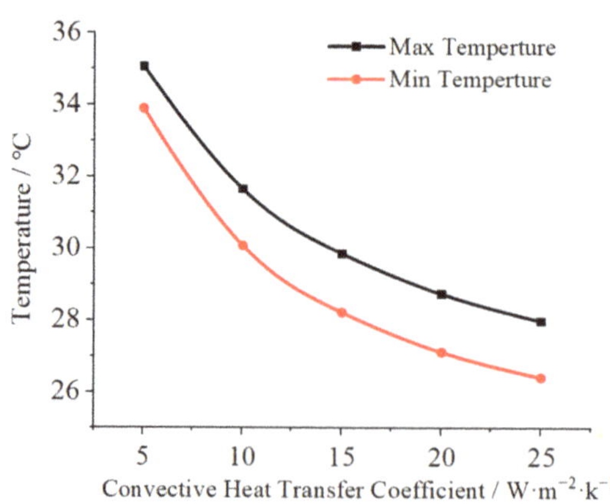

Figure 4. Sensitivity analysis of convective heat transfer coefficient.

2.2. Thermal Management System

The structure of the BTMS with immersion cooling is shown in Figure 5. The coolant is pumped into the battery module, where it passes through the cells and takes the heat away. The temperature of the coolant rises after it absorbs the heat. Then, the coolant flows into the heat exchanger and exchanges heat with the cooling system, and the air before it flows into the battery module again. The coolant circulates in the system and keeps the cells working under suitable temperature conditions.

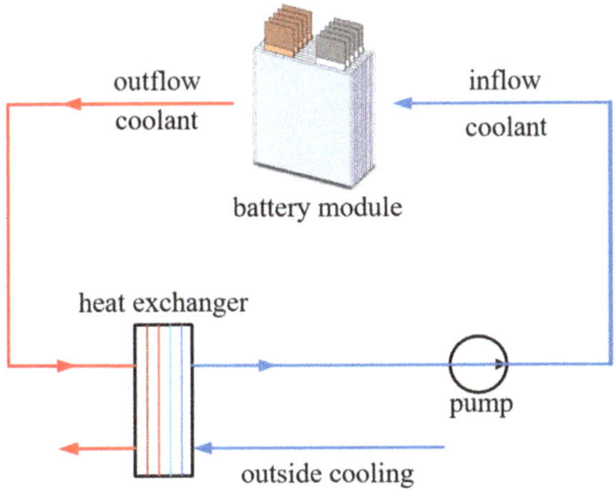

Figure 5. Structure of the BTMS with immersion cooling.

The structure of the battery module is shown in Figure 6a. The cells and the coolant channels were arranged vertically in the battery module, and separated from each other. The width of each coolant channel was 7.6 mm, and its length and height had the same dimensions with those of the cell, as shown in Table 6. Compared with the front or back surface area of the cell, the side surface area of the cell was very small. When designing the immersion cooling system, the inlets and outlets of the coolant channels are set on the planes of the side surfaces of the cells. So, the heat transfer between the side surfaces of the cells and the coolant can be ignored. Only the heat transfer between the front and back surfaces of the cells and the coolant was considered. Since the coolant was in direct contact with the cells, and the water–glycol solution commonly used in plate cooling was electrically conductive, it was necessary to use a non-conductive coolant. Silicone oil has excellent insulation resistivity, and its resistance and viscosity changes little in a wide temperature range, so it is often used as a heat carrier. At a suitable flow rate, silicone oil can provide good cooling effects for the cells [34]. Moreover, silicone oil has good shear stability. When the vehicle runs on a bumpy road, the impact of the oil on the cells is very small. So dimethyl silicone oil was selected as the coolant in this research. Its material parameters are shown in Table 7.

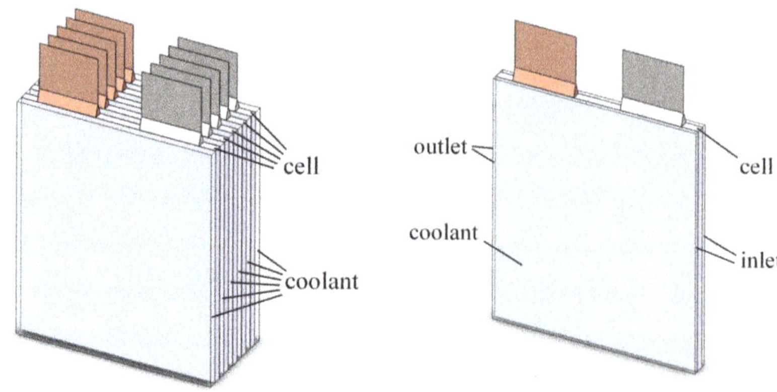

Figure 6. (**a**) Structure of the battery module; (**b**) model used for calculation.

Table 6. Dimensions of the coolant channel.

Dimension	Value
Length (mm)	166
Width (mm)	7.2
Height (mm)	205

Table 7. Material parameters of dimethyl silicone oil.

Material Parameter	Value
Density ρ_c (kg·m^{-3})	968
Viscosity μ_c (N·s·m^{-2})	1.452
Specific heat capacity $C_{p,c}$ (J·kg^{-1}·K^{-1})	1630
Thermal conductivity λ_c(W·m^{-1}·K^{-1})	0.16

In the model, the dimethyl silicone oil was assumed to be an ideal incompressible fluid, and no phase change occurred during the cooling process. The initial flow rate of the coolant at the inlet was 1 mm/s. According to the structure and material parameters, as well as the flow rates involved in the following investigation, the Reynolds numbers calculated were all less than 2300, so the flow pattern was a laminar flow.

The continuity equation of the coolant is:

$$\frac{\partial \rho_c}{\partial t} + \nabla(\rho_c v) = 0 \tag{13}$$

The energy conservation equation of the coolant is:

$$\rho_c \frac{\partial T_c}{\partial t} + \nabla \cdot (\rho_c v T_c) = \nabla \cdot \left(\frac{\lambda_c}{C_{p,c}} \nabla T_c \right) \tag{14}$$

The momentum conservation equation of the coolant is:

$$\rho_c \frac{dv}{dt} = -\nabla P + \mu_c \nabla^2 v \tag{15}$$

where ρ_c, μ_c, λ_c, $C_{p,c}$ and T_c are the density, viscosity, thermal conductivity, specific heat capacity and temperature of the coolant, respectively. v and p are the velocity and pressure of the coolant, respectively.

The front and the back surfaces of each cell were both in contact with the coolant, that is, one cell was in contact with two fluid domains. To simplify the model, we took out one cell in the center of the battery module and the two fluid domains attached to it. We cut each fluid domain from its middle plane in the spanwise direction, and applied symmetrical boundary conditions on the middle planes. The final model used for calculation is shown in Figure 6b. The boundary conditions and initial conditions are listed in Table 8. Since the heat production rate of the positive tab was higher than that of the negative tab, the coolant flowed in from the positive tab side of the cell and flowed out from the other side.

Table 8. Boundary conditions and initial conditions.

Name	Condition
Inlet flow rate (mm/s)	1
Initial temperature of coolant (°C)	23
Flow pattern	Laminar
Outlet pressure	Atmosphere
Wall condition	Non-slip

Five different grid numbers were used for the grid independence verification, and the average grid qualities were all above 0.9. The maximum temperature (MAT) and minimum temperature (MIT) of the cell (excluding the tabs) at the end of 3 C discharge were used as indicators. As shown in Figure 7a, when the number of grids increased from 86,488 to 129,868, the MAT and MIT remained basically unchanged. Five different time steps were used for the time step independence verification. As shown in Figure 7b, when the time step decreased from 3 s to 1 s, the MAT and MIT remained basically unchanged. Considering both the amount of computation and the accuracy of results, the grid number of 86,488 and the time step of 3 s were finally adopted for the subsequent calculations. The final grids are shown in Figure 2b.

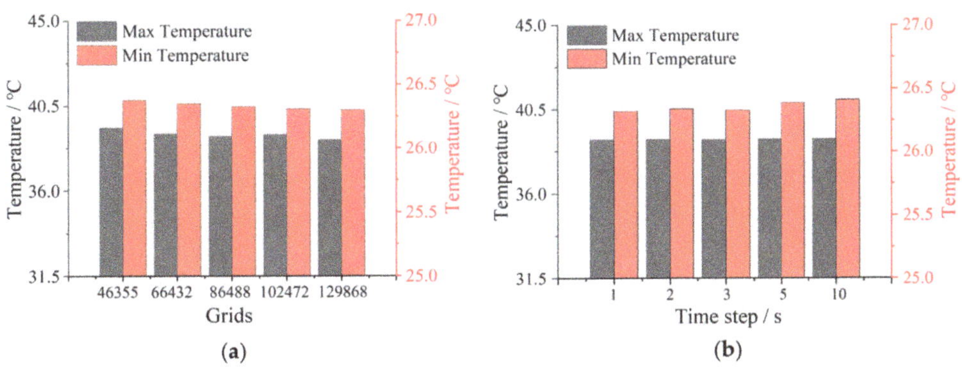

Figure 7. (**a**) Grid independence verification; (**b**) time step independence verification.

3. Performance of BTMS with Immersion Cooling

According to whether the coolant experiences phase transition during the convective heat transfer process [35], immersion cooling can be classified into single-phase immersion cooling and two-phase immersion cooling [20]. In this research, due to the thermal characteristics of the dimethyl silicone oil, the coolant did not experience phase transition, so it belongs to single-phase immersion cooling. Compared with two-phase immersion cooling, single-phase immersion cooling usually has a simpler structure and lower cost., The MAT and the maximum temperature difference (MATD) of the cell (excluding the tabs) with immersion cooling during the 3 C discharging process are shown in Figure 8. The temperature distribution on the cell at the end of the 3 C discharge is shown in Figure 9. There are two conditions considered in immersion cooling: stationary immersion cooling (the flow rate of the coolant is 0 mm/s), and flowing immersion cooling (the flow rate of the coolant is 1 mm/s). For comparison, the temperature information on the cell with natural cooling is also provided in the figures. Natural cooling is to put one single cell in the air, cooled passively by natural convective heat transfer. The model is the same with the one in Section 2.1.

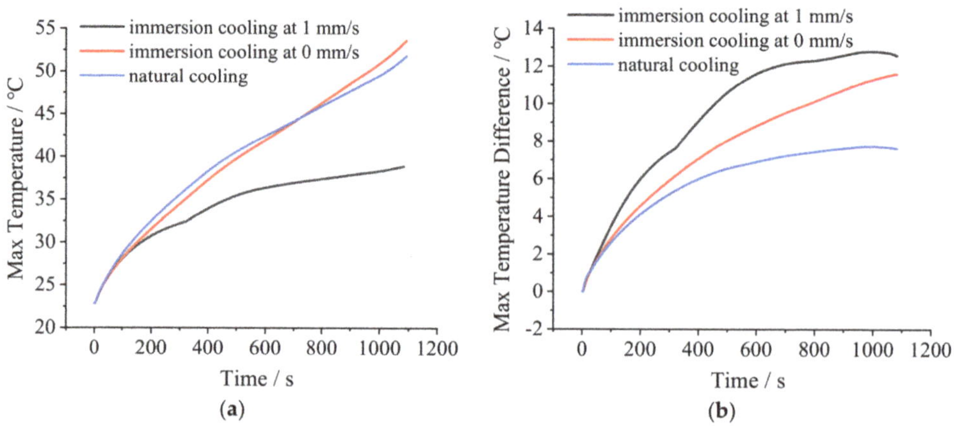

Figure 8. (**a**) MAT of the cell during 3 C discharging; (**b**) MATD of the cell during 3 C discharging.

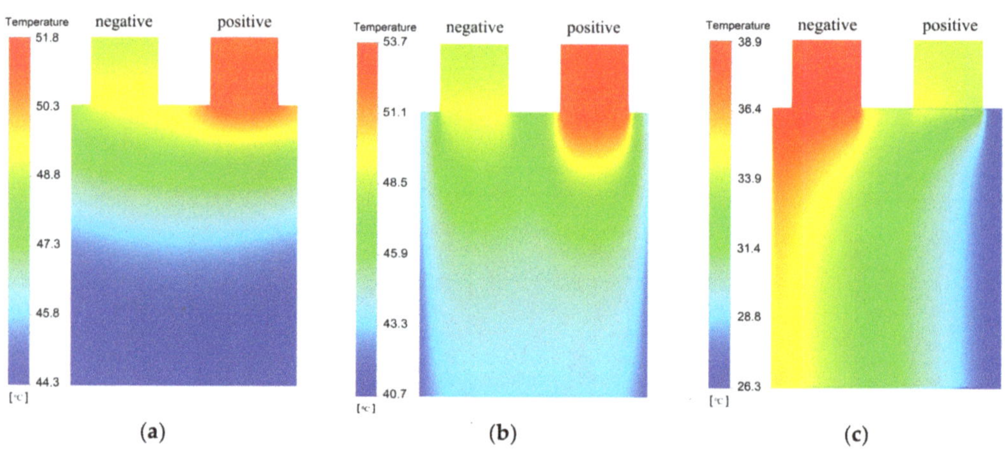

Figure 9. Temperature distributions on the cell at the end of 3 C discharge: (**a**) natural cooling; (**b**) immersion cooling at 0 mm/s; (**c**) immersion cooling at 1 mm/s.

It can be seen from Figure 8a that the MAT of the cell with natural cooling demonstrated a rapid upward trend during discharging. At the end of discharge, the MAT of the cell reached 51.8 °C, which exceeded the suitable temperature range of lithium-ion batteries. However, with stationary immersion cooling, the MAT of the cell also demonstrated a rapid upward trend. At the end of discharge, the MAT reached 53.6 °C, which even exceeded that with natural cooling. This was caused by several reasons. First, the specific heat capacity of dimethyl silicone oil is relatively small (1630 $J{\cdot}kg^{-1}{\cdot}k^{-1}$), and is much smaller than that of water (4200 $J{\cdot}kg^{-1}{\cdot}k^{-1}$). So, its heat absorption and storage capability are relatively weak. Second, the model of the cell with immersion cooling is part of the battery module. The tight arrangement of the cells and the heat generated by the surrounding cells both affect the heat dissipation of the cell. Third, the stationary coolant only exchanges heat with the cell. Other thermal boundary conditions of the coolant are adiabatic. Therefore, during the discharging process, the temperature of the coolant increases continuously, and the heat exchange between the coolant and the cell decreases continuously. Eventually, the temperature of the cell rises rapidly. However, natural cooling involves only putting one cell in the air at 23 °C. The higher the temperature of the cell, the stronger the heat exchange between the air and the cell. So, at the end of discharge, the MAT of the cell with natural cooling is slightly lower than that with stationary immersion cooling. However, if the cell is put in a tightly packed battery module, the MAT at the end of discharge could be much higher.

With flowing immersion cooling, the MAT of the cell was significantly lower than those with natural cooling and stationery immersion cooling during the entire discharging process. At the end of discharge, the MAT was greatly reduced to 38.9 °C, within the suitable temperature range of lithium-ion batteries. This is because the flowing coolant can absorb and take the heat away from the cell continuously, thus greatly improving the cooling effect.

However, as shown in Figure 8b, the MATDs of the cells with stationary immersion cooling and flowing immersion cooling were both significantly greater than that with natural cooling during the entire discharging process. At the end of discharge, the MATD of the cell with stationary immersion cooling reached 11.6 °C, and that with flowing immersion cooling reached 12.6 °C. Both of them exceeded the limit of 5 °C. The reason is related to the temperature distribution of the cell. As shown in Figure 9a,b, with natural cooling and stationary immersion cooling, the MATs both appear at the junction between the positive tab and the cell. The MITs both appear at the lower part of the cell. Therefore, the MATDs of the cells were determined by the temperature near the positive tab and the temperature of the lower part of the cell under these two conditions.

With natural cooling, the temperature near the tab is higher, where the convective heat exchange between the cell and the air is stronger. So, the increase in the MAT is inhibited. The temperature of the lower part of the cell is lower, where the convective heat exchange between the cell and the air is weaker. So, the increase in the MIT is accelerated. The combined action of the two factors results in a smaller MATD of the cell. With stationary immersion cooling, the temperature of the coolant close to the junction between the positive tab and the cell is higher after the coolant absorbs heat, so its cooling capability is weakened, and the increase in the MAT is accelerated. The temperature of the coolant far from the tab is lower, so its cooling capability is stronger, and the increase in the MIT is inhibited. The coolant with higher temperature is in the upper part, and coolant with lower temperature is in the lower part. It inhibits the convective heat transfer in the coolant, thereby further increasing the MATD of the cell.

As shown in Figure 9c, with flowing immersion cooling, the MAT appears at the left side of the cell, and the MIT appears at the right side of the cell. The temperature gradually increases from the right to the left. This is because the coolant with lower temperature was in direct contact with the cell at the inlet, so the cooling effect on the cell close to the inlet was very good. However, in the initial settings, the flow rate of the coolant was only 1 mm/s. During the flowing process, its temperature gradually increased, and the convective heat exchange between the coolant and the cell was weaker. So, the closer to the outlet, the greater the temperature rise of the cell, resulting in the increase in MATD. It is also noted that the MAT was transferred from the junction of the positive tab and the cell to the junction of the negative tab and the cell. Although the positive tab has greater heat production, and is not in direct contact with the coolant, a good cooling effect can also be achieved through heat conduction since it is closer to the inlet of the coolant. The negative tab is far away from the inlet, so its temperature is the highest, but still far below the corresponding temperatures with natural cooling and stationary immersion cooling.

Through the above analysis it can be seen that, compared with natural cooling and stationary immersion cooling, flowing immersion cooling demonstrates strong temperature suppression capability. The MAT of the cell can be controlled within 40 °C during the 3 C discharging process. The temperature of the tabs is also much lower. However, due to the small flow rate of the coolant, the MATD of the cell cannot be controlled within 5 °C, which needs to be further improved.

4. Effects of Different Parameters on the Performance of Immersion Cooling

The BTMS with immersion cooling has many parameters that could affect its performance, such as the flow rate, viscosity, specific heat capacity, and thermal conductivity of the coolant. By changing these parameters, their effects on the performance of immersion cooling were investigated, so as to provide references for the design and optimization of the BTMS. The method was to change one parameter at a time, while keeping the other parameters constant: the single variable method. The MAT and MATD of the cell (excluding the tabs) at the end of 3 C discharge, as well as the pressure drop of the coolant, were used as indicators. To facilitate analysis, the MIT of the cell was also supplemented as an indicator.

4.1. Flow Rate

The flow rate determines the mass flow of the coolant passing through the cell, which in turn determines whether it can quickly remove heat from the cell. However, the flow of the coolant requires power. The lower the flow rate of the coolant, the lower the energy consumption of the system. In the initial settings, the flow rate of the coolant was 1 mm/s. To investigate its effect on the performance of immersion cooling, the flow rate was set to 2 mm/s, 4 mm/s, 6 mm/s, 8 mm/s, 10 mm/s, and 12 mm/s, with the other parameters remaining unchanged. The MAT, MIT, and MATD of the cell at the end of discharge, as well as the pressure drop of the coolant, are shown in Figure 10.

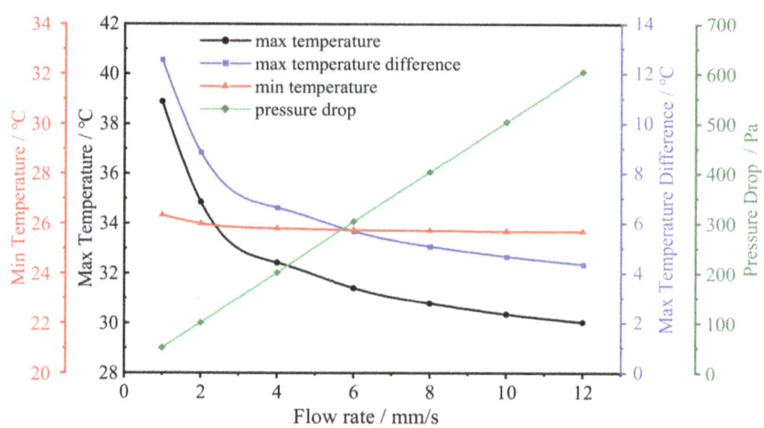

Figure 10. Relationship between the flow rate of the coolant and the MAT, MIT, and MATD of the cell, as well as the pressure drop of the coolant.

It can be seen that, with the increase in the flow rate, the MAT of the cell decreased from 38.9 °C to 29.9 °C, a reduction of 9.0 °C. The MATD decreased from 12.5 °C to 4.3 °C, a reduction of 8.2 °C. Both of them demonstrate significant downward trends. This is because, as the flow rate increases, the coolant can take more and more heat away from the cell. However, as shown in Figure 9c, when the flow rate is 1 mm/s, the temperature of the cell near the inlet of the coolant is already low, but the temperature of the cell near the outlet of the coolant is much higher. When the flow rate increases, the convective heat exchange between the coolant and the cell near the inlet of the coolant and the cell near the outlet of the coolant will both increases. However, the increase in the latter is greater than that of the former. So, the decline in the MAT will be much greater than that of the MIT, which reduces the MATD of the cell. When the flow rate reaches 8 mm/s, the MATD decreases to 5 °C.

It can also be seen that the flow rate influences the pressure drop significantly. The higher the flow rate, the greater the pressure drop, and they are in direct proportion. This is because the flow pattern of the coolant is laminar, and the pressure drop is mainly composed of the frictional pressure drop in the channel:

$$\Delta P = f \frac{l}{D} \frac{\rho_c v^2}{2} \tag{16}$$

where $f = \frac{64}{Re}$, after substitution:

$$\Delta P = \frac{32 l v \mu_c}{D^2} \tag{17}$$

where v is the flow rate of the coolant, μ_c is the dynamic viscosity of the coolant, l is the length of the channel, and D is the hydraulic diameter of the channel.

The pressure drop of the coolant determines the pumping power of the circulating system. The pumping power required by the coolant in the flow channel is:

$$P = \Delta P \cdot A \cdot v \tag{18}$$

where A is the cross-sectional area of the flow channel. Therefore, the pumping power increases with the pressure drop. In practical applications, there is a trade-off between cooling performance and pumping cost.

Therefore, by increasing the flow rate of the coolant, the MAT and MATD of the cell can be effectively reduced. However, the pump providing the power will consume more energy. The flow rate should be adjusted in real time according to the actual needs of the BTMS.

4.2. Viscosity

The viscosity of the coolant is one of the material properties. The viscosity of water is about 0.001 $N \cdot s \cdot m^{-2}$, while that of high-viscosity silicone oil can reach tens of thousands of $N \cdot s \cdot m^{-2}$. In the initial settings, the coolant is dimethyl silicone oil, and its viscosity is 1.456 $N \cdot s \cdot m^{-2}$. To investigate its effect on the performance of immersion cooling, the viscosity was set to 0.1 $N \cdot s \cdot m^{-2}$, 0.8 $N \cdot s \cdot m^{-2}$, 1.5 $N \cdot s \cdot m^{-2}$, 2.2 $N \cdot s \cdot m^{-2}$, and 2.9 $N \cdot s \cdot m^{-2}$, with the other parameters remaining unchanged. The MAT, MIT, and MATD of the cell at the end of discharge, as well as the pressure drop of the coolant, are shown in Figure 11.

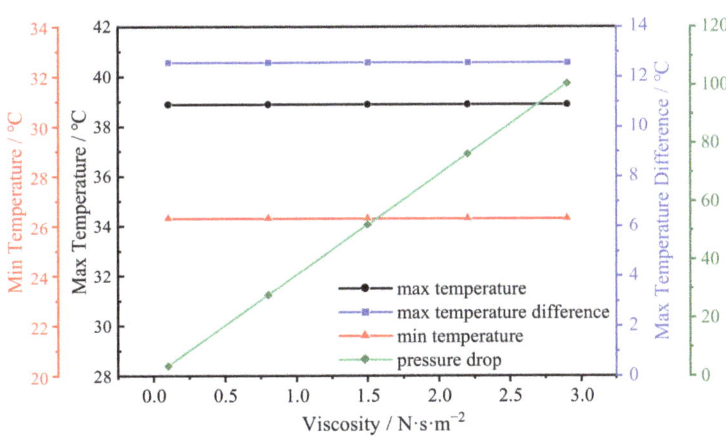

Figure 11. Relationship between the viscosity of the coolant and the MAT, MIT, MATD of the cell, as well as the pressure drop of the coolant.

It can be seen that the viscosity of the coolant basically does not influence the MAT and MATD of the cell. The MAT is maintained at 38.9 °C stably, and the MATD is maintained at 12.5 °C stably. After calculating the Reynolds numbers of the coolant with different viscosities, it was found that they were all below 2100, so the flow patterns were all laminar. Therefore, the change in the viscosity has little effect on the flowing process of the coolant, and the convective heat transfer between the coolant and cell basically does not change.

However, the viscosity influences the pressure drop significantly. The greater the viscosity, the greater the pressure drop—in direct proportion—which is consistent with Equation (17). For the BTMS with immersion cooling in this research, shown in Figure 5, the coolant needs to circulate in the system and exchange heat in the heat exchanger. Higher viscosity is not conducive to the flow of coolant, and will also increase the energy consumption of the system. Therefore, the coolant with lower viscosity is better.

4.3. *Specific Heat Capacity*

Specific heat capacity refers to the heat absorbed or released by a substance of unit mass when it rises or falls to unit temperature. The specific heat capacity of the coolant is also one of the material properties. At room temperature, the specific heat capacity of water is about 4200 $J{\cdot}kg^{-1}{\cdot}k^{-1}$. While in the initial settings, the specific heat capacity of dimethyl silicone oil is 1630 $J{\cdot}kg^{-1}{\cdot}k^{-1}$. To investigate its effect on the performance of immersion cooling, the specific heat capacity was set to 1700 $J{\cdot}kg^{-1}{\cdot}k^{-1}$, 2300 $J{\cdot}kg^{-1}{\cdot}k^{-1}$, 2900 $J{\cdot}kg^{-1}{\cdot}k^{-1}$, 3500 $J{\cdot}kg^{-1}{\cdot}k^{-1}$, 4100 $J{\cdot}kg^{-1}{\cdot}k^{-1}$, with the other parameters remaining unchanged. The MAT, MIT, MATD of the cell at the end of discharge, as well as the pressure drop of the coolant are shown in Figure 12.

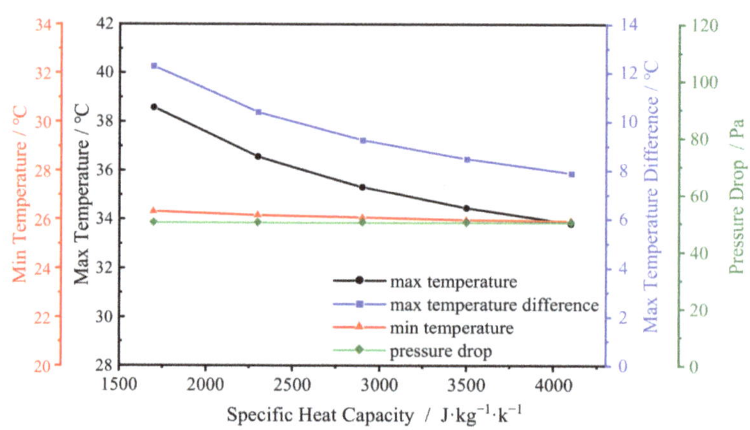

Figure 12. Relationship between the specific heat capacity of the coolant and the MAT, MIT, MATD of the cell, as well as the pressure drop of the coolant.

It can be seen that with the increase in the specific heat capacity of the coolant, both the MAT and the MATD of the cell demonstrate downward trends. The MAT of the cell decreases from 38.6 °C to 33.8 °C, a reduction of 4.8 °C. The MATD decreased from 12.2 °C to 7.8 °C, a reduction of 4.4 °C. However, the MIT decreased from 26.3 °C to 25.9 °C, a reduction of only 0.4 °C. The reason for this phenomenon is that specific heat capacity reflects the heat capacity per unit mass of the coolant. In the convective heat transfer process between the coolant and the cell, the greater its value is, the more heat can be transferred through convection. So, the convective heat exchange is stronger, significantly reducing the temperature of the cell. However, since the temperature of the cell near the inlet of the coolant is already low, the change in the specific heat capacity has a greater effect on the MAT than on the MIT. So, the MATD also decreases with the MAT.

It can also be seen that the pressure drop of the coolant basically does not change with the specific heat capacity. This is because the specific heat capacity is a thermal parameter of the material. It mainly affects the heat exchange between the coolant and the cell, not the flow pattern of the coolant.

Therefore, when designing the BTMS with immersion cooling, the coolant with greater specific heat capacity can effectively enhance its performance. Adding phase change material (PCM) into the coolant is an effective way to increase its specific heat capacity, such as the PCM emulsion and encapsuled PCM [9,36].

4.4. Thermal Conductivity

Thermal conductivity refers to the heat transmitted directly by the material of unit section and unit length under unit temperature difference. The thermal conductivity of the coolant can directly affect the convective heat transfer between the coolant and cell. At room temperature, the thermal conductivity of water is 0.6009 $W{\cdot}m^{-1}{\cdot}k^{-1}$, and that of the nanofluid prepared by Huang et al. [37] increased by 4.3% compared with water. In the initial settings, the thermal conductivity of dimethyl silicone oil was 0.16 $W{\cdot}m^{-1}{\cdot}k^{-1}$. To investigate its effect on the performance of immersion cooling, the thermal conductivity was set to 0.1 $W{\cdot}m^{-1}{\cdot}k^{-1}$, 0.2 $W{\cdot}m^{-1}{\cdot}k^{-1}$, 0.3 $W{\cdot}m^{-1}{\cdot}k^{-1}$, 0.4 $W{\cdot}m^{-1}{\cdot}k^{-1}$, 0.5 $W{\cdot}m^{-1}{\cdot}k^{-1}$, 0.6 $W{\cdot}m^{-1}{\cdot}k^{-1}$, with the other parameters remaining unchanged. The MAT, MIT, and MATD of the cell at the end of discharge, as well as the pressure drop of the coolant, are shown in Figure 13.

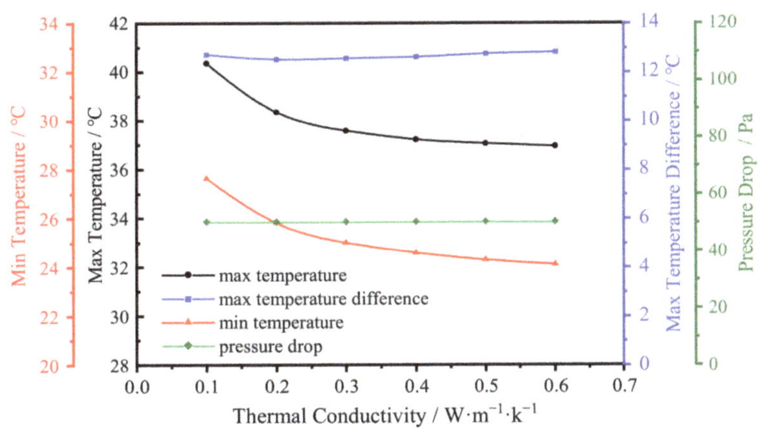

Figure 13. Relationship between the thermal conductivity of the coolant and the MAT, MIT, and MATD of the cell, as well as the pressure drop of the coolant.

It can be seen that, with the increase in the thermal conductivity of the coolant, both the MAT and the MIT of the cell demonstrate downward trends. The MAT decreased from 40.4 °C to 37.0 °C, a reduction of 3.4 °C. The MIT decreased from 27.6 °C to 24.1 °C, a reduction of 3.5 °C. When the fluid with uniform temperature passes through a solid wall with another temperature, a fluid layer with a great temperature change will be formed near the solid wall, which is called the thermal boundary layer. It can be regarded as a fluid layer with a temperature gradient, which is also the main area where heat exchange occurs. In the thermal boundary layer of laminar flow, the heat transfer in the direction perpendicular to the wall mainly depends on thermal conduction. Therefore, the thermal conductivity plays an important role in the convection heat exchange between the coolant and the cell. Increasing thermal conductivity can better promote the convective heat exchange between them, thereby enhancing the cooling effect. Since the convective heat exchange between the coolant and the entire cell is enhanced, the MAT and the MIT are reduced synchronously, resulting in insignificant changes in the MATD of the cell.

It was also noticed that, in the previous analysis, the MIT of the cell was always around 26 °C. However, after increasing the thermal conductivity, the MIT demonstrated a significant downward trend. The reason is that the convective heat exchange between the coolant and the cell at the inlet determines the MIT. In the initial setting, the smaller thermal conductivity of dimethyl silicone oil functioned as a bottleneck, restricting the convective heat exchange between the coolant and the cell at the inlet.

It can also be seen that the pressure drop of the coolant basically did not change with thermal conductivity. This is also because the thermal conductivity was a thermal parameter of the material. It mainly affects the heat exchange between the coolant and the cell, not the flow pattern of the coolant.

Therefore, when designing the BTMS with immersion cooling, the coolant with greater thermal conductivity can effectively enhance its performance. Adding thermal conductive micro-particles into the coolant is an effective way to improve its thermal conductivity [13,38].

5. Conclusions

In this research, a BTMS with immersion cooling was designed based on the electric–thermal coupled model of lithium-ion battery and the coolant of dimethyl silicone oil. The performance of the BTMS with immersion cooling was obtained through numerical calculations and compared with that with natural cooling. The effects of different parameters on the performance of immersion cooling were investigated. The following conclusions were obtained:

1. Compared with natural cooling and stationary immersion cooling, flowing immersion cooling can significantly reduce the MAT of the cell and the temperature of the tabs during the 3 C discharging process, even at a flow rate of 1 mm/s. However, both the stationary immersion cooling and the flowing immersion cooling at 1 mm/s increase the MATD of the cell significantly;
2. Within a certain range, the MAT and MATD of the cell can be greatly reduced by increasing the flow rate and specific heat capacity of the coolant. Increasing the thermal conductivity of the coolant can reduce the MAT and MIT of the cell, but has little effect on the MATD. The viscosity of the coolant basically does not affect the temperature of the cell;
3. The pressure drop of the coolant varies with the flow rate and viscosity of the coolant proportionally. Lower flow rate and smaller viscosity can reduce the energy consumption of the system. The specific heat capacity and thermal conductivity of the coolant basically do not affect the pressure drop;
4. Compared with other BTMS technologies, the structure design of the BTMS with immersion cooling is simple and compact. Its pressure drop is lower, and it can achieve a better cooling effect with smaller power consumption. The temperature uniformity can also be improved through the adjustments of design and working parameters. So, the BTMS with immersion cooling has good application prospects in electric vehicles.

Author Contributions: Conceptualization, Y.W. and Y.Z.; methodology, Y.Z.; software, Y.Z. and Z.W.; validation, Y.W. and Y.Z.; formal analysis, Y.Z.; investigation, Y.Z.; resources, Z.X.; data curation, Z.W.; writing—original draft preparation, Y.Z.; writing—review and editing, Z.X. and Y.W.; visualization, Z.W.; supervision, Y.W.; project administration, Z.X.; funding acquisition, Z.X. and Y.W. All authors have read and agreed to the published version of the manuscript.

Funding: This work is supported by Shandong Provincial Natural Science Foundation, China (ZR2020ME138); Guangdong Provincial Natural Science Foundation, China (2022A1515010240); National Key R&D Program of China (SQ2020YFF0416411); Key R&D Program of Shandong Province (2019JZZY010914); Foundation of State Key Laboratory of Automotive Simulation and Control (20181102).

Data Availability Statement: The data that support the findings of this study are available from the corresponding author upon reasonable request.

Conflicts of Interest: There are no relevant financial or non-financial competing interests to report.

References

1. Takami, N.; Ise, K.; Harada, Y.; Iwasaki, T.; Kishi, T.; Hoshina, K. High-energy, fast-charging, long-life lithium-ion batteries using $TiNb_2O_7$ anodes for automotive applications. *J. Power Sources* **2018**, *396*, 429–436. [CrossRef]
2. Diouf, B.; Pode, R. Potential of lithium-ion batteries in renewable energy. *Renew. Energy* **2015**, *76*, 375–380. [CrossRef]
3. Foss, C.E.L.; Svensson, A.M.; Gullbrekken, Ø.; Sunde, S.; Vullum-Bruer, F. Temperature effects on performance of graphite anodes in carbonate-based electrolytes for lithium-ion batteries. *J. Energy Storage* **2018**, *17*, 395–402. [CrossRef]
4. Li, H.; Wang, Y.; He, X.; Peng, W.; Yu, K.; Xie, Z. Effects of Material Parameters on the Thermal Characteristics of a Lithium Ion Battery under Fast Charging Conditions. *Energy Fuels* **2021**, *35*, 3426–3437. [CrossRef]
5. Li, W.; Chen, S.; Peng, X.; Xiao, M.; Gao, L.; Grag, A.; Bao, N. A Comprehensive Approach for the Clustering of Similar-Performance Cells for the Design of a Lithium-Ion Battery Module for Electric Vehicles. *Engineering* **2019**, *5*, 595–812. [CrossRef]
6. Mohammed, A.H.; Esmaeeli, R.; Aliniagerdroudbari, H.; Alhadri, M.; Hashemi, S.R.; Nadkarni, G.; Farhad, S. Dual-purpose cooling plate for thermal management of prismatic lithium-ion batteries during normal operation and thermal runaway. *Appl. Therm. Eng.* **2019**, *160*, 114106. [CrossRef]
7. Wei, L.; Lu, Z.; Cao, F.; Zhang, L.; Yang, X.; Yu, L.; Jin, L. A comprehensive study on thermal conductivity of the lithium-ion battery. *Int. J. Energy Res.* **2020**, *44*, 9466–9478. [CrossRef]
8. Alipour, M.; Hassanpouryouzband, A.; Kizilel, R. Investigation of the Applicability of Helium-Based Cooling System for Li-Ion Batteries. *Electrochem* **2021**, *2*, 135–148. [CrossRef]
9. Li, H.; Xiao, X.; Wang, Y.; Lian, C.; Li, Q.; Wang, Z. Performance investigation of a battery thermal management system with microencapsulated phase change material suspension. *Appl. Therm. Eng.* **2020**, *180*, 115795. [CrossRef]
10. Huang, Y.; Mei, P.; Lu, Y.; Huang, R.; Yu, X.; Chen, Z.; Anthony, R.P. A novel approach for Lithium-ion battery thermal management with streamline shape mini channel cooling plates. *Appl. Therm. Eng.* **2019**, *157*, 113623. [CrossRef]
11. Sheng, L.; Zhang, H.; Su, L.; Zhang, Z.; Zhang, H.; Li, K.; Fang, Y.; Ye, W. Effect analysis on thermal profile management of a cylindrical lithium-ion battery utilizing a cellular liquid cooling jacket. *Energy* **2020**, *220*, 119725. [CrossRef]
12. Al-Zareer, M.; Dincer, I.; Rosen, M.A. Performance assessment of a new hydrogen cooled prismatic battery pack arrangement for hydrogen hybrid electric vehicles. *Energy Convers. Manag.* **2018**, *173*, 303–319. [CrossRef]
13. Jilte, R.D.; Kumar, R.; Ahmadi, M.H. Cooling Performance of Nanofluid submerged vs. Nanofluid Circulated Battery Thermal Management Systems. *J. Clean. Prod.* **2019**, *240*, 118131. [CrossRef]
14. Wang, Y.; Wang, Z.; Min, H.; Li, Q. Performance investigation of a passive battery thermal management system applied with phase change material. *J. Energy Storage* **2021**, *35*, 102279. [CrossRef]
15. Li, Q.; Wang, C.; Wang, Y.; Wang, Z.; Li, H.; Lian, C. Study on the effect of the adiabatic section parameters on the performance of pulsating heat pipes. *Appl. Therm. Eng.* **2020**, *180*, 115813. [CrossRef]
16. Lu, M.; Zhang, X.; Ji, J.; Xu, X.; Zhang, Y. Research progress on power battery cooling technology for electric vehicles. *J. Energy Storage* **2020**, *27*, 101155. [CrossRef]
17. Lin, J.; Liu, X.; Li, S.; Zhang, Z.; Yang, S. A review on recent progress, challenges and perspective of battery thermal management system. *Int. J. Heat Mass Transf.* **2021**, *167*, 120834. [CrossRef]
18. Kim, J.; Oh, J.; Lee, H. Review on battery thermal management system for electric vehicles. *Appl. Therm. Eng.* **2019**, *149*, 192–212. [CrossRef]
19. Yang, S.; Ling, C.; Fan, Y.; Yang, Y.; Tan, X.; Dong, H. A Review of Lithium-Ion Battery Thermal Management System Strategies and the Evaluate Criteria. *Int. J. Electrochem. Sci.* **2019**, *14*, 6077–6170. [CrossRef]
20. Roe, C.; Feng, X.; White, G.; Li, R.; Wang, H.; Rui, X.; Li, C.; Zhang, F.; Null, V.; Parkes, M.; et al. Immersion cooling for lithium-ion batteries—A review. *J. Power Sources* **2022**, *525*, 231094. [CrossRef]
21. Chen, D.; Jiang, J.; Kim, G.-H.; Yang, C.; Pesaran, A. Comparison of different cooling methods for lithium-ion battery cells. *Appl. Therm. Eng.* **2016**, *94*, 846–854. [CrossRef]
22. Dubey, P.; Pulugundla, G.; Srouji, A.K. Direct Comparison of Immersion and Cold-Plate Based Cooling for Automotive Li-Ion Battery Modules. *Energies* **2021**, *14*, 1259. [CrossRef]
23. Sundin, D.W.; Sponholtz, S. Thermal Management of Li-Ion Batteries with Single-Phase Liquid Immersion Cooling. *IEEE Open J. Veh. Technol.* **2020**, *1*, 82–92. [CrossRef]
24. Bhattacharjee, A.; Mohanty, R.K.; Ghosh, A. Design of an Optimized Thermal Management System for Li-Ion Batteries under Different Discharging Conditions. *Energies* **2020**, *13*, 5695. [CrossRef]
25. Jithin, K.V.; Rajesh, P.K. Numerical analysis of single-phase liquid immersion cooling for lithium-ion battery thermal management using different dielectric fluids. *Int. J. Heat Mass Transf.* **2022**, *188*, 122608. [CrossRef]
26. Tian, X.; Lyu, P.; Fan, Y.; Rao, J.; Ouyang, K. Numerical investigation of the direct liquid cooling of a fast-charging lithium-ion battery pack in hydrofluoroether. *Appl. Therm. Eng.* **2021**, *196*, 117279.
27. Wang, H.; Tao, T.; Xu, J.; Shi, H.; Mei, X.; Gou, P. Thermal performance of a liquid-immersed battery thermal management system for lithium-ion pouch batteries. *J. Energy Storage* **2022**, *46*, 103835. [CrossRef]
28. Patil, M.S.; Seo, J.-H.; Lee, M.-Y. A novel dielectric fluid immersion cooling technology for Li-ion battery thermal management. *Energy Convers. Manag.* **2021**, *229*, 113715. [CrossRef]
29. Luo, M.; Cao, J.; Liu, N.; Zhang, Z.; Fang, X. Experimental and Simulative Investigations on a Water Immersion Cooling System for Cylindrical Battery Cells. *Front. Energy Res.* **2022**, *10*, 803882. [CrossRef]

30. Chen, Z.; Zhang, B.; Xiong, R.; Shen, W.; Yu, Q. Electro-thermal coupling model of lithium-ion batteries under external short circuit. *Appl. Energy* **2021**, *193*, 116910. [CrossRef]
31. Wu, S.; Lao, L.; Wu, L.; Liu, L.; Lin, C.; Zhang, Q. Effect analysis on integration efficiency and safety performance of a battery thermal management system based on direct contact liquid cooling. *Appl. Therm. Eng.* **2022**, *201*, 117788. [CrossRef]
32. Wu, B.; Li, Z.; Zhang, J. Thermal Design for the Pouch-Type Large-Format Lithium-Ion Batteries: I. Thermo-Electrical Modeling and Origins of Temperature Non-Uniformity. *J. Electrochem. Soc.* **2015**, *162*, A181–A191. [CrossRef]
33. Madani, S.S.; Swierczynski, M.; Kær, S.K. Cooling simulation and thermal abuse modeling of lithium-ion batteries using the Newman, Tiedemann, Gu, and Kim (NTGK) model. *Electrochem. Soc. Trans.* **2017**, *81*, 261–270. [CrossRef]
34. Karimi, G.; Dehghan, A.R. Thermal analysis of high-power lithium-ion battery packs using flow network approach: Thermal management of a high-power Li-ion battery pack. *Int. J. Energy Res.* **2014**, *38*, 1793–1811. [CrossRef]
35. Wang, Z.; Wang, Y.; Xie, Z.; Li, H.; Peng, W. Parametric investigation on the performance of a direct evaporation cooling battery thermal management system. *Int. J. Heat Mass Transf.* **2022**, *189*, 122685. [CrossRef]
36. Li, Q.; Wang, Y.; Lian, C.; Li, H.; He, X. Effect of micro encapsulated phase change material on the anti-dry-out ability of pulsating heat pipes. *Appl. Therm. Eng.* **2019**, *159*, 113854. [CrossRef]
37. Huang, D.; Wu, Z.; Sunden, B. Effects of hybrid nanofluid mixture in plate heat exchangers. *Exp. Therm. Fluid Sci.* **2016**, *72*, 190–196. [CrossRef]
38. Jilte, R.; Afzal, A.; Panchal, S. A novel battery thermal management system using nano-enhanced phase change materials. *Energy* **2020**, *219*, 119564. [CrossRef]

energies

MDPI

Article

Luenberger Observer-Based Microgrid Control Strategy for Mixed Load Conditions

Yong Cheng * and Cong Li

College of Electrical and Control Engineering, Xi'an University of Science and Technology, Xi'an 710054, China; 19206029021@stu.xust.edu.cn
* Correspondence: chengbati@163.com

Abstract: In this paper, a Luenberger observer-based microgrid control strategy is proposed to enhance the power quality of microgrids, when the grid loads are mixed and strongly non-linear. To improve performance under this condition, a Luenberger observer is designed for three phase power grids. On the basis of the observer, the components of different frequencies and sequences of voltages and currents are obtained accurately. The virtual impedance of different frequencies and sequences is designed, which makes the equivalent line impedance meet the power-sharing condition, reducing the fundamental negative sequence voltages and harmonic voltages. The active power droop equation, meanwhile, is proposed, where the bus voltage is modified. The value range of virtual impedance is discussed in the complex frequency domain. The proposed control strategy does not require any communication lines, so the hardware structure is simplified. The simulations and experiments are provided to verify the effectiveness of the proposed method.

Keywords: low-voltage microgrid; droop control; harmonic power-sharing; virtual impedance; Luenberger observer; power quality

Citation: Cheng, Y.; Li, C. Luenberger Observer-Based Microgrid Control Strategy for Mixed Load Conditions. *Energies* **2022**, *15*, 3655. https://doi.org/10.3390/en15103655

Academic Editor: André Madureira

Received: 11 March 2022
Accepted: 5 May 2022
Published: 16 May 2022

1. Introduction

With the impact of the energy crisis and concerns about carbon emissions, new energy power generations have developed rapidly [1,2]. The microgrid is composed of loads, distributed generators (DG), energy storage equipment, power electronic devices, measurements, monitoring and protection devices. It has been built and put into the application in islands, remote areas, commercial enterprises, campuses, and industrial power [3–5]. It can be connected to the grid or operated on an island. The power distribution of the distributed power sources of the microgrid under island operation is particularly important, and because there is no frequency and voltage support of the large grid, its electricity quality has many problems [6,7].

Nowadays, droop control has been widely employed in microgrid inverter control [8]. Because the line impedance is difficult to meet the power-sharing condition, the output power distribution of traditional droop control is disproportion [9,10]. Some scholars introduce virtual impedance technology into droop control and conduct a lot of research on DG power-sharing [11–22]. An intermediate instantaneous droop control loop is adopted to fix the output impedance in [11], hence, achieving excellent power balance when sharing linear or non-linear loads. A novel virtual impedance implementation method in terms of a second-order general-integrator (SOGI) algorithm is proposed for parallel inverter systems without communication signals in [12]. For the sake of achieving accurate power-sharing and improving voltage quality without the impact of hardware parameter variations, a novel voltage stabilization and power-sharing control method is studied in [13], which is on the basis of the virtual complex impedance. Paper [14] considers the impact of current harmonics in islanded and grid-connected microgrids. The virtual impedance is applied to the equal distribution of the harmonic power of the island microgrid and improved the output current when connected to the grid. An adaptive virtual impedance control method is proposed based on the injection of an extra small AC signal (SACS) in the output voltage of each inverter in [15], which achieves power-sharing in a steady-state. The virtual impedance proposed in [16] is modified by the feed-forward terms of DG line current and the point of common coupling (PCC) voltage at the fundamental frequency and harmonic frequency respectively, and the communication system is introduced simultaneously. In [17], a control strategy that does not require recognition of line parameters and the employment of conventional droop is proposed. Nevertheless, this strategy still requires communication between DG and the PCC. Paper [18] proposes an enhanced virtual harmonic impedance control scheme to compensate for the PCC voltage harmonics without harmonic power-sharing error for islanded microgrids. Whereas it is aimed at single-phase systems and requires a central controller and its communication with the DG. In [19], an adaptive virtual impedance control method is applied to DG units in islanded microgrids. However, this control strategy still requires the central controller and its communication with the DG. Paper [20] introduces a kind of harmonic droop control to realize harmonic power-sharing. In [21], the paper proposes an adaptive Virtual Synchronous Generator (VSG) control method based on virtual impedance. The virtual resistance is adaptively designed according to the operating point of the microgrid. However, the control performance under mixed load is not considered. Paper [22] adopts a virtual impedance tuning method based on a successive approximation to accurately compensate for the mismatch among line impedances. To realize this strategy, a common triggering signal from the microgrid central controller is indispensable to start the internal time sequence of each DG unit. [16,22] requires a communication system, which increases the complexity and cost of the system.

This paper proposes a control strategy for island microgrid under mixed load, which is different from other methods as follows. Firstly, the mathematical model of three-phase current is established, the Luenberger observer of three-phase current is designed and its stability is proved. Based on the observer configuration of virtual impedances of different frequencies and sequences, the inverter output power is equally divided and the unbalanced voltage and harmonic voltage of the PCC are reduced. Secondly, the value range of the virtual impedance is discussed. At the h-order frequency, with harmonic voltage as input and harmonic current as output, the root locus of the virtual resistance is obtained. Analyzing the root locus, it is concluded that the critical value of the virtual resistance should be its corresponding line resistance value. Finally, the active power droop control with voltage compensation is employed to increase the PCC voltage. The effectiveness of the control strategy is verified under different load conditions in both simulation and experiment.

The paper is arranged as below. In Section 2, they are introduced, which are the structure of the system and analyzed the employment of virtual impedance to achieve power-sharing and improve power quality. In Section 3, it is established that the three-phase current mathematical model, its Luenberger observer, and the stability are derived. In Section 4, the control strategy structure is discussed. The value range of the harmonic virtual impedance is analyzed and the droop control with voltage compensation is deducted. The simulation and experiment of two inverters without communication in parallel are performed in Sections 5 and 6, respectively, and conclusions are given in Section 7.

2. Power Distribution and Electricity Quality Analysis

Figure 1 shows a simplified diagram of an islanded microgrid, where n DG units are interfaced to the PCC through different DG feeders. Each DG unit consists of a DC source, an inverter, and an LC filter. r is the equivalent resistance of the filter inductor. Z_{li} is the line impedance from DGi to PCC. P_i + jQ_i is the power transferred from DGi to PCC. The microgrid also includes several unbalanced and nonlinear loads placed at the PCC.

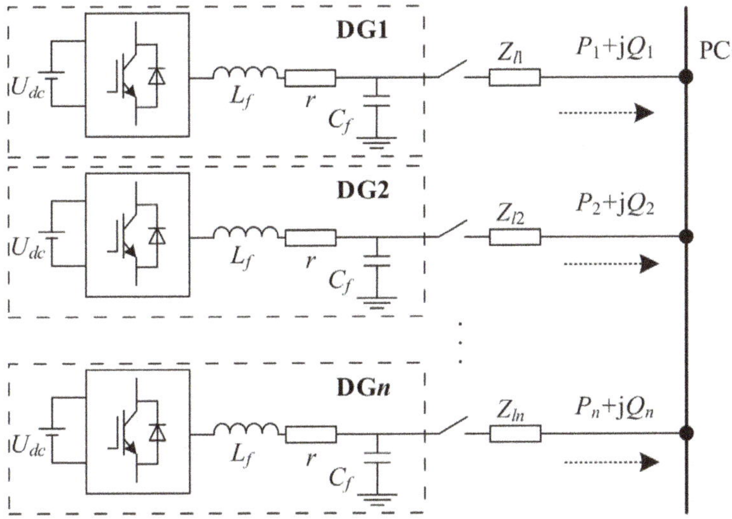

Figure 1. Configuration of the islanded microgrid.

2.1. Power Sharing

Unbalanced loads will produce negative sequence currents, and nonlinear loads produce harmonic currents. The voltage generated by these current components on the equivalent line impedance causes the system voltage to be unbalanced and the harmonic content is too large, and at the same time affects the power-sharing performance of the parallel inverters.

Using virtual impedance to reconstruct the equivalent line impedance at different frequencies and sequences can improve the power-sharing accuracy and PCC power quality. The equivalent circuits under fundamental positive sequence, negative sequence and h-order harmonics are shown in Figure 2.

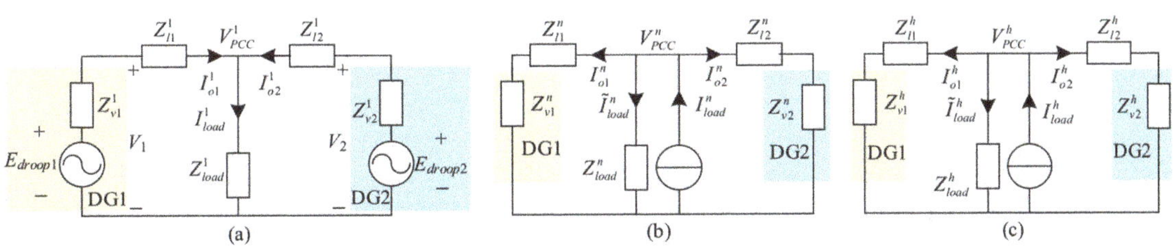

Figure 2. Equivalent circuits of a microgrid at different frequencies and sequences: (**a**) Equivalent circuit at fundamental positive sequence; (**b**) Equivalent circuit at fundamental negative sequence; (**c**) Equivalent circuit at harmonic frequencies.

The superscripts 1, n, and h in Figure 2 represent the variables under the fundamental positive sequence, fundamental negative sequence, and h-order harmonic network, respectively. Z_{vi} and Z_{li} are virtual impedance and line impedance, respectively. E_{droopi} is the given voltage for traditional active droop control. I_{o1}, I_{on} and I_{oh} are fundamental positive sequence components, fundamental negative sequence components and h-order harmonic components of inverter output current respectively. Moreover, the load is much larger than the line impedance, where the negative sequence current $\tilde{I}^n_{load}$ flowing through the balanced load under the negative sequence network and the harmonic current $\tilde{I}^h_{load}$ flowing through the balanced load under the h-order harmonic network can be ignored.

Parallel inverters without communication usually use resistive droop control as follows

$$\begin{cases} E_i = E^* - k_{pi}P_i, k_{pi} > 0 \\ f_i = f^* + k_{qi}Q_i, k_{qi} > 0 \end{cases} \tag{1}$$

where, E^* and f^* are the voltage amplitude and frequency at no-load; P_i, Q_i are the active power and reactive power output by the inverter, V_i, f_i are the voltage amplitude and frequency in the corresponding state; k_{pi}, k_{qi} are the droop coefficients of active and reactive, respectively.

The power-sharing of parallel inverters needs to satisfy the following relationship

$$\frac{Z_{l1}}{Z_{l2}} = \frac{k_{p1}}{k_{p2}} \tag{2}$$

To achieve power-sharing, the setting of virtual impedance should satisfy

$$\begin{cases} \frac{Z_{v1}^1 + Z_{l1}^1}{Z_{v2}^1 + Z_{l2}^1} = \frac{k_{p1}}{k_{p2}} \\ \frac{Z_{v1}^n + Z_{l1}^n}{Z_{v2}^n + Z_{l2}^n} = \frac{k_{p1}}{k_{p2}} \\ \frac{Z_{v1}^h + Z_{l1}^h}{Z_{v2}^h + Z_{l2}^h} = \frac{k_{p1}}{k_{p2}} \end{cases} \tag{3}$$

2.2. *PCC Voltage Quality Analysis*

According to Figure 2a, it can be seen that the PCC fundamental positive sequence voltage is calculated as

$$V_{PCC}^1 = E_{droopi} - \left(Z_{v1}^1 + Z_{l1}^1\right) I_{o1}^1 \tag{4}$$

It can be known from Equation (4) that when the circuit reaches a steady-state, the PCC voltage is lower than the reference voltage. This is mainly because a part of the voltage drop occurs on the equivalent line impedance, which causes the PCC voltage amplitude to drop. The greater the output current, the greater the voltage drop.

According to Figure 2b,c, the PCC fundamental negative sequence voltage and the h-order harmonic voltage are

$$\begin{cases} V_{PCC}^n = (Z_{vi}^n + Z_{li}^n) I_{oi}^n \\ V_{PCC}^h = (Z_{vi}^h + Z_{li}^h) I_{oi}^h \end{cases} \tag{5}$$

According to Equation (5), to reduce the fundamental negative sequence voltage and harmonic voltage of PCC, the virtual impedances should take a negative value.

According to the IEEE 1459-2010 standard [23], the DG harmonic power H_i^h is calculated as

$$H_i^h = \frac{3}{2} E_0 I_i^h \tag{6}$$

where E_0 is the amplitude of the fundamental positive sequence voltage. I_i^h is the amplitude of the harmonic current.

The three-phase voltage unbalance of PCC is calculated as

$$\varepsilon_{UVF} = \frac{\max\left[\left|V_a - V_{Pavg}\right|, \left|V_b - V_{Pavg}\right|, \left|V_c - V_{Pavg}\right|\right]}{V_{Pavg}} \times 100\% \tag{7}$$

where V_a, V_b and V_c are the effective value of phase voltage, and V_{pavg} is the average value of phase voltage.

3. Luenberger Observer for Three-Phase Power Grid

The calculation of the virtual impedance voltage at different frequencies and sequences requires the corresponding inverter output current components. For this reason, a Luenberger observer for the three phase grid current is designed in this paper.

3.1. Three Phase Grid Current Model

The 3P3W system does not contain zero sequence components. The grid current can be divided into the positive and negative sequence components of each harmonic [24], namely

$$i_{abc} = \begin{bmatrix} i_a \\ i_b \\ i_c \end{bmatrix} = \sum_{h=1}^{m} (i_{abc\_p}^h + i_{abc\_n}^h) \tag{8}$$

where the subscripts p and n represent positive and negative sequence components respectively.

The positive sequence component $i_{abc\_p}^h$ and the negative sequence component $i_{abc\_n}^h$ of the current in the abc coordinate system are

$$i_{abc\_p}^h = I_p^h \begin{bmatrix} \cos\left(h\omega t + \varphi_p^h\right) \\ \cos\left(h\omega t - \frac{2\pi}{3} + \varphi_p^h\right) \\ \cos\left(h\omega t + \frac{2\pi}{3} + \varphi_p^h\right) \end{bmatrix}; i_{abc\_n}^h = I_n^h \begin{bmatrix} \cos\left(h\omega t + \varphi_n^h\right) \\ \cos\left(h\omega t + \frac{2\pi}{3} + \varphi_n^h\right) \\ \cos\left(h\omega t - \frac{2\pi}{3} + \varphi_n^h\right) \end{bmatrix}$$

where, I_p^h and I_n^h are the amplitudes of h-order harmonic positive and negative sequence current respectively; φ_p^h and φ_n^h are the initial phase angles of positive and negative sequence current, respectively. ω is the grid current angular frequency.

After Clarke transformation, the grid current in the $\alpha\beta$ coordinate system can be obtained as

$$i_{\alpha\beta} = \begin{bmatrix} i_\alpha \\ i_\beta \end{bmatrix} = \begin{bmatrix} \sum_{h=1}^{m}\left(i_{\alpha\_p}^h + i_{\alpha\_n}^h\right) \\ \sum_{h=1}^{m}\left(i_{\beta\_p}^h + i_{\beta\_n}^h\right) \end{bmatrix} \tag{9}$$

where, $i_{\alpha\_p}^h$ and $i_{\beta\_p}^h$ are the positive sequence currents in the $\alpha\beta$ coordinate system; $i_{\alpha\_n}^h$ and $i_{\beta\_n}^h$ are the negative sequence currents in the $\alpha\beta$ coordinate system.

According to (9), the positive sequence current and negative sequence current in the coordinate system $\alpha\beta$ are respectively

$$i_{\alpha\_p}^h = I_p^h \cos\left(h\omega t + \varphi_p^h\right) \tag{10}$$

$$i_{\alpha\_n}^h = I_n^h \cos\left(h\omega t + \varphi_n^h\right) \tag{11}$$

$$i_{\beta\_p}^h = I_p^h \sin\left(h\omega t + \varphi_p^h\right) \tag{12}$$

$$i_{\beta\_n}^h = -I_n^h \sin\left(h\omega t + \varphi_n^h\right) \tag{13}$$

From Equations (9)–(11), a dynamic equation from the fundamental wave to the m-order harmonics can be established, and the current i_α and its derivative satisfy

$$\begin{cases} \dot{x}_\alpha = Ax_\alpha \\ y_\alpha = i_\alpha = Cx_\alpha \end{cases} \tag{14}$$

where,

$$x_\alpha = \begin{bmatrix} x_{\alpha,1} \\ x_{\alpha,2} \\ \vdots \\ x_{\alpha,2m-1} \\ x_{\alpha,2m} \end{bmatrix} = \begin{bmatrix} I_p^1 \cos(\omega t + \varphi_p^1) + I_n^1 \cos(\omega t + \varphi_n^1) \\ -I_p^1 \sin(\omega t + \varphi_p^1) - I_n^1 \sin(\omega t + \varphi_n^1) \\ \vdots \\ I_p^m \cos(m\omega t + \varphi_p^m) + I_n^m \cos(m\omega t + \varphi_n^m) \\ -I_p^m \sin(m\omega t + \varphi_p^m) - I_n^m \sin(m\omega t + \varphi_n^m) \end{bmatrix};$$

$$A = \begin{bmatrix} 0 & \omega & \cdots & 0 & 0 \\ -\omega & 0 & \cdots & 0 & 0 \\ \vdots & \vdots & \ddots & \vdots & \vdots \\ 0 & 0 & \cdots & 0 & m\omega \\ 0 & 0 & \cdots & -m\omega & 0 \end{bmatrix}_{2m \times 2m}$$

$$C = \begin{bmatrix} 1 & 0 & 1 & 0 & \cdots & 1 & 0 \end{bmatrix}_{1 \times 2m}$$

For system (14), there are

$$rank\begin{bmatrix} C & CA & \cdots & CA^{2m-1} \end{bmatrix}^T = 2m \tag{15}$$

According to (15), the system (14) is observable.

Similarly, according to (9), (12) and (13), the current i_β and its derivative satisfy the dynamic equation

$$\begin{cases} \dot{x}_\beta = A x_\beta \\ y_\beta = i_\beta = C x_\beta \end{cases} \tag{16}$$

where,

$$x_\beta = \begin{bmatrix} x_{\beta,1} \\ x_{\beta,2} \\ \vdots \\ x_{\beta,2m-1} \\ x_{\beta,2m} \end{bmatrix} = \begin{bmatrix} I_p^1 \sin(\omega t + \varphi_p^1) - I_n^1 \sin(\omega t + \varphi_n^1) \\ I_p^1 \cos(\omega t + \varphi_p^1) - I_n^1 \cos(\omega t + \varphi_n^1) \\ \vdots \\ I_p^m \sin(m\omega t + \varphi_p^m) - I_n^m \sin(m\omega t + \varphi_n^m) \\ I_p^m \cos(m\omega t + \varphi_p^m) - I_n^m \cos(m\omega t + \varphi_n^m) \end{bmatrix}$$

The system (16) is also observable.

3.2. Luenberger Observer of Three Phase Grid Current and Its Stability Analysis

The Luenberger observer for system (14) is as follows

$$\begin{cases} \dot{\hat{x}}_\alpha = A\hat{x}_\alpha + G(y_\alpha - \hat{y}_\alpha) \\ \hat{y}_\alpha = C\hat{x}_\alpha \end{cases} \tag{17}$$

where the feedback coefficient $G = \begin{bmatrix} g_1 & g_2 & \cdots & g_{2m} \end{bmatrix}^T$, "^" means the estimated value of the variable.

Then the h-order harmonic current is estimated as

$$\begin{cases} \hat{i}^h_{\alpha\_p} = \frac{\hat{x}_{\alpha,2h-1} + \hat{x}_{\beta,2h}}{2}, \ \hat{i}^h_{\alpha\_n} = \frac{\hat{x}_{\alpha,2h-1} - \hat{x}_{\beta,2h}}{2} \\ \hat{i}^h_{\beta\_p} = \frac{-\hat{x}_{\alpha,2h} + \hat{x}_{\beta,2h-1}}{2}, \ \hat{i}^h_{\beta\_n} = \frac{\hat{x}_{\alpha,2h} + \hat{x}_{\beta,2h-1}}{2} \end{cases} \tag{18}$$

The 5th and 7th harmonic content in the system is large enough to total current. The designed observer only estimates the fundamental wave, the 5th harmonic and the 7th harmonic. The following stability analysis is performed only for three-phase current Luenberger observers with 5th and 7th harmonics.

3. Luenberger Observer for Three-Phase Power Grid

The calculation of the virtual impedance voltage at different frequencies and sequences requires the corresponding inverter output current components. For this reason, a Luenberger observer for the three phase grid current is designed in this paper.

3.1. Three Phase Grid Current Model

The 3P3W system does not contain zero sequence components. The grid current can be divided into the positive and negative sequence components of each harmonic [24], namely

$$i_{abc} = \begin{bmatrix} i_a \\ i_b \\ i_c \end{bmatrix} = \sum_{h=1}^{m} (i^h_{abc\_p} + i^h_{abc\_n}) \tag{8}$$

where the subscripts p and n represent positive and negative sequence components respectively.

The positive sequence component $i^h_{abc\_p}$ and the negative sequence component $i^h_{abc\_n}$ of the current in the abc coordinate system are

$$i^h_{abc\_p} = I^h_p \begin{bmatrix} \cos\left(h\omega t + \varphi^h_p\right) \\ \cos\left(h\omega t - \frac{2\pi}{3} + \varphi^h_p\right) \\ \cos\left(h\omega t + \frac{2\pi}{3} + \varphi^h_p\right) \end{bmatrix}; i^h_{abc\_n} = I^h_n \begin{bmatrix} \cos\left(h\omega t + \varphi^h_n\right) \\ \cos\left(h\omega t + \frac{2\pi}{3} + \varphi^h_n\right) \\ \cos\left(h\omega t - \frac{2\pi}{3} + \varphi^h_n\right) \end{bmatrix}$$

where, I^h_p and I^h_n are the amplitudes of h-order harmonic positive and negative sequence current respectively; φ^h_p and φ^h_n are the initial phase angles of positive and negative sequence current, respectively. ω is the grid current angular frequency.

After Clarke transformation, the grid current in the $\alpha\beta$ coordinate system can be obtained as

$$i_{\alpha\beta} = \begin{bmatrix} i_\alpha \\ i_\beta \end{bmatrix} = \begin{bmatrix} \sum_{h=1}^{m}\left(i^h_{\alpha\_p} + i^h_{\alpha\_n}\right) \\ \sum_{h=1}^{m}\left(i^h_{\beta\_p} + i^h_{\beta\_n}\right) \end{bmatrix} \tag{9}$$

where, $i^h_{\alpha\_p}$ and $i^h_{\beta\_p}$ are the positive sequence currents in the $\alpha\beta$ coordinate system; $i^h_{\alpha\_n}$ and $i^h_{\beta\_n}$ are the negative sequence currents in the $\alpha\beta$ coordinate system.

According to (9), the positive sequence current and negative sequence current in the coordinate system $\alpha\beta$ are respectively

$$i^h_{\alpha\_p} = I^h_p \cos\left(h\omega t + \varphi^h_p\right) \tag{10}$$

$$i^h_{\alpha\_n} = I^h_n \cos\left(h\omega t + \varphi^h_n\right) \tag{11}$$

$$i^h_{\beta\_p} = I^h_p \sin\left(h\omega t + \varphi^h_p\right) \tag{12}$$

$$i^h_{\beta\_n} = -I^h_n \sin\left(h\omega t + \varphi^h_n\right) \tag{13}$$

From Equations (9)–(11), a dynamic equation from the fundamental wave to the m-order harmonics can be established, and the current i_α and its derivative satisfy

$$\begin{cases} \dot{x}_\alpha = Ax_\alpha \\ y_\alpha = i_\alpha = Cx_\alpha \end{cases} \tag{14}$$

where,

$$x_\alpha = \begin{bmatrix} x_{\alpha,1} \\ x_{\alpha,2} \\ \vdots \\ x_{\alpha,2m-1} \\ x_{\alpha,2m} \end{bmatrix} = \begin{bmatrix} I_p^1 \cos(\omega t + \varphi_p^1) + I_n^1 \cos(\omega t + \varphi_n^1) \\ -I_p^1 \sin(\omega t + \varphi_p^1) - I_n^1 \sin(\omega t + \varphi_n^1) \\ \vdots \\ I_p^m \cos(m\omega t + \varphi_p^m) + I_n^m \cos(m\omega t + \varphi_n^m) \\ -I_p^m \sin(m\omega t + \varphi_p^m) - I_n^m \sin(m\omega t + \varphi_n^m) \end{bmatrix};$$

$$A = \begin{bmatrix} 0 & \omega & \cdots & 0 & 0 \\ -\omega & 0 & \cdots & 0 & 0 \\ \vdots & \vdots & \ddots & \vdots & \vdots \\ 0 & 0 & \cdots & 0 & m\omega \\ 0 & 0 & \cdots & -m\omega & 0 \end{bmatrix}_{2m \times 2m}$$

$$C = \begin{bmatrix} 1 & 0 & 1 & 0 & \cdots & 1 & 0 \end{bmatrix}_{1 \times 2m}$$

For system (14), there are

$$rank\begin{bmatrix} C & CA & \cdots & CA^{2m-1} \end{bmatrix}^T = 2m \tag{15}$$

According to (15), the system (14) is observable.

Similarly, according to (9), (12) and (13), the current i_β and its derivative satisfy the dynamic equation

$$\begin{cases} \dot{x}_\beta = Ax_\beta \\ y_\beta = i_\beta = Cx_\beta \end{cases} \tag{16}$$

where,

$$x_\beta = \begin{bmatrix} x_{\beta,1} \\ x_{\beta,2} \\ \vdots \\ x_{\beta,2m-1} \\ x_{\beta,2m} \end{bmatrix} = \begin{bmatrix} I_p^1 \sin(\omega t + \varphi_p^1) - I_n^1 \sin(\omega t + \varphi_n^1) \\ I_p^1 \cos(\omega t + \varphi_p^1) - I_n^1 \cos(\omega t + \varphi_n^1) \\ \vdots \\ I_p^m \sin(m\omega t + \varphi_p^m) - I_n^m \sin(m\omega t + \varphi_n^m) \\ I_p^m \cos(m\omega t + \varphi_p^m) - I_n^m \cos(m\omega t + \varphi_n^m) \end{bmatrix}$$

The system (16) is also observable.

3.2. Luenberger Observer of Three Phase Grid Current and Its Stability Analysis

The Luenberger observer for system (14) is as follows

$$\begin{cases} \dot{\hat{x}}_\alpha = A\hat{x}_\alpha + G(y_\alpha - \hat{y}_\alpha) \\ \hat{y}_\alpha = C\hat{x}_\alpha \end{cases} \tag{17}$$

where the feedback coefficient $G = \begin{bmatrix} g_1 & g_2 & \cdots & g_{2m} \end{bmatrix}^T$, "^" means the estimated value of the variable.

Then the h-order harmonic current is estimated as

$$\begin{cases} \hat{i}_{\alpha\_p}^h = \frac{\hat{x}_{\alpha,2h-1} + \hat{x}_{\beta,2h}}{2}, \ \hat{i}_{\alpha\_n}^h = \frac{\hat{x}_{\alpha,2h-1} - \hat{x}_{\beta,2h}}{2} \\ \hat{i}_{\beta\_p}^h = \frac{-\hat{x}_{\alpha,2h} + \hat{x}_{\beta,2h-1}}{2}, \ \hat{i}_{\beta\_n}^h = \frac{\hat{x}_{\alpha,2h} + \hat{x}_{\beta,2h-1}}{2} \end{cases} \tag{18}$$

The 5th and 7th harmonic content in the system is large enough to total current. The designed observer only estimates the fundamental wave, the 5th harmonic and the 7th harmonic. The following stability analysis is performed only for three-phase current Luenberger observers with 5th and 7th harmonics.

In Equation (17), replacing $\hat{y}_\alpha$ in the first equation with the second equation, the following equation can be obtained

$$\dot{\hat{x}}_\alpha = A\hat{x}_\alpha + G(y_\alpha - C\hat{x}_\alpha) \tag{19}$$

where,

$$x_\alpha = \begin{bmatrix} x_{\alpha,1} \\ x_{\alpha,2} \\ x_{\alpha,3} \\ x_{\alpha,4} \\ x_{\alpha,5} \\ x_{\alpha,6} \end{bmatrix} = \begin{bmatrix} I_1^+ \cos(\omega t + \varphi_1^+) + I_1^- \cos(\omega t + \varphi_1^-) \\ -I_1^+ \sin(\omega t + \varphi_1^+) - I_1^- \sin(\omega t + \varphi_1^-) \\ I_5^+ \cos(5\omega t + \varphi_5^+) + I_5^- \cos(5\omega t + \varphi_5^-) \\ -I_5^+ \sin(5\omega t + \varphi_5^+) - I_5^- \sin(5\omega t + \varphi_5^-) \\ I_7^+ \cos(7\omega t + \varphi_7^+) + I_7^- \cos(7\omega t + \varphi_7^-) \\ -I_7^+ \sin(7\omega t + \varphi_7^+) - I_7^- \sin(7\omega t + \varphi_7^-) \end{bmatrix};$$

$$A = \begin{bmatrix} 0 & \omega & 0 & 0 & 0 & 0 \\ -\omega & 0 & 0 & 0 & 0 & 0 \\ 0 & 0 & 0 & 5\omega & 0 & 0 \\ 0 & 0 & -5\omega & 0 & 0 & 0 \\ 0 & 0 & 0 & 0 & 0 & 7\omega \\ 0 & 0 & 0 & 0 & -7\omega & 0 \end{bmatrix}$$

$$G = \begin{bmatrix} g_1 & g_2 & \cdots & g_6 \end{bmatrix}^T$$

$$C = \begin{bmatrix} 1 & 0 & 1 & 0 & 1 & 0 \end{bmatrix}$$

To simplify the analysis, each element in the feedback matrix G is taken as g. Equation (19) is discretized and simplified to

$$\hat{x}_\alpha(k+1) = A_e\hat{x}_\alpha(k) + GTy_\alpha(k) \tag{20}$$

where T is the sampling period, A_e is given in the Appendix A.

The dynamic response of the Luenberger state observer depends on the eigenvalues of the state equation coefficient matrix A_e. When the eigenvalues of A_e are all distributed within the unit circle of the Z domain, the discrete Luenberger disturbance observer is stable.

Let $\omega = 314$, $T = 1 \times 10^{-5}$s. When the feedback coefficient g increases from small, the pole change of the coefficient matrix A_e is shown in Figure 3. It can be concluded from Figure 3 that when $46 < g < 67{,}127$, the modulus value of all eigenvalues of the coefficient matrix A_e is less than 1, so the observer is stable; when $g > 67{,}127$ or $g < 46$, there is one eigenvalue whose modulus value is greater than 1, so the observer is unstable.

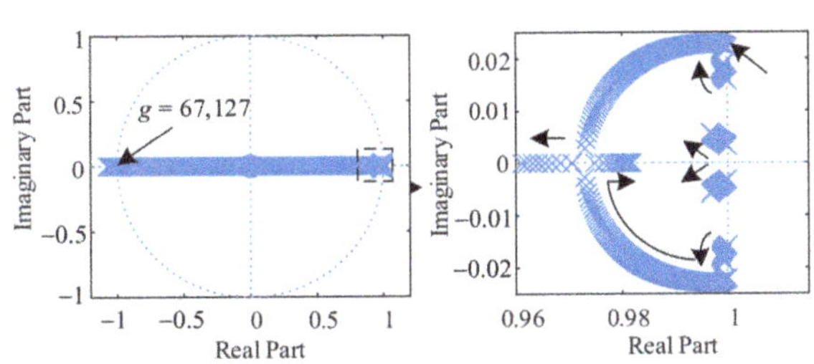

Figure 3. Pole distribution diagram of A_e when g changes.

4. Proposed Control Strategy

4.1. Basic Idea

The overall control structure of the inverter is shown in Figure 4. v_{oabc} and i_{oabc} are the output voltage and current of DG*i* respectively. i_{Labc} is the filter inductor current. ω_{ref} and V_{ref} are the reference values of the frequency and voltage amplitude of the improved droop control output respectively.

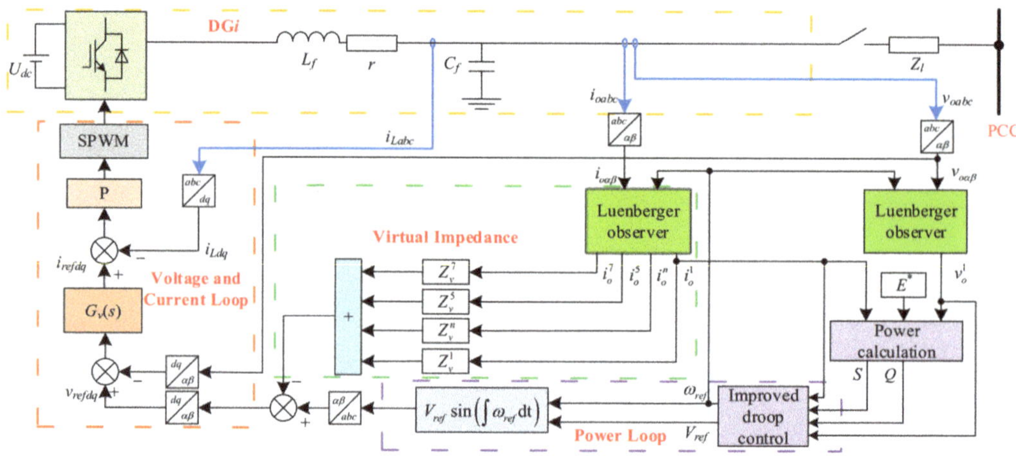

Figure 4. Control system structure block diagram.

The positive fundamental sequence component of the current is extracted by the three phase grid Luenberger observer designed in the previous section. The apparent power S calculated from the positive sequence component of the fundamental wave of the output current and the no-load voltage E^* is fed back to the active droop control, which can reduce the output current difference of the inverter. The voltage controller $G_v(s)$ is a parallel connection of PI and quasi-proportional resonance controllers, which realizes the no steady-state error control of fundamental positive sequence, fundamental negative sequence and harmonics. In this paper, the virtual impedance feeds back the currents of different frequencies and sequences, and configures each virtual impedance independently, which not only satisfies the power-sharing condition, but also reduces the equivalent fundamental negative sequence impedance and harmonic impedance.

4.2. Analysis of Virtual Impedance Value Range

The voltage and current double closed-loop control with *h*-order virtual harmonic impedance are shown in Figure 5. i_{Lf}, i_{Cf} and i_o are the filter inductor current, filter capacitor current and inverter output current, respectively. v_o is the inverter output voltage.

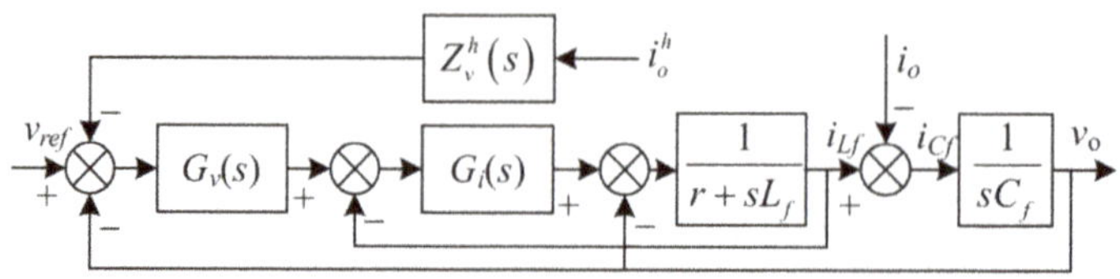

Figure 5. Voltage and current double closed-loop control with *h*-order virtual harmonic impedance.

In Figure 5, G_i is the proportional controller, $G_v(s)$ is the parallel voltage controller of PI and QPR, which is expressed as

$$G_v(s) = \left(k_{vp} + \frac{k_{vi}}{s}\right) + \frac{2K_{r1}\omega_c s}{s^2 + 2\omega_c s + (2\omega_0)^2} + \frac{2K_{r2}\omega_c s}{s^2 + 2\omega_c s + (6\omega_0)^2} \quad (21)$$

where, k_{vp} and k_{vi} are the proportional coefficients and integral coefficients of the PI controller, respectively; K_{r1} and K_{r2} are the gain of the QPR controller, ω_c is the shear frequency, and ω_0 is the fundamental frequency.

The output voltage of the inverter can be obtained from Figure 5 as

$$v_o(s) = G_s(s)\left[v_{ref}(s) + Z_v^h(s)i_o^h(s)\right] - Z_o(s)i_o(s) \quad (22)$$

where,

$$G_s(s) = \frac{G_v(s)G_i(s)}{1 + sC_f(sL_f + r) + G_v(s)G_i(s) + sC_fG_i(s)}$$

$$Z_o(s) = \frac{sL_f + r + G_i(s)}{1 + sC_f(sL_f + r) + G_v(s)G_i(s) + sC_fG_i(s)}$$

The equivalent harmonic impedance from the inverter to the PCC can be described as

$$Z^{h*}(s) = \left[G_s(s)Z_v^h(s) + Z_o(s)\right] + R_l + sL_l \quad (23)$$

where, $Z_v^h(s) = (-R_v^h - sL_v^h)$.

To reduce power coupling, there is $L_{v1}^h = L_1$. DG1 and the line impedance can be equivalent to impedance $Z_1^{h*}(s)$ under the h-order harmonic. With PCC harmonic voltage $E_h(s)$ as input and $I_{o1}^h(s)$ as output, $1/Z_1^{h*}(s)$ is the closed-loop transfer function of the system. Let k_{vp} = 0.8, k_{vi} = 20, k_{ip} = 0.6, ω_0 = 314, ω_c = 6.28, K_{r1} = 5, K_{r2} = 30. The filter parameters are shown in Table 1. The root locus of the system gain R_{v1}^h can be obtained by substituting the parameters as shown in Figure 6. The root locus equation has been given in the Appendix A.

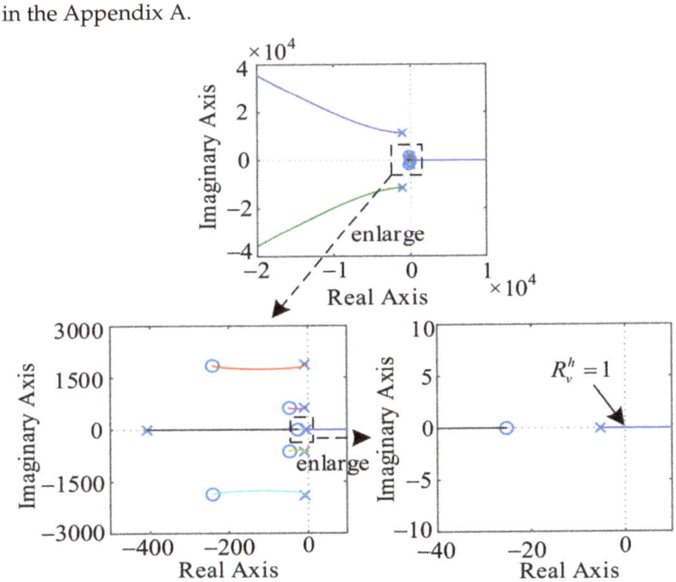

Figure 6. Root locus of system.

According to Figure 6, when the system gain R_{v1}^{h} is greater than the line resistance R_{l1} = 1 Ω, there is a pole in the right half-plane, and the system is unstable. Similarly, the critical value of other virtual resistors in the parallel system should be the corresponding line resistance. To sum up, for the sake of making the system stable, the value of virtual negative resistance should be less than the value of line resistance as far as possible, and the value of virtual inductance should ensure the resistance of the whole system.

4.3. Droop Control with Voltage Compensation

According to Equation (4), it can be known that the PCC voltage is lower than the reference voltage, so this section compensates for the PCC voltage.

According to Figure 7, the output reference voltage amplitude of the active droop equation is

$$E_{droop} = E^* - k_{pi}S_i \tag{24}$$

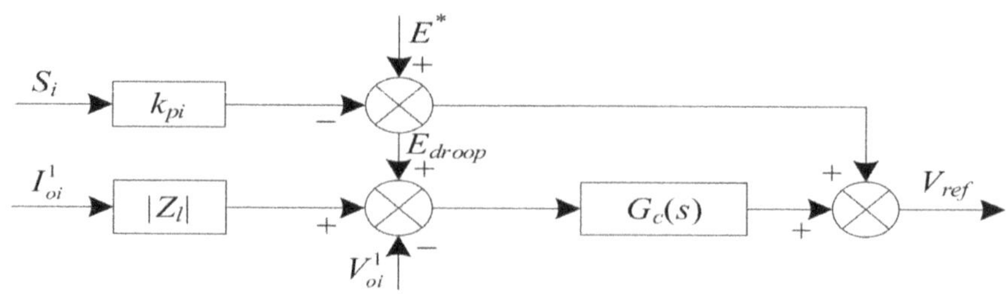

Figure 7. Improved active droop control block diagram.

The output voltage amplitude reference value is

$$V_{ref} = V_{droop} + G_c(s)\left[E_{droop} - V_{oi}^1 + I_{oi}^1|Z_l|\right] \tag{25}$$

where V_{oi}^1 and I_{oi}^1 are the fundamental amplitudes of the inverter output voltage and current, $|Z_l|$ is the line impedance modulus, and $G_c(s)$ adopt proportional-integral controller ($G_c(s)$ = k_{cp} + k_{ci}/s).

When the system reaches a steady-state, there are

$$V_{PCC} = V_{oi}^1 - I_{oi}^1|Z_l| = E_{droop} \tag{26}$$

Equation (26) shows that the PCC voltage is equal to the output reference voltage of the traditional active droop controller in the steady-state.

5. Simulation Results

To verify the effectiveness of the proposed control strategy, this paper builds a simulation model based on Matlab/Simulink. In the model, two voltage source three-phase inverters are operated in parallel in islanded mode. The droop control parameters of the two inverters have the same value, and the impedance of the line connected to the PCC in parallel is not equal. The inverter output voltage is a sine wave with a frequency of 50 Hz and a peak value of 311 V. The three phase uncontrolled rectifier module is used as a non-linear load. The low-voltage microgrid line parameters are shown in Table 1.

The simulation parameters of the two control strategies are shown in Tables 2 and 3, respectively. The common load of PCC connections is shown in Table 4. The simulation results are shown in Figures 7 and 8.

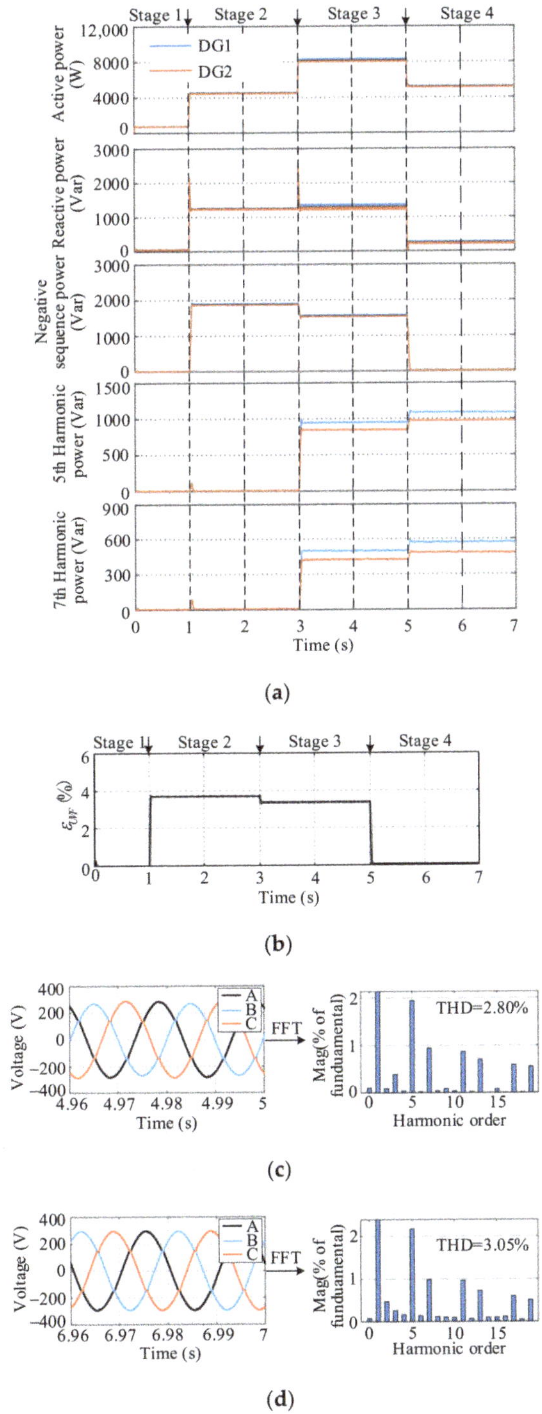

Figure 8. Traditional virtual impedance droop control: (**a**) power-sharing performance; (**b**) PCC three-phase voltage unbalance ε_{UVF}; (**c**) PCC voltage and A-phase voltage FFT analysis in stage 3; (**d**) PCC voltage and A-phase voltage FFT analysis in stage 4.

Figures 8a and 9a show the power-sharing performance of comparison, which are the fundamental positive sequence active power, fundamental positive sequence reactive power, fundamental negative sequence power, 5th harmonic power and 7th harmonic power from top to bottom. In stage 1 and stage 2, both strategies achieve power-sharing. When the nonlinear load 3 is connected at 3 s, the harmonic power of the two DGs is not equal in tradition, but in the proposed strategy. It can be seen that in the proposed control strategy, the virtual impedance is used to compensate for the harmonic impedance, so that the harmonic power can also be equally divided.

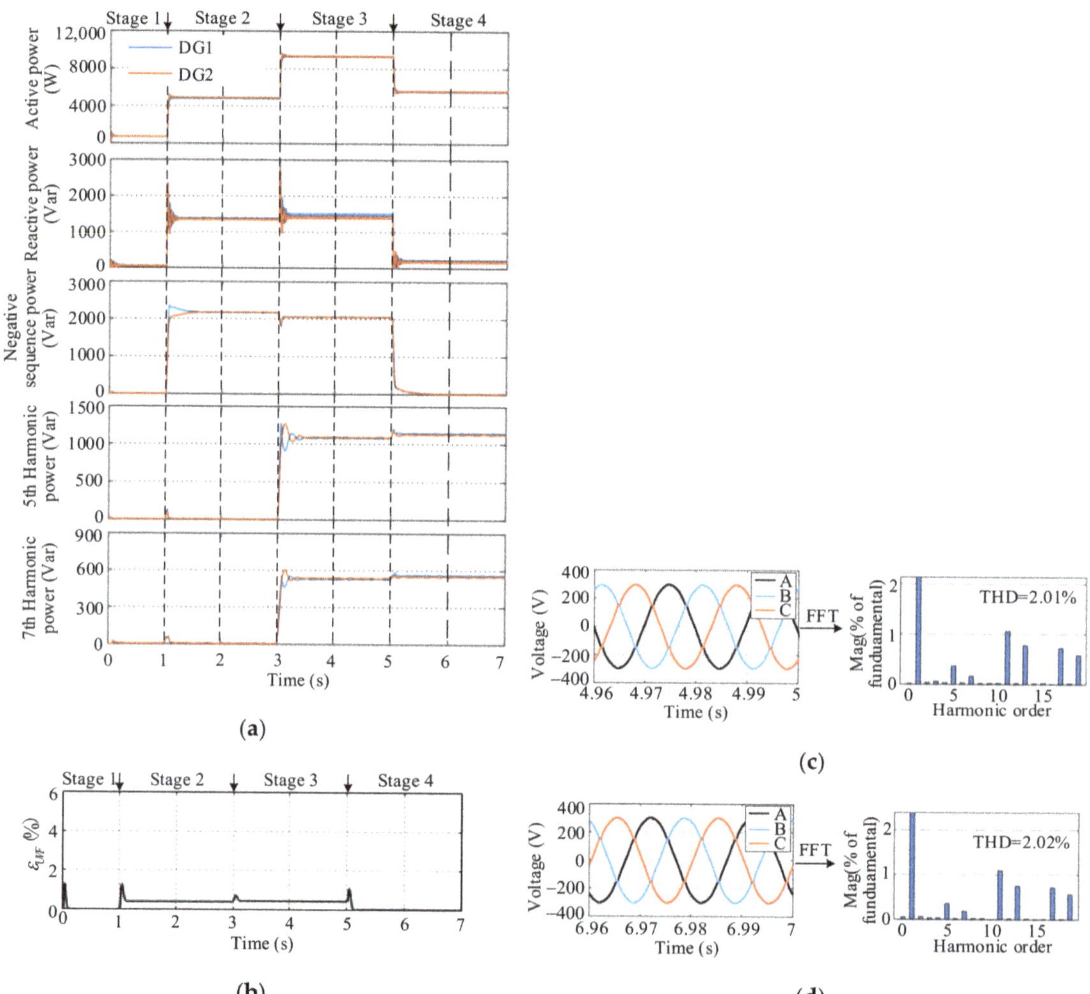

Figure 9. The proposed control strategy: (**a**) power-sharing performance; (**b**) PCC three-phase voltage unbalance ε_{UVF}; (**c**) PCC voltage and A-phase voltage FFT analysis in stage 3; (**d**) PCC voltage and A-phase voltage FFT analysis in stage 4.

Figures 8b and 9b shows the three phase voltage unbalance at each stage. In stages 2 and 3, there is an unbalanced load 2 connected to the PCC, the three phase voltage unbalance is 3.7% at stage 2 in tradition, which is 0.4% in the proposed. Due to the reduction of the fundamental negative sequence impedance and the use of parallel voltage controllers of PI and QPR, the three phase voltage unbalance of the PCC is greatly reduced.

Figures 8c,d and 9c,d show the PCC voltage waveforms and phase A voltage FFT analysis of stages 3 and 4. The PCC voltage is distorted due to the nonlinear load 3 being connected to the PCC in stages 3 and 4. In traditional virtual impedance droop control, the 5th and 7th harmonic components of the PCC voltage are higher. The proposed control strategy reduces the impedance of the 5th and 7th harmonics, and significantly reduces the content of the 5th and 7th harmonics of the PCC voltage.

Table 1. Line Parameters of Low Voltage Microgrid.

Parameters	Values	Parameters	Values
U_{dc}	700 V	L_{l1}	0.413 mH
L_f	5 mH	R_{l2}	2 Ω
r	0.2 Ω	L_{l2}	0.826 mH
C_f	20 μF	Balance load 1	100 Ω + 20 mH
f_s	10 kHz	Unbalance load 2	Phase A 30 Ω + 30 mH Phase B 0 Ω Phase C 30 Ω + 30 mH
R_{l1}	1 Ω	Non-linear load 3	25 Ω

Table 2. Simulation Parameters of Traditional Virtual Impedance Droop Control.

Parameters	Values	Parameters	Values
k_{pi}	1.555×10^{-3}	R_{v1}	0 Ω
k_{qi}	2.5×10^{-4}	L_{v1}	−0.413 mH
k_{vp}, k_{vi}	0.8, 50	R_{v2}	−1 Ω
k_{ip}	0.6	L_{v2}	−0.826 mH

Table 3. The Simulation Parameters of the Control Strategy Proposed.

Parameters	Values	Parameters	Values
k_{vp}, k_{vi}	0.8, 20	R^5_{v1} L^5_{v1}	−0.6 Ω, −0.413 mH
ω_c	6.28	R^7_{v1} L^7_{v1}	−0.6 Ω, −0.413 mH
k_{cp}, k_{ci}	0.4, 50	R^1_{v2} L^1_{v2}	−1 Ω, −0.826 mH
g	260	R^n_{v2} L^n_{v2}	−2 Ω, −0.826 mH
R^1_{v1} L^1_{v1}	0 Ω, −0.413 mH	R^5_{v2} L^5_{v2}	−1.6 Ω, −0.826 mH
R^n_{v1} L^n_{v1}	−1 Ω, −0.413 mH	R^7_{v2} L^7_{v2}	−1.6 Ω, −0.826 mH

Table 4. Connected common load.

	Stage1	Stage2	Stage3	Stage4
time of duration	0–1 s	1–3 s	3–5 s	5–7 s
Common load	load 1	load 1 and 2	load 1, 2 and 3	load 2 and 3

6. Experimental Verification

To further verify the effectiveness of the proposed method, a parallel system of two inverters is built with TMS320F28335 as the controller, which is shown in Figure 10. The experimental parameters are shown in Table 5, and the two inverters have the same droop control coefficient.

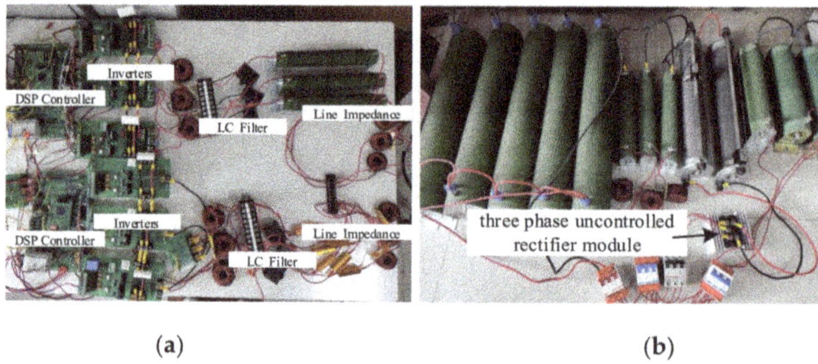

Figure 10. Experimental hardware: (**a**) Parallel inverters; (**b**) loads.

Table 5. Experiment parameters.

Parameters	Values	Parameters	Values
U_{dc}	90 V	R_{l1}	0.5 Ω
L_f	5 mH	L_{l1}	1 mH
r	0.2 Ω	R_{l2}	1.5 Ω
C_f	20 μF	L_{l2}	2 mH
f_s	20 kHz	Balance load 1	10 Ω
E^*	30 V	Unbalance load between phase AB 2	20 Ω
f^*	50 Hz	non-linear load 3	15 Ω

6.1. PCC Connected to Unbalanced Load

PCC is connected to balanced load 1 and unbalanced load 2. The PCC three phase voltage and DG output current phase A under the two control strategies are shown in Figures 11 and 12.

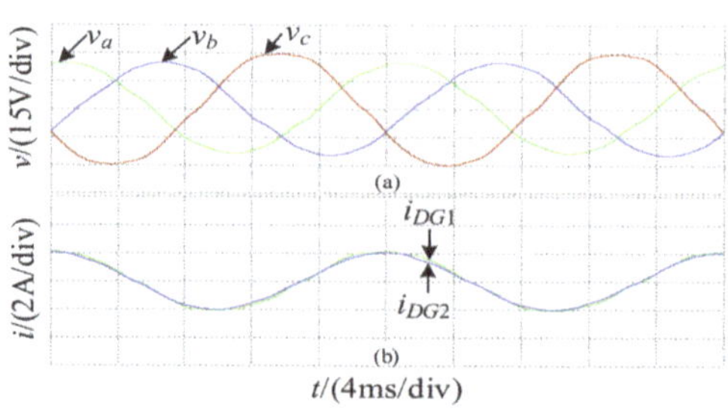

Figure 11. The experimental results of the traditional control method under the unbalanced load connected to the PCC: (**a**) PCC voltage; (**b**) Phase A output current of DGs.

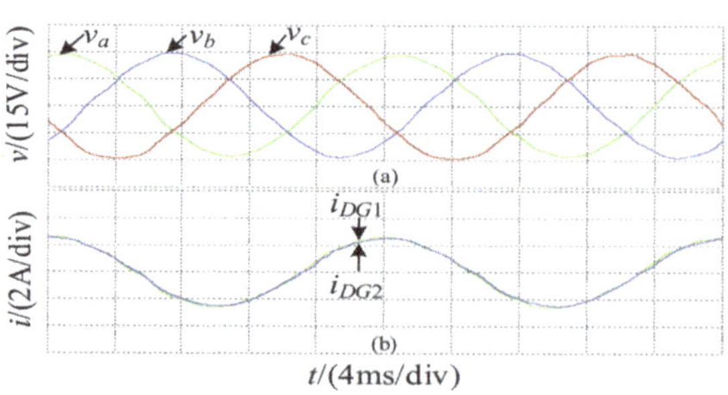

Figure 12. The experimental results of the method proposed under unbalanced load connected to PCC: (**a**) PCC voltage; (**b**) Phase A output current of DGs.

The DGi output currents under the two control strategies are almost equal. The three phase voltage unbalance is reduced from 14.66% under the traditional control strategy to 1.70% under the proposed control strategy.

6.2. PCC Connected to Non-Linear Load

In Figure 13a,b, because the PCC is connected with a balanced load 1 and a non-linear load 3, the PCC voltage is severely distorted, and the FFT analysis of phase A voltage is shown in Figure 13c. THD has reached 8.17%, and the 5th and 7th harmonic content are relatively high. The FFT analysis of DG phase A output current is shown in Figure 13d. Due to the small impedance of the DG1 line, the harmonic content of its output current is relatively high. The experimental results show that the traditional virtual impedance droop control can not realize the equal distribution of harmonic power, and the PCC voltage is severely distorted.

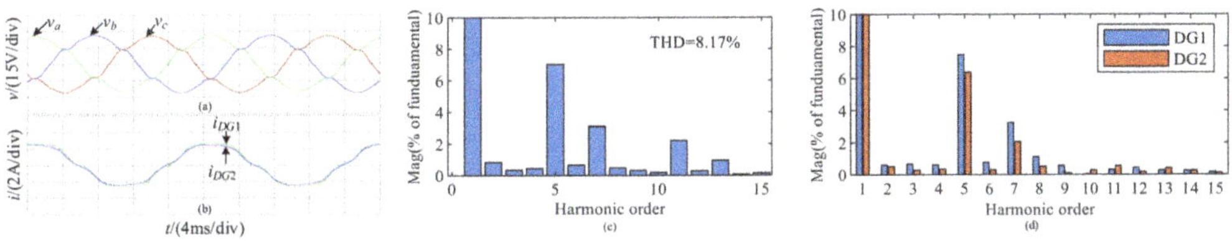

Figure 13. The experimental results of the traditional control method under the non-linear load connected to the PCC: (**a**) PCC voltage; (**b**) Phase A output current of DGs; (**c**) FFT analysis of phase A voltage of PCC; (**d**) FFT analysis of phase A output current of DGs.

After adopting the control method proposed, the PCC voltage and Phase A output current of DGs are shown in Figure 14a,b. The FFT analysis of the PCC phase A voltage is shown in Figure 14c, the THD is reduced to 4.41%, and the 5th and 7th harmonic content of the control are significantly reduced. The FFT analysis of DG phase A output current is shown in Figure 14d, and the 5th and 7th harmonic content are almost equal. The experimental results show that the method proposed can realize the equal distribution of harmonic power, reduce the PCC voltage THD, and improve the electricity quality. Since the equivalent harmonic impedance is reduced, the voltage harmonic content of the PCC is reduced, and the inverter also gets more harmonic current. But at the same time, the harmonic current of the linear load is reduced. Compared with the traditional virtual impedance control, the harmonic content of the output currents of inverter 1 and inverter 2 under the proposed control strategy is increased. This phenomenon is more pronounced due to the lower experimental voltage level and the smaller impedance of the load.

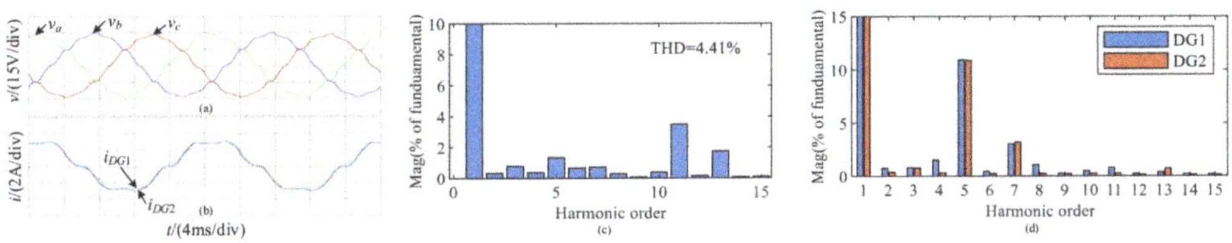

Figure 14. The experimental results of the method proposed under the non-linear load connected to the PCC: (**a**) PCC voltage; (**b**) Phase A output current of DGs; (**c**) FFT analysis of phase A voltage of PCC; (**d**) FFT analysis of phase A output current of DGs.

6.3. PCC Connected to Unbalanced and Non-Linear Mixed Load

In Figure 15a,b, because the PCC is connected with a balanced load 1, an unbalanced load 2 and a non-linear load 3, the PCC voltage quality is seriously degraded. The three-phase voltage imbalance reached 11.04%. The FFT analysis of the phase A voltage is shown in Figure 15c. THD reaches 7.87%, so part of the harmonic current flows to the balanced load

1. The FFT analysis of DG phase A output current is shown in Figure 15d. The harmonic content of the DG1 output current is relatively high. The experimental results show that the traditional virtual impedance droop control can not realize the equal distribution of harmonic power under the condition of a mixed load, the PCC voltage is severely distorted and the three-phase voltage unbalance is large.

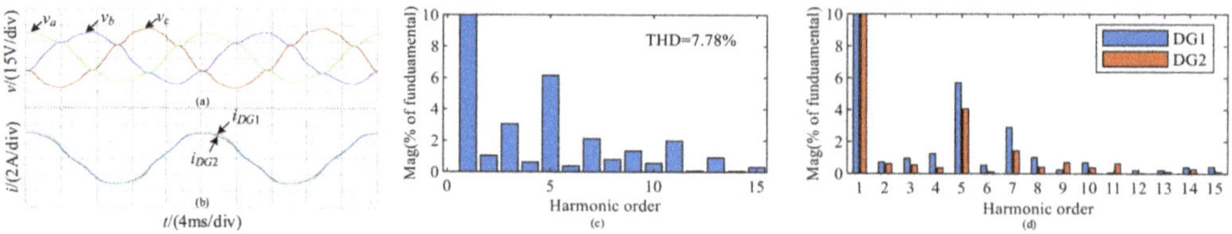

Figure 15. The experimental results of the traditional control method under the mixed load connected to the PCC: (**a**) PCC voltage; (**b**) Phase A output current of DGs; (**c**) FFT analysis of phase A voltage of PCC; (**d**) FFT analysis of phase A output current of DGs.

After adopting the control method proposed, the PCC three phase voltage is shown in Figure 16a. Its three phase voltage unbalance has dropped to 1.69%. The FFT analysis of the PCC phase A voltage is shown in Figure 16c. The THD is reduced to 4.48%, and the 5th and 7th harmonic content of the control are significantly reduced. The FFT analysis of DG phase A output current is shown in Figure 16d, and the 5th and 7th harmonic content are almost equal. The experimental results show that the method proposed in this paper can realize the equal distribution of harmonic power under the mixed load, reduce the PCC voltage THD and the three-phase voltage unbalance, and improve the electricity quality.

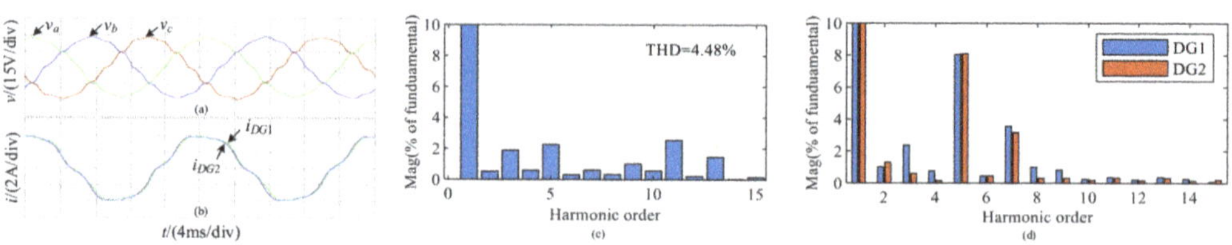

Figure 16. The experimental results of the method proposed under the mixed load connected to the PCC: (**a**) PCC voltage; (**b**) Phase A output current of DGs; (**c**) FFT analysis of phase A voltage of PCC; (**d**) FFT analysis of phase A output current of DGs.

6.4. Transient Performance of the Proposed Method

The inverter adopts the proposed control strategy, and the output current of the inverter when the load changes are shown in Figure 17. Figure 17a,b are the experimental waveforms of connecting and disconnecting the nonlinear load 3 when the PCC is connected with the three-phase balanced load 1, respectively. It can be seen that the proposed control strategy has a faster response.

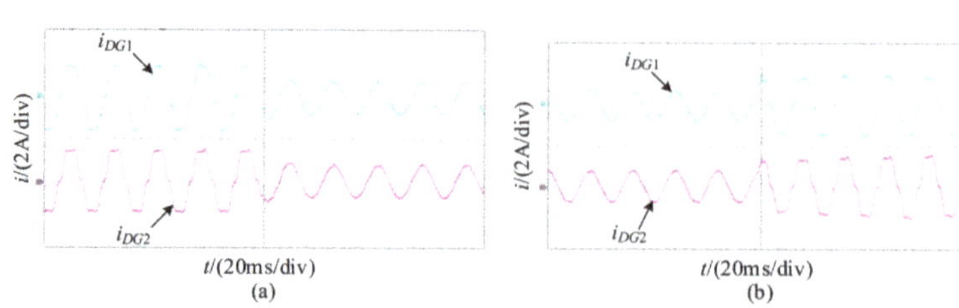

Figure 17. Experimental results of the proposed method under load changing conditions: (**a**) nonlinear load access to PCC and (**b**) nonlinear loads disconnected from PCC.

7. Conclusions

To solve the distribution problem of negative sequences and harmonics in microgrids, and the issue of the degradation of PCC voltage quality when the low-voltage microgrid is connected with unbalanced and non-linear loads, a control method based on the Luenberger observer is proposed. The conclusions are as follows:

(1) Analyze the generation mechanism of fundamental negative sequence voltage and harmonic voltage, and it is pointed out that the fundamental negative sequence voltage and harmonic voltage can be reduced based on power-sharing by using virtual impedance with a negative value.
(2) Design a Luenberger observer for three phase current and prove its stability. Use this observer to get the fundamental positive sequence, fundamental negative sequence and harmonic components. Configure each sequence of virtual impedances to achieve power-sharing and improve electricity quality. And analyzed the value range of virtual impedance.
(3) Design the active power droop equation with voltage compensation to increase the voltage amplitude.

Author Contributions: Methodology, Y.C.; Software, C.L. All authors have read and agreed to the published version of the manuscript.

Funding: This work was supported by the Science and Technology Department Industrial Public Relations of Shaanxi Province (No. 2016GY-064).

Institutional Review Board Statement: Not applicable.

Informed Consent Statement: Not applicable.

Data Availability Statement: Not applicable.

Conflicts of Interest: The authors declare no conflict of interest.

Appendix A

$$A_e = AT - GTC + I$$
$$= \begin{bmatrix} 1-gT & \omega T & -gT & 0 & -gT & 0 \\ -(gT+\omega T) & 1 & -gT & 0 & -gT & 0 \\ -gT & 0 & 1-gT & 5\omega T & -gT & 0 \\ -gT & 0 & -(gT+5\omega T) & 1 & -gT & 0 \\ -gT & 0 & -gT & 0 & 1-gT & 7\omega T \\ -gT & 0 & -gT & 0 & -(gT+7\omega T) & 1 \end{bmatrix}$$

where, I is the identity matrix.

Root locus equation:

$$\frac{R_{v1}^{h}\left(a_5s^5 + a_4s^4 + a_3s^3 + a_2s^2 + a_1s + a_0\right)}{A_8s^8 + A_7s^7 + A_6s^6 + A_5s^5 + A_4s^4 + A_3s^3 + A_2s^2 + A_1s + A_0} = -1$$

where,

$a_5 = -0.48$ $a_4 = -287.8$ $a_3 = -1.9 \times 10^6$ $a_2 = -2.94 \times 10^8$
$a_1 = -6.72 \times 10^{11}$ $a_0 = -1.68 \times 10^{13}$ $A_8 = 4.13 \times 10^{-11}$ $A_7 = 1.08 \times 10^{-7}$
$A_6 = 5.6 \times 10^{-3}$ $A_5 = 2.84$ $A_4 = 2.18 \times 10^4$ $A_3 = 9.41 \times 10^6$
$A_2 = 7.98 \times 10^9$ $A_1 = 3.19 \times 10^{12}$ $A_0 = 1.68 \times 10^{13}$

References

1. Kohn, W.; Zabinsky, Z.B.; Nerode, A. A Micro-Grid Distributed Intelligent Control and Management System. *IEEE Trans. Smart Grid* **2015**, *6*, 2964–2974. [CrossRef]
2. Nasser, N.; Fazeli, M. Buffered-Microgrid Structure for Future Power Networks; a Seamless Microgrid Control. *IEEE Trans. Smart Grid* **2021**, *12*, 131–140. [CrossRef]
3. Lee, K.; Yun, S.; Hwang, P.; Seo, D.; Pak, Y. Demonstration project of distribution level microgrid in Penetanguishene of Canada. In Proceedings of the CIRED Workshop 2016, Helsinki, Finland, 14–15 June 2016; pp. 1–3.
4. Cho, J.; Kim, H.; Cho, Y.; Kim, H.; Kim, J. Demonstration of a DC Microgrid with Central Operation Strategies on an Island. In Proceedings of the 2019 IEEE Third International Conference on DC Microgrids (ICDCM), Matsue, Japan, 20–23 May 2019; pp. 1–5.
5. Nishita, Y.; Izui, Y.; Honda, M.; Mizuochi, M.; Natsuume, D.; Tabata, H. DC Microgrid Experimental System at KIT and its Autonomous Distributed DC Voltage Control Method. In Proceedings of the 2021 IEEE Fourth International Conference on DC Microgrids (ICDCM), Arlington, VA, USA, 18–21 July 2021; pp. 1–6.
6. Fujimoto, S.; Obara, S. Electricity Quality Analysis of Teuri-Yagishiri Island Microgrid. In Proceedings of the 2019 IEEE 10th International Symposium on Power Electronics for Distributed Generation Systems (PEDG), Xi'an, China, 3–6 June 2019; pp. 353–358.
7. Ul Hassan, S.; ul Abideen, Z.; Izhar, T. Advanced control techniques for micro-grids power quality improvement. In Proceedings of the 2017 Asian Conference on Energy, Power and Transportation Electrification (ACEPT), Singapore, 24–26 October 2017; pp. 1–6.
8. De Brabandere, K.; Bolsens, B.; van den Keybus, J.; Woyte, A.; Driesen, J.; Belmans, R. A Voltage and Frequency Droop Control Method for Parallel Inverters. *IEEE Trans. Power Electron.* **2007**, *22*, 1107–1115. [CrossRef]
9. Zhao, J.; Li, J.; Nian, H.; Yang, L.; Gai, B.; Pan, Z.; Xu, Y. A Feeder Impedance Identification Based Droop Control Method for Accurate Reactive Power Sharing in Islanded Microgrids. In Proceedings of the 2019 10th International Conference on Power Electronics and ECCE Asia (ICPE 2019—ECCE Asia), Busan, Korea, 27–30 May 2019; pp. 1–5.
10. Han, Y.; Li, H.; Shen, P.; Coelho, E.A.A.; Guerrero, J.M. Review of Active and Reactive Power Sharing Strategies in Hierarchical Controlled Microgrids. *IEEE Trans. Power Electron.* **2017**, *32*, 2427–2451. [CrossRef]
11. Guerrero, J.M.; De Vicuna, L.G.; Matas, J.; Castilla, M.; Miret, J. Output impedance design of parallel-connected UPS inverters with wireless load-sharing control. *IEEE Trans. Ind. Electron.* **2005**, *52*, 1126–1135. [CrossRef]
12. Alcala, J.M.; Castilla, M.; De Vicuña, L.G.; Miret, J.; Vasquez, J.C. Virtual impedance loop for droop-controlled single-phase parallel inverters using a second-order general-integrator scheme. *IEEE Trans. Power Electron.* **2010**, *25*, 2993–3002. [CrossRef]
13. Wai, R.; Zhang, Q.; Wang, Y. A Novel Voltage Stabilization and Power Sharing Control Method Based on Virtual Complex Impedance for an Off-Grid Microgrid. *IEEE Trans. Power Electron.* **2019**, *34*, 1863–1880. [CrossRef]
14. Micallef, A.; Apap, M.; Spiteri-Staines, C.; Guerrero, J.M. Mitigation of Harmonics in Grid-Connected and Islanded Microgrids Via Virtual Admittances and Impedances. *IEEE Trans. Smart Grid* **2017**, *8*, 651–661. [CrossRef]
15. Liu, B.; Liu, Z.; Liu, J.; An, R.; Zheng, H.; Shi, Y. An Adaptive Virtual Impedance Control Scheme Based on Small-AC-Signal Injection for Unbalanced and Harmonic Power Sharing in Islanded Microgrids. *IEEE Trans. Power Electron.* **2019**, *34*, 12333–12355. [CrossRef]
16. He, J.; Li, Y.W.; Guerrero, J.M.; Blaabjerg, F.; Vasquez, J.C. An islanding microgrid power sharing approach using enhanced virtual impedance control scheme. *IEEE Trans. Power Electron.* **2013**, *28*, 5272–5282. [CrossRef]
17. Razi, R.; Iman-Eini, H.; Hamzeh, M. An Impedance-Power Droop Method for Accurate Power Sharing in Islanded Resistive Microgrids. *IEEE J. Emerg. Sel. Top. Power Electron.* **2020**, *8*, 3763–3771. [CrossRef]
18. Hoang, T.V.; Lee, H.-H. Virtual Impedance Control Scheme to Compensate for Voltage Harmonics with Accurate Harmonic Power Sharing in Islanded Microgrids. *IEEE J. Emerg. Sel. Top. Power Electron.* **2021**, *9*, 1682–1695. [CrossRef]
19. He, J.; Li, Y.W.; Blaabjerg, F. An Enhanced Islanding Microgrid Reactive Power, Imbalance Power, and Harmonic Power Sharing Scheme. *IEEE Trans. Power Electron.* **2015**, *30*, 3389–3401. [CrossRef]
20. Moussa, H.; Shahin, A.; Martin, J.; Nahid-Mobarakeh, B.; Pierfederici, S.; Moubayed, N. Harmonic Power Sharing with Voltage Distortion Compensation of Droop Controlled Islanded Microgrids. *IEEE Trans. Smart Grid* **2018**, *9*, 5335–5347. [CrossRef]
21. Liang, X.; Andalib-Bin-Karim, C.; Li, W.; Mitolo, M.; Shabbir, M.N.S.K. Adaptive Virtual Impedance-Based Reactive Power Sharing in Virtual Synchronous Generator Controlled Microgrids. *IEEE Trans. Ind. Appl.* **2021**, *57*, 46–60. [CrossRef]
22. An, R.; Liu, Z.; Liu, J. Successive-Approximation-Based Virtual Impedance Tuning Method for Accurate Reactive Power Sharing in Islanded Microgrids. *IEEE Trans. Power Electron.* **2021**, *36*, 87–102. [CrossRef]
23. *IEEE Std 1459–2010 (Revision of IEEE Std 1459–2000)*; IEEE Standard Definitions for the Measurement of Electric Power Quantities Under Sinusoidal, Nonsinusoidal, Balanced, or Unbalanced Conditions. IEEE: New York, NY, USA, 2010; pp. 1–50.
24. Dai, Z.; Lin, W. Adaptive Estimation of Three-Phase Grid Voltage Parameters under Unbalanced Faults and Harmonic Disturbances. *IEEE Trans. Power Electron.* **2017**, *32*, 5613–5627. [CrossRef]

energies

MDPI

Article

Lithium-Ion Battery Health State Prediction Based on VMD and DBO-SVR

Chunling Wu [1], Juncheng Fu [1,*], Xinrong Huang [1], Xianfeng Xu [1] and Jinhao Meng [2,*]

[1] School of Energy and Electrical Engineering, Chang'an University, Xi'an 710064, China; wuchl@chd.edu.cn (C.W.)

[2] School of Electrical Engineering, Xi'an Jiaotong University, Xi'an 710049, China

* Correspondence: 2021132007@chd.edu.cn (J.F.); jinhao@xjtu.edu.cn (J.M.)

Abstract: Accurate estimation of the state-of-health (SOH) of lithium-ion batteries is a crucial reference for energy management of battery packs for electric vehicles. It is of great significance in ensuring safe and reliable battery operation while reducing maintenance costs of the battery system. To eliminate the nonlinear effects caused by factors such as capacity regeneration on the SOH sequence of batteries and improve the prediction accuracy and stability of lithium-ion battery SOH, a prediction model based on Variational Modal Decomposition (VMD) and Dung Beetle Optimization -Support Vector Regression (DBO-SVR) is proposed. Firstly, the VMD algorithm is used to decompose the SOH sequence of lithium-ion batteries into a series of stationary mode components. Then, each mode component is treated as a separate subsequence and modeled and predicted directly using SVR. To address the problem of difficult parameter selection for SVR, the DBO algorithm is used to optimize the parameters of the SVR model before training. Finally, the predicted values of each subsequence are added and reconstructed to obtain the final SOH prediction. In order to verify the effectiveness of the proposed method, the VMD-DBO-SVR model was compared with SVR, Empirical Mode Decomposition-Support Vector Regression (EMD-SVR), and VMD-SVR methods for SOH prediction of batteries based on the NASA dataset. Experimental results show that the proposed model has higher prediction accuracy and fitting degree, with prediction errors all within 1% and better robustness.

Keywords: lithium-ion battery; state of health; variational mode decomposition; dung beetle optimization algorithm; support vector regression

Citation: Wu, C.; Fu, J.; Huang, X.; Xu, X.; Meng, J. Lithium-Ion Battery Health State Prediction Based on VMD and DBO-SVR. *Energies* **2023**, *16*, 3993. https://doi.org/10.3390/en16103993

Academic Editor: Djaffar Ould-Abdeslam

Received: 12 April 2023
Revised: 8 May 2023
Accepted: 8 May 2023
Published: 9 May 2023

1. Introduction

With the acceleration of economic globalization and the massive use of fossil fuels, environmental pollution and energy shortage have become increasingly prominent issues. Lithium-ion batteries for energy storage have found extensive applications across various facets of daily life and industrial production, owing to their substantial energy-storage capacity and excellent cycling performance [1]. With the growing number of applications for lithium-ion batteries, the battery will gradually age, leading to performance degradation. If used improperly, it may cause more serious accidents. Therefore, accurately and quickly estimating the SOH of the battery can provide necessary information for decision makers, enabling them to plan ahead, extend battery life, and enhance the safety of utilizing lithium-ion batteries. This has important practical significance [2]. Currently, although many methods exist for predicting the SOH, they are primarily categorized into two groups: model-based and data-driven approaches [3–6].

Equivalent circuit models or electrochemical models are the primary approaches utilized in model-based methods. Reference [7] proposed a method based on Electrochemical Impedance Spectroscopy (EIS) to estimate the SOH of lithium-ion batteries. Reference [8] estimated the SOH by constructing a nonlinear equivalent circuit battery model, where the incremental state of charge (ΔSOC) is proportional to the SOH within a suitable voltage range. Reference [9] utilized the Forgetting Factor Recursive Least Squares (FFRLS) algorithm to identify the equivalent circuit model of Thevenin for battery parameters. In this model, the electrical resistance is used as the characteristic factor for battery health status and is estimated to determine the battery's SOH. The equivalent circuit model has a simple structure and low computational complexity, but its robustness is poor, and it is easy to produce estimation errors under different operating conditions. The electrochemical model establishes a dynamic system model based on the battery's electrochemical reaction mechanism and has higher accuracy, but it is difficult to identify parameters.

Data-driven methods essentially create a black-box model, where the internal structure of the battery does not need to be explicitly constructed. To construct a prediction model for SOH, it is only necessary to extract and analyze external parameters of the battery that are extremely correlated with SOH, and use them as training data. Reference [10] uses Convolutional Neural Network (CNN) to extract features and reduce data dimensionality of model input factors. Then, these factors are used as inputs of the Bidirectional Long and Short Term Memory (BiLSTM) network to predict the SOH. Compared with other neural networks, this approach provides higher prediction accuracy. Literature [11] proposed an SOH estimation method based on an improved Ant Lion Optimizer algorithm and Support Vector Regression (IALO-SVR). This method uses the IALO algorithm to optimize the kernel parameters of SVR, thereby improving the accuracy of SOH prediction. However, the accuracy of prediction results in the SVR model is directly affected by the penalty factor and kernel function parameters. Therefore, selecting appropriate model parameters is a critical issue that requires immediate attention when using the SVR method for estimating the SOH of lithium-ion batteries [12–14]. Literature [15] utilizes the Ensemble Empirical Mode Decomposition (EEMD) algorithm to decompose the original sequence signal into a trend signal and low-frequency residual signal in order to reduce the influence of various noises. Then, LSTM and CNN models are used to predict the SOH of the two types of signals separately. However, EEMD is prone to model mixing, which may affect the overall prediction accuracy.

To address the above issues, we extracted the available capacity of each charge-discharge cycle of the battery and calculated the corresponding SOH data. We used this SOH data as a health indicator and proposed a lithium-ion battery SOH prediction method based on the VMD-DBO-SVR model. Firstly, the VMD method is employed to decompose the original SOH sequence into a series of Intrinsic Mode Function (IMF) components that represent local features at multiple scales. Then, SVR is employed to model and predict each IMF element directly, and to address the difficulty in selecting SVR parameters, the DBO algorithm is utilized to optimize the parameters of the SVR model before model training. Finally, the predicted values of each sub-sequence are combined and reconstructed to derive the ultimate SOH prediction value. The proposed method is evaluated using the NASA dataset.

We compared the accuracy and practicality of our method with other methods using the same dataset in the literature. In reference [16], a Deep Neural Network (DNN) method was used to predict the SOH of lithium-ion batteries, which has better predictive performance compared to other machine learning models. However, due to the use of a single prediction model, the prediction accuracy is still not high enough. Reference [17] used a new physics-informed machine learning prediction model PIDDA, which includes three parts: autoencoder, physical information model training, and physics-based prediction adjustment, to achieve more accurate SOH prediction with less training data, but it did not consider the influence of dataset noise. Compared with the above two methods, our method considers the influence of noise and compensates for the problem of single model parameter selection, which has certain advantages. In summary, the main objective of this study is to eliminate the impact of battery capacity regeneration and various noises on capacity data, thereby eliminating the impact on SOH data, and solving the problem of difficult selection of SVR model parameters, thereby improving the accuracy of SOH prediction.

2. Basic Theory

2.1. Definition of Battery SOH

Battery SOH refers to the current health status of a battery, which is an important indicator of battery performance and service life, as the health status of a battery gradually deteriorates over time. SOH is typically expressed as a percentage and is defined as follows [18–20]:

$$\mathrm{SOH} = \frac{C_i}{C_0} \times 100\% \tag{1}$$

where, C_i is the available capacity of the i-th charge-discharge cycle, C_0 is the rated capacity.

2.2. VDM Decomposition

VMD is a variational method-based technique used to decompose nonlinear and non-stationary signals into multiple Intrinsic Mode Functions (IMF) [21,22]. The central concept of VMD is minimizing the interference among each IMF and other frequency bands through iterative optimization, which avoids information overlap and makes it robust to noise and interference. Since lithium-ion batteries are subject to capacity recovery and random interference during use, it is essential to employ VMD to remove noise interference. Here are the key stages of the algorithm:

Step 1: Construct the variational model. The original SOH signal is decomposed into K IMF, and the variational constraint function is formulated as follows:

$$\begin{gathered}\min_{\{u_k\},\{\omega_k\}}\left\{\sum_{k=1}^{K}\left\|\partial_t\left[\left(\delta(t)+\frac{j}{\pi t}\right)*u_k(t)\right]e^{-j\omega_k t}\right\|_2^2\right\}\\ \text{s.t. } \sum_{k=1}^{k}u_k(t)=f(t)\end{gathered} \tag{2}$$

where, $f(t)$ is the SOH data, $\{u_k\}$ is a set of K IMF components that have been decomposed, $\{\omega_k\}$ is a collection of central frequencies corresponding to each IMF component, $\delta(t)$ is the impulse function, $*$ is the convolution function.

Step 2: Introduce a penalty factor α and the Lagrange multiplier λ transforming a constrained variational problem into an unconstrained variational problem. The augmented Lagrange expression is derived as:

$$\begin{aligned}L=(\{u_k\},\{\omega_k\},\lambda)&=\alpha\sum_{k=1}^{K}\|\partial_t[(\delta(t)+j/\pi t)*u_k(t)]e^{-j\omega_k t}\|_2^2\\&+\left\|f(t)-\sum_{k=1}^{K}u_k(t)\right\|_2^2+\left[\lambda(t),f(t)-\sum_{k=1}^{K}u_k(t)\right]\end{aligned} \tag{3}$$

Step 3: Initialize $\{u_k^1\}$, $\{\omega_k^1\}$, $\{\lambda^1\}$ and upper limit on the number of iterations n. In the Fourier transform domain, continuously iterate and update $\left\{\hat{u}_k^{n+1}\right\}$, $\left\{\omega_k^{n+1}\right\}$ and $\{\hat{\lambda}^{n+1}\}$. The update formula is as follows:

$$\hat{u}_k^{n+1}(\omega)=\frac{\hat{f}(\omega)-\sum_{i\neq k}^{K}\hat{u}_i(\omega)+\hat{\lambda}(\omega)/2}{1+2\alpha(\omega-\omega_k)^2} \tag{4}$$

$$\omega_k^{n+1}=\frac{\int_0^\infty \omega\left|\hat{u}_k^{n+1}(\omega)\right|^2 d\omega}{\int_0^\infty\left|\hat{u}_k^{n+1}(\omega)\right|^2 d\omega} \tag{5}$$

$$\hat{\lambda}^{n+1}(\omega)=\hat{\lambda}^n(\omega)+\gamma\left[\hat{f}(\omega)-\sum_{k=1}^{K}\hat{u}_k^{n+1}(\omega)\right] \tag{6}$$

where, γ is the noise tolerance; ω is the frequency.

Step 4: Stop the iterative updates until the stopping criteria are satisfied, which are as follows:

$$\sum_{k=1}^{K} \frac{\|\hat{u}_k^{n+1} - \hat{u}_k^n\|_2^2}{\|\hat{u}_k^n\|_2^2} < \varepsilon \quad (7)$$

where, ε is the discriminant accuracy, $\varepsilon > 0$.

2.3. Dung Beetle Optimization Algorithm (DBO)

The Dung Beetle Optimization (DBO) algorithm is a population-based intelligent optimization algorithm proposed by Jiankai Xue et al. [23], inspired by the rolling, dancing, foraging, stealing, and breeding behaviors of dung beetles. Specifically, the DBO algorithm achieves the traversal and search of the search space by simulating the behavior of dung beetles rolling dung balls. This algorithm also introduces some strategies, such as dance behavior to determine the direction of advancement, and grabbing behavior to jump out of local optimal solutions. The algorithm exhibits comparable competitiveness to the latest optimization strategies regarding the speed of convergence and accuracy of solutions.

The DBO algorithm mainly includes four types of behavior: rolling, breeding, foraging, and stealing, corresponding to four types of dung beetles: rolling dung beetle, breeding dung beetle, foraging dung beetle, and thief dung beetle. The algorithm achieves parameter optimization by having each type of dung beetle perform its corresponding operation. The specific four behaviors of the DBO algorithm are as follows:

(1) Rolling ball

The dung beetle rolls a much larger dung ball than itself and usually uses celestial cues such as the sun to navigate in order to maintain the dung ball's motion in a linear path. During the rolling process, the position of the rolling dung beetle is updated according to the following formula:

$$\begin{gathered} x_i(t+1) = x_i(t) + \alpha \times k \times x_i(t-1) + b \times \Delta x \\ \Delta x = |x_i(t) - X^w| \end{gathered} \quad (8)$$

where, t is the iteration count, $x_i(t)$ is the position information of the i-th beetle during the t-th iteration, $k \in (0, 0.2]$ is a fixed parameter representing the deviation factor, $b \in (0, 1)$ is a fixed parameter, α is the natural coefficient, which is allocated a value of either 1 or -1, X^w is the global worst position, Δx simulates the variation of light brightness.

When a dung beetle confronts an obstruction that blocks its path, it needs to reposition itself by dancing in order to find a new route. To simulate this dance behavior, a tangent function is used to obtain a new rolling direction. After the dung beetle determines a different direction, it continues to roll the ball backward. The position of the dung beetle is updated as follows:

$$x_i(t+1) = x_i(t) + \tan(\theta)|x_i(t) - x_i(t-1)| \quad (9)$$

where, θ is the deflection angle, which takes a value of $[0, \pi]$.

(2) Reproduction

In nature, female dung beetles roll their dung balls to a safe place suitable for laying eggs and hide them. Inspired by this behavior, a strategy for selecting boundaries is chosen to mimic the oviposition area of female dung beetles, which is defined as follows:

$$\begin{gathered} L_b^* = \max(X^* \times (1 - R), L_b) \\ U_b^* = \min(X^* \times (1 + R), U_b) \end{gathered} \quad (10)$$

where, X^* represents the current optimal position, $L_b{}^*$ and $U_b{}^*$, respectively, represent the lower and upper bounds of the oviposition area, $R = 1 - t/T_{max}$, T_{max} is the maximum number of iterations, L_b and U_b, respectively, represent the lower and upper bounds of the optimization problem.

Once the oviposition area is determined, the female beetle will select an egg in this area for laying. The boundary range of the oviposition area will dynamically change, which is mainly determined by the value of R, as can be clearly seen from Equation (10). Therefore, the position of the egg is also dynamic during the iteration process, defined as:

$$B_i(t+1) = X^* + b_1 \times (B_i(t) - Lb^*) + b_2 \times (B_i(t) - Ub^*) \quad (11)$$

where, $B_i(t)$ is the position of the i-th egg in the t-th iteration, b_1 and b_2 are two uncorrelated stochastic vectors of size $1 \times D$, D is the dimension of the optimization problem.

(3) Foraging

The eggs laid by female beetles will gradually grow. Some matured small beetles will come out of the ground to search for food. The optimal foraging area of small beetles is modeled as follows:

$$\begin{aligned} L_b{}^b &= \max\left(X^b \times (1-R), L_b\right) \\ U_b{}^b &= \min\left(X^b \times (1+R), U_b\right) \end{aligned} \quad (12)$$

where, X^b is the global optimal position, $L_b{}^b$ and $U_b{}^b$ represent the lower and upper bounds of the optimal foraging area. Therefore, the position update of the small dung beetles is as follows:

$$x_i(t+1) = x_i(t) + C_1 \times \left(x_i(t) - L_b{}^b\right) + C_2 \times \left(x_i(t) - U_b{}^b\right) \quad (13)$$

where, $x_i(t)$ represents the position of the i-th small dung beetle in the t-th iteration, C_1 represents a random variable that follows a normal distribution, C_2 is a random vector within the range of (0, 1).

(4) Stealing

There are some beetles, called thieves, that steal dung balls from other beetles. From Equation (12), it can be seen that X^b is the optimal food source, so it can be assumed that the area around X^b is the optimal location for competing food. During the iteration process, the position of the thief is updated as follows:

$$x_i(t+1) = X^b + S \times g \times \left(|x_i(t) - X^*| + \left|x_i(t) - X^b\right|\right) \quad (14)$$

where, $x_i(t)$ represents the position of the i-th thief at the t-th iteration, g is a random vector of size $1 \times D$ following a normal distribution, S is a constant.

2.4. Support Vector Regression (SVR)

The SVR method finds the best hyperplane that fits the data by continuously reducing the error between predicted and actual values. Its advantage is strong generalization ability and good performance in handling nonlinear problems [24,25].

Assuming a given sample set $S = \{x_i, y_i\}_{i=1}^{n} (x_i \in X = R^n, y_i \in Y = R)$. Where x_i is the i-th input vector, y_i is the corresponding output vector, and n is the total number of all samples. By utilizing the SVR method, non-linear mapping is applied to map the sample set from a low-dimensional space to a high-dimensional space. This non-linear mapping can be defined as:

$$f(x) = \omega \cdot \phi(x) + b \quad (15)$$

where, x, b, ω represent input data, intercept, and weights, respectively. Then introduce the slack variables $\{\xi_i\}_{i=1}^{n}$ and $\{\xi_i^*\}_{i=1}^{n}$, we can get:

$$\min R(\omega, b, \xi) = \frac{1}{2}\|\omega\|^2 + C\sum_{i}^{n}(\xi_i + \xi_i^*) \quad (16)$$

$$\text{s.t.} \begin{cases} y_i - \omega \cdot \phi(x) - b \leq \varepsilon + \xi_i \\ \omega \cdot \phi(x) + b - y_i \leq \varepsilon + \xi_i^* \\ \xi_i, \xi_i^* \geq 0. \end{cases} \tag{17}$$

where, C is the penalty factor, $\varepsilon(\varepsilon > 0)$ is the highest acceptable error of the regression. Introducing Lagrange multipliers and kernel function, Equation (16) can be converted into the following equation:

$$\max R(\alpha_i^*, \alpha_i) = -\frac{1}{2}\sum_{i,j}^{n}(\alpha_i^* - \alpha_i)\left(\alpha_j^* - \alpha_j\right)\phi(x_i)\phi(x_j) - \sum_{i}^{n}\alpha_i(y_i + \varepsilon) + \sum_{i}^{n}\alpha_i^*(y_i - \varepsilon) \tag{18}$$

Subject to:

$$\text{s.t.} \begin{cases} \sum_{i}^{n}(\alpha_i - \alpha_i^*) = 0, \\ 0 \leq \alpha_i, \alpha_i^* \leq C, \ i = 1, 2, \ldots n \end{cases} \tag{19}$$

where, α_i and α_i^* are Lagrange multipliers. After minimizing the Lagrangian function, the SVR expression for the non-linear mapping can be obtained. This expression can be formulated as:

$$f(x) = \sum_{i}^{n}(\alpha_i - \alpha_i^*)K(x_i, x) + b \tag{20}$$

where, $K(x_i, x) = \phi(x_i)\phi(x_j)$ is the kernel function. The Radial Basis Function (RBF) is a widely adopted nonlinear function and is also frequently used in SVR [26]. RBF can be defined as Equation (21).

$$K_{RBF}(x_i, x) = \exp\left(-\frac{1}{2\sigma^2}\|x_i - x\|^2\right) \tag{21}$$

where, σ is the kernel parameter.

3. SOH Prediction Based on VMD-DBO-SVR Combination Model

3.1. SVR Method Based on DBO Optimization

This paper employs the SVR method to establish the SOH prediction model. However, the selection of the penalty factor C and kernel parameter σ has a crucial impact on the forecasting precision of the SVR model. Larger C values and smaller σ values may lead to overfitting, while smaller C values and larger σ values may lead to underfitting. Due to the strong global optimization ability and fast convergence speed, as well as robustness, of the DBO algorithm, this paper uses the DBO algorithm to tune the penalty factor C and kernel parameter σ, aiming to improve the prediction accuracy of the SVR method.

The process of DBO for optimizing SVR parameters is illustrated in Figure 1, with the main procedures as follows:

(1) Initialization parameters for DBO algorithm: population size of dung beetles (pop), the proportion of four types of dung beetles in the population including rollers, breeders, foragers, and thieves, the dimension of variable parameters (dim), the maximum number of iterations (T_{max}), and lower bound (L_b) and upper bound (U_b).

(2) Randomly initialize the positions of all dung beetles.

(3) Calculate the fitness values of all dung beetles and record the global optimum position. In this study, the Mean Square Error (MSE) between actual and predicted values is applied to establish the fitness function, namely:

$$MSE = \frac{1}{n}\sum_{i}^{n}(\hat{y}_i - y_i)^2 \tag{22}$$

where, $\hat{y}_i$ is the i-th decomposition component of predicted SOH, y_i is the i-th decomposition component of actual SOH.

(4) Update the positions of all dung beetles: if it is a rolling dung beetle, update its position by rolling action according to Equation (8) in obstacle-free mode or by dancing action according to Equation (9) in obstacle mode; if it is a breeding dung beetle, update its position by breeding action according to Equations (10) and (11); if it is a foraging dung beetle, update its position by foraging action according to Equations (12) and (13); if it is a thief dung beetle, update its position by theft action according to Equation (14).

(5) After the update, determine if the position of each individual dung beetle exceeds the boundaries L_b and U_b. If it does, return to step (3). Otherwise, continue the execution.

(6) Update the present optimal solution and its fitness value.

(7) Repeat the above steps (3) to (6) until the iteration limit T_{max} is attained, and output the optimal parameters to the SVR model.

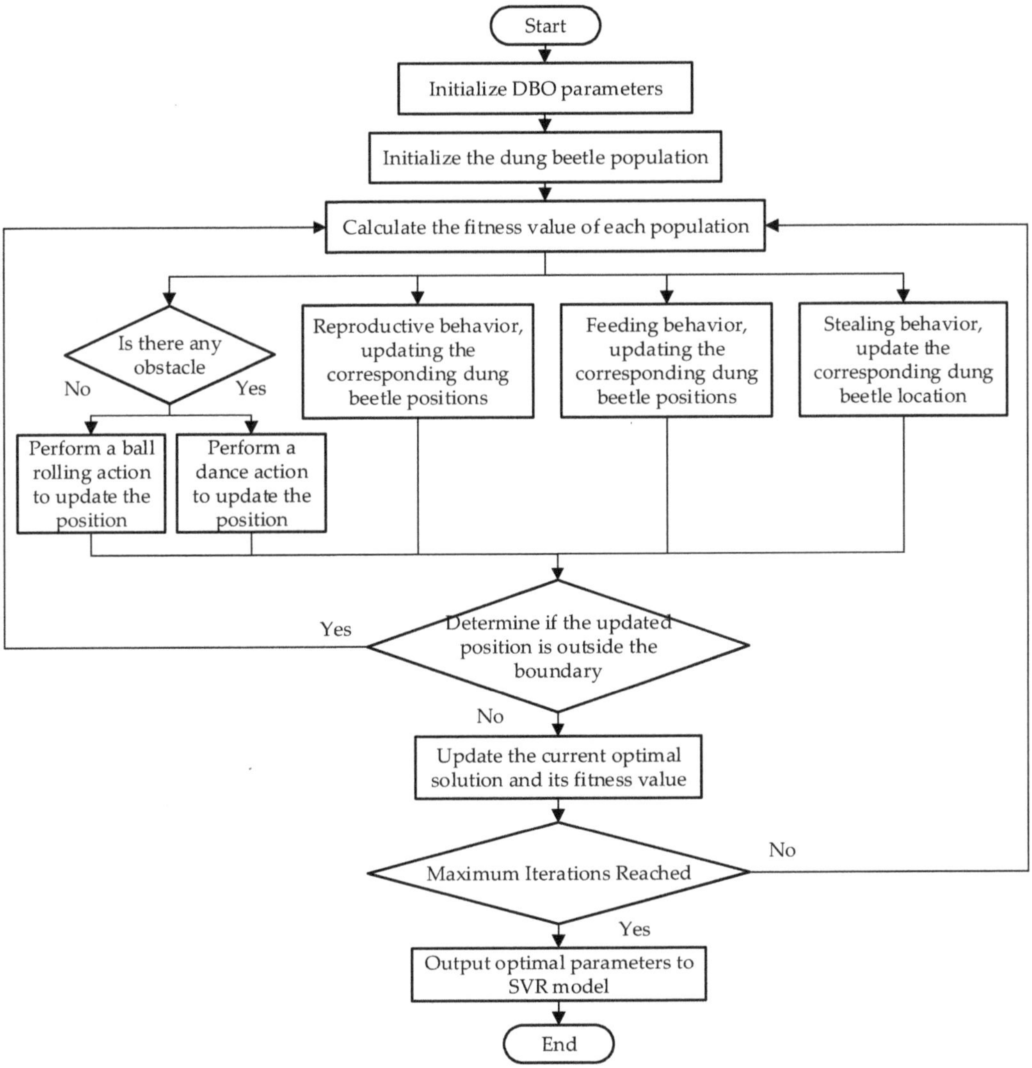

Figure 1. Optimization process of SVR parameters using DBO algorithm.

3.2. Combined Forecasting Model Framework Based on VMD-DBO-SVR

The flowchart of the combined prediction model based on VMD-DBO-SVR is illustrated in Figure 2.

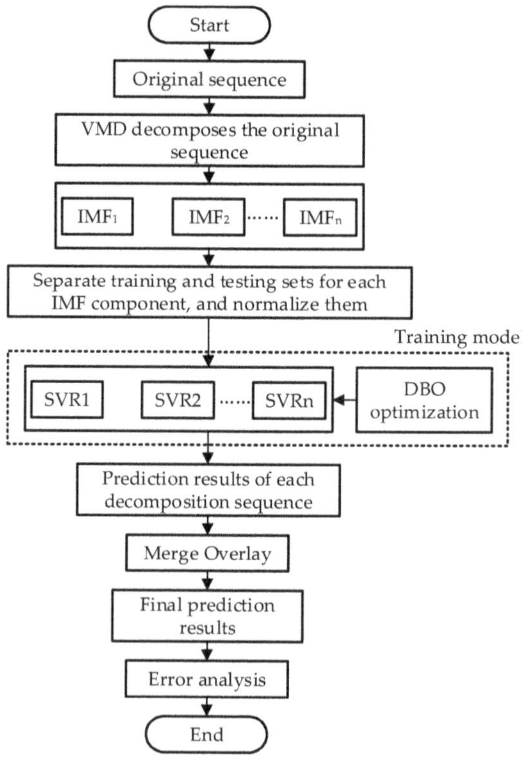

Figure 2. VMD-DBO-SVR forecasting model process.

The lithium-ion battery prediction model in this paper consists of four steps:

(1) Using VMD decomposition to decompose the original sequence of battery SOH into various modal components with different frequency bands.

(2) Preprocess the data of each mode component obtained by VMD decomposition, normalize it, and divide the preprocessed data into training data and testing data. Use the mode component data of the k-th iteration as input and the mode component data of the k + 1-th iteration as output.

(3) Build an SVR prediction model for each component separately and update the optimal SVR parameters using the DBO optimization algorithm.

(4) Validate the existing model with the test data, perform reverse normalization on the predicted values of each component, combine and superimpose the components, obtain the final SOH prediction result, and conduct error analysis.

4. Experimental Results and Comparative Analysis

4.1. Experimental Data and Parameter Settings

The dataset used in this experiment is from the Prognostics Center of Excellence (PCoE) at NASA. The dataset includes aging test data for three 18,650 lithium-ion batteries with a rated capacity of 2 Ah each, labeled B5, B6, and B7. All data were collected at a temperature of 24 °C using the CC-CV cycle test method for aging testing. Firstly, the battery is charged with a steady current of 1.5 A. When the battery voltage hits 4.2 V, the steady voltage mode is applied to continue charging the battery until the charging current drops below 20 mA. Then, the battery is discharged at a steady current of 2 A until the voltages of B5, B6, and B7 drop to 2.7 V, 2.5 V, and 2.2 V, respectively. B5–B7 batteries have undergone 168 charge-discharge cycles, and the data includes measured voltage, current, temperature, and the available capacity for each cycle. Therefore, this dataset is mainly used for predicting the SOH and RUL, and estimating the State of Charge (SOC) of lithium-ion batteries. We extract the available capacity from each charge-discharge cycle and process it as SOH degradation data according to the SOH definition, enabling us to estimate the SOH of lithium-ion batteries. The detailed parameters of the 3 batteries are shown in Table 1, and the SOH degradation curves are shown in Figure 3.

Table 1. Detailed parameters of the experimental dataset.

Number	Temperature/°C	Discharge Current	Capacity/Ah	Shutdown Voltage/V
B5	24	2A/CC	2	2.7
B6	24	2A/CC	2	2.5
B7	24	2A/CC	2	2.2

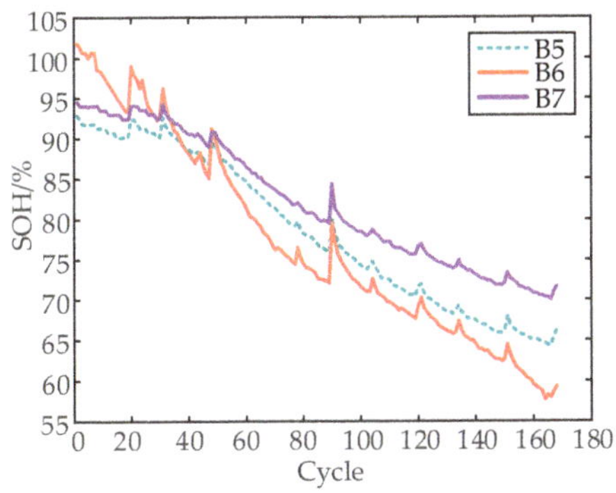

Figure 3. SOH degradation curves of NASA dataset battery B5, B6, and B7.

To more comprehensively demonstrate the adaptability of the VMD-DBO-SVR prediction method, in this experiment, 50% and 60% of the SOH data were, respectively, selected as the training set, and the remaining 50% and 40% of the data were applied as the test set to test the performance of the model.

When using the VMD algorithm for signal decomposition, it is necessary to select the mode number K beforehand. If the chosen number of modes is insufficient, certain significant information in the initial signal may be lost. On the other hand, if the selected mode number is too large, it may lead to frequency aliasing. Therefore, in this study, the mode number K is determined by examining the arrangement of center frequencies under various decomposition mode numbers. Taking the B5 battery as an example, the center frequencies under different K values are shown in Table 2.

Table 2. Center frequency of B5 battery under different K values.

K	Center Frequency/Hz					
2	1.97×10^{-5}	0.233	-	-	-	-
3	1.97×10^{-5}	0.166	0.328	-	-	-
4	1.96×10^{-5}	0.095	0.233	0.357	-	-
5	1.96×10^{-5}	0.093	0.167	0.292	0.401	-
6	1.96×10^{-5}	0.066	0.1664	0.224	0.330	0.402

As shown in Table 2, when K = 6, the central frequencies of the third and fourth mode components are close, indicating an over-decomposition phenomenon. Therefore, K = 5 is determined. The time-domain and corresponding frequency-domain plots of the SOH signal of battery B5 after VMD decomposition are shown in Figures 4 and 5, respectively.

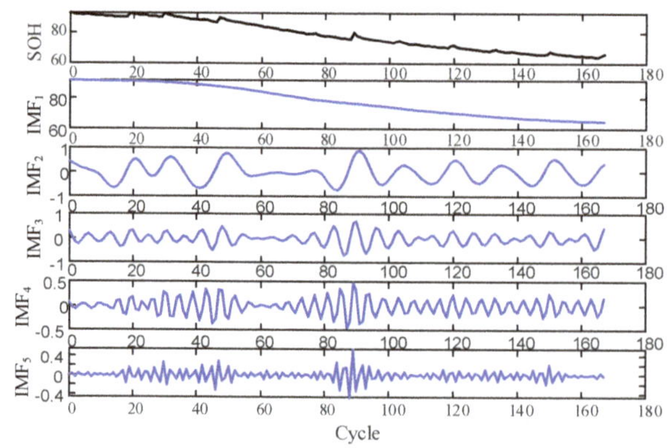

Figure 4. Time-domain diagram of VMD decomposition for B5 battery.

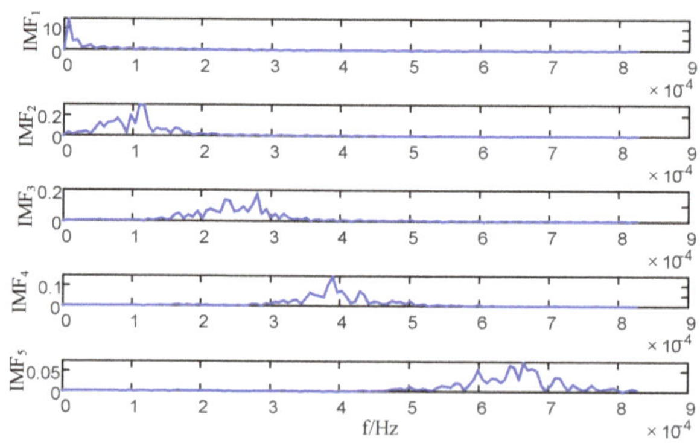

Figure 5. Spectrum diagram of B5 battery after VMD decomposition.

In Figure 4, IMF_1 represents the main trend of the original signal, IMF_2 and IMF_3 represent the periodic small fluctuations of the original signal within a shorter time period, and IMF_4 and IMF_5 represent the signal variations in the higher frequency band. Therefore, VMD can effectively decompose the different components of lithium battery SOH aging data. As shown in Figure 5, the mode mixing phenomenon of the original signal is well suppressed.

The parameter settings used in this paper for optimizing with the DBO algorithm are as follows: the number of dung beetle populations is pop = 30; the proportion of roller dung beetles, breeding dung beetles, foraging dung beetles, and thief dung beetles in the dung beetle population is 0.2, 0.2, 0.2, and 0.4, respectively; the dimension of variable parameters is dim = 2; the upper limit for the number of iterations is $T_{max} = 50$; the lower boundary is $L_b = 0.01$; and the upper boundary is $U_b = 100$.

4.2. Evaluation Index

(1) Mean Absolute Percentage Error (MAPE):

$$\mathrm{MAPE} = \frac{1}{n}\sum_{i}^{n}\left|\frac{\hat{y}_i - y_i}{y_i}\right| \times 100\% \tag{23}$$

(2) Root Mean Square Error (RMSE):

$$\mathrm{RMSE} = \sqrt{\frac{1}{n}\sum_{i}^{n}(\hat{y}_i - y_i)^2} \tag{24}$$

(3) Relative Accuracy (RA):

$$\mathrm{RA} = \frac{1}{n}\sum_{i}^{n}\left[1 - \left|\frac{\hat{y}_i - y_i}{y_i}\right|\right] \tag{25}$$

where, $\hat{y}_i$ is the predicted SOH; y_i is the actual SOH.

The performance of the proposed model is evaluated using the above indicators in this paper. A smaller value of MAPE and RMSE indicates a more accurate prediction result, while a value closer to 1 for RA demonstrates a better prediction performance of the model.

4.3. *Experimental Verification and Analysis of SOH Prediction Based on VMD-DBO-SVR Model*

When 60% of the data from three lithium-ion batteries were taken as the training set (B5, B6, and B7 with 100 cycles each), the fitting of the SOH prediction results based on the VMD-DBO-SVR model and the actual test results of the lithium-ion batteries are shown in Figure 6. The corresponding prediction errors are shown in Table 3, where ST represents the starting point of the prediction.

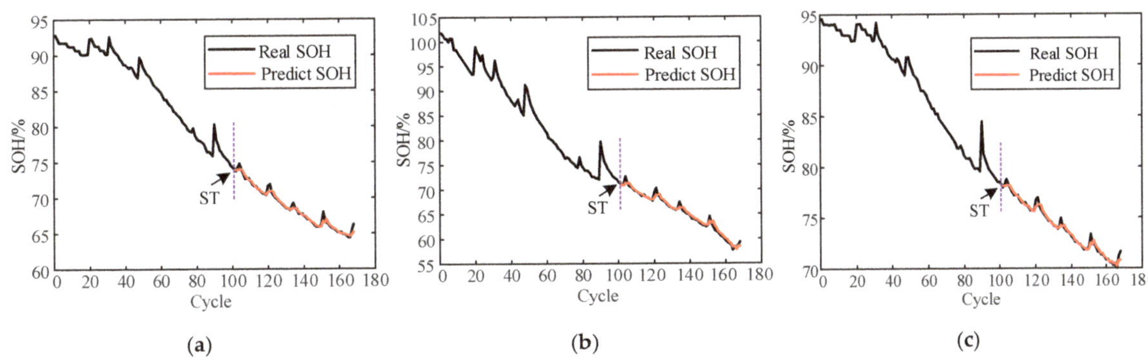

Figure 6. SOH prediction results with 60% data set of three batteries as training set. (**a**) B5; (**b**) B6; (**c**) B7.

Table 3. Prediction results of the VMD-DBO-SVR model based on the training dataset of 60%.

Battery	MAPE/%	RMSE	RA
B5	0.3511	0.3488	0.9964
B6	0.5863	0.5019	0.9941
B7	0.2594	0.2765	0.9974

As shown in Figure 6, the predicted SOH values for the three batteries are very close to the true values, indicating that the proposed method can effectively predict the trend of battery SOH and has good prediction accuracy. As shown in Table 3, the best prediction performance among the three batteries is B7, with an RA of 99.74% and RMSE and MAPE of 0.5019 and 0.2594%, respectively. The worst prediction performance is for B6, but its RA value is also as high as 99.41%, and RMSE and MAPE are only 0.2765 and 0.5863%, respectively. These results demonstrate that the VMD-DBO-SVR model proposed in this paper has good applicability to different lithium-ion batteries and can maintain high prediction accuracy.

To further verify the SOH prediction accuracy of the proposed model with insufficient training data, 50% of the data from three batteries were selected as the training set (B5, B6, and B7 with 84 cycles each), and the remaining 50% were used as the test set for SOH prediction. When 50% of the data was used for training, the fitting of the lithium-ion battery SOH prediction results based on the VMD-DBO-SVR model to the true test results is shown in Figure 7, and the corresponding prediction errors are shown in Table 4.

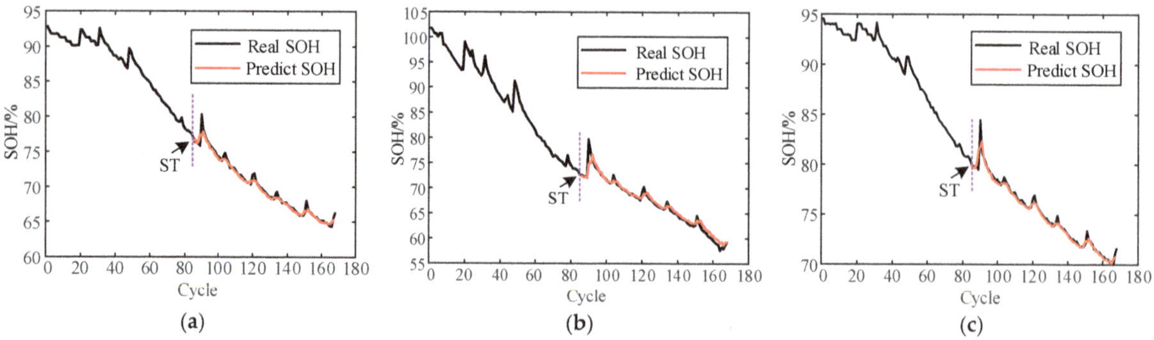

Figure 7. SOH prediction results of three batteries with 50% dataset as training set. (**a**) B5; (**b**) B6; (**c**) B7.

Table 4. Prediction results of the VMD-DBO-SVR model based on a 50% dataset as training set.

Battery	MAPE/%	RMSE	RA
B5	0.3906	0.4771	0.9961
B6	0.7892	0.8227	0.9921
B7	0.3318	0.4828	0.9966

When using fewer data as the training set, the prediction accuracy will decrease as shown in Figure 7, indicating that the less effective information provided during modeling from the early stage of prediction, the greater the error in the prediction results. However, the proposed model in this paper still has a good predictive effect. As shown in Table 4, even for battery B6 with the worst prediction accuracy, its RA value still reaches 99.21%, and the RMSE and MAPE values are only 0.8227 and 0.7892%, respectively. Compared with 60% training data, the corresponding RA value only decreases by 0.2%, while the RMSE and MAPE values increase by only 0.3227 and 0.2029%, respectively, indicating that the proposed prediction model has good generalization ability.

4.4. Comparative Analysis of VMD-DBO-SVR Model with Other Models

To validate the effectiveness and superiority of the VMD-DBO-SVR model proposed in this paper for predicting the SOH, three batteries were selected with 50% of their data employed as the training set, and compared with three other prediction models: SVR, EMD-SVR, and VMD-SVR. Figure 8 shows the comparison of the prediction results under different algorithms, and the corresponding prediction errors are shown in Table 5.

Figure 8 reveals that the single model SVR has the worst prediction performance, and its estimation error gradually increases in the later stage, indicating poor prediction accuracy. Although the EMD-SVR model reduces the error, the prediction accuracy is still poor and has not stabilized. The VMD-SVR model overcomes the end effect and modal aliasing phenomenon of EMD decomposition, and compared with the EMD-SVR model, the prediction performance is significantly improved, but the prediction accuracy is still not high enough. The VMD-DBO-SVR model proposed in this paper preprocesses the initial SOH data using the VMD decomposition method to reduce noise in the original data. After optimizing the SVR model parameters using the DBO algorithm, the prediction model is trained to accurately predict the overall degradation trend of the battery and has good tracking ability for capacity regeneration, resulting in the best prediction performance.

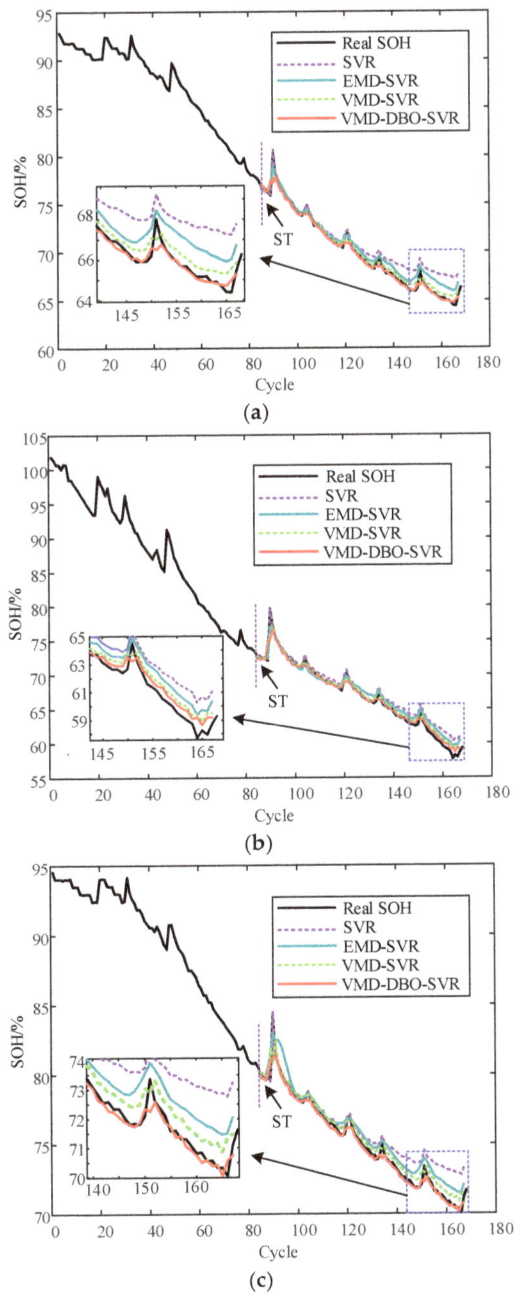

Figure 8. Comparison of SOH prediction results under different models for three batteries. (**a**) B5; (**b**) B6; (**c**) B7.

Table 5. SOH Prediction Errors of Different Models for three Batteries.

Battery	Model	MAPE/%	RMSE	RA	Prediction Time/s
B5	SVR	1.7833	1.5486	0.9821	0.6875
	EMD-SVR	1.1607	0.9775	0.9883	3.2114
	VMD-SVR	0.6467	0.5797	0.9935	3.4732
	VMD-DBO-SVR	0.3906	0.4771	0.9961	3.7297
B6	SVR	1.9822	1.6458	0.9801	0.7751
	EMD-SVR	1.3974	1.3233	0.9860	3.2046
	VMD-SVR	1.0090	0.9082	0.9899	3.6104
	VMD-DBO-SVR	0.7892	0.8148	0.9921	3.9867
B7	SVR	1.4489	1.4325	0.9855	0.6658
	EMD-SVR	1.1933	1.0556	0.9880	3.1699
	VMD-SVR	0.7364	0.6431	0.9926	3.1105
	VMD-DBO-SVR	0.3318	0.4828	0.9966	3.4405

From Table 5, it can be seen that the SVR model has a shorter prediction time, but its error is too large, which may result in unsatisfactory prediction results in practical applications. Compared with the EMD-SVR and VMD-SVR methods, the proposed model has a significant improvement in prediction accuracy with a relatively small increase in prediction time. Taking the B5 battery as an example, the RMSE of SVR, EMD-SVR, and VMD-SVR models are 1.5486, 0.9775, and 0.5797, and their prediction time are 0.6875 s, 3.2114 s, and 3.4732 s, respectively, while the RMSE and prediction time of the proposed model are 0.4771 and 3.7297 s, proving that the proposed model has higher accuracy than the other three models with only a small increase in prediction time.

To further demonstrate the superiority of the proposed VMD-DBO-SVR prediction method, still using battery B5 as an example and under the same initial conditions with the same training set, we compared the prediction results of our method with those of recently published models in related literature. The comparison results are shown in Table 6.

Table 6. Comparison of prediction results between proposed method and other method in literature.

Battery	Model	MAPE/%	RMSE
B5	IALO-SVR [11]	0.7400	0.6841
	ABMS-CEEMDAN-LSTM [27]	1.3145	1.0948
	VMD-DBO-SVR	0.3906	0.4771

From Table 6, it can be seen that although Reference [11] uses an improved ant lion optimizer to optimize the SVR parameters, it ignores the noise effect of the initial capacity data of batteries, resulting in lower prediction accuracy. Reference [27] reduces the impact of battery capacity regeneration by using an adaptive double exponential model smoothing method and denoising the lithium battery capacity data using CEEMDAN. However, the modal aliasing effect caused by CEEMDAN is unavoidable, resulting in inaccurate prediction. In general, the VMD-DBO-SVR model proposed in this work achieves higher prediction accuracy.

In this section, we first conducted simulations with 60% of the training data to validate the precision of the proposed method on three batteries. Then, we reduced the training data to 50% to demonstrate that the prediction accuracy decreases with the decrease of training data, but the proposed model still has good prediction performance. Furthermore, we compared the proposed model with SVR, EMD-SVR, and VMD-SVR models from prediction accuracy and running time, the experimental results showed that the proposed model had significantly improved prediction accuracy with a small increase in computation time. Finally, we compared the proposed model with relevant methods in recent literature to confirm its superior predictive performance.

5. Conclusions

Accurately predicting the SOH of lithium-ion batteries can improve their safety during operation and prevent accidents. We propose a SOH prediction model based on Variational Mode Decomposition (VMD) and Dung Beetle Optimization-Support Vector Regression (DBO-SVR). Through verification and analysis, the primary conclusions of this paper are as follows:

(1) The VMD algorithm can decompose the battery SOH sequence into multiple stationary mode components, which can effectively reduce noise interference, such as capacity regeneration and testing errors, and minimize prediction errors.
(2) The selection of kernel parameters in the SVR method directly affects the accuracy of SOH prediction. To address this issue, we proposed a DBO optimization algorithm to provide the optimal parameters for the SVR method. The combination of the two methods can improve the prediction accuracy and stability of SOH.
(3) NASA battery dataset was employed to validate the prediction performance of the proposed VMD-DBO-SVR model. The results showed that the VMD-DBO-SVR model had good prediction accuracy and stability, and the prediction error was maintained within 1%.

The above conclusion indicates that the model solves various noise interference problems through the VMD algorithm, and solves the problem of difficult SVR parameter selection through the DBO optimization algorithm, thereby improving prediction accuracy and achieving the preliminary research objectives of this article.

In actual SOH prediction for lithium-ion batteries, it is sometimes difficult to directly measure the available capacity of the battery, making the approach proposed in this paper unsuitable. Therefore, the next research direction of this paper is to use easily measurable feature factors that characterize the degradation pattern of SOH.

Author Contributions: Conceptualization, C.W. and J.F.; methodology, C.W. and J.F.; software, X.H. and X.X.; validation, J.M. and C.W.; formal analysis, J.M.; investigation, J.F.; resources, C.W. and X.H.; data curation, X.X.; writing—original draft preparation, C.W. and J.F.; writing—review and editing, C.W.; visualization, X.X.; supervision, J.F.; project administration, C.W.; funding acquisition, C.W. and X.H. All authors have read and agreed to the published version of the manuscript.

Funding: This research was funded by the National Key R&D Program of China and the Key R&D Plan of Shaanxi Province, grant number 2021YFB2601300 and 2022GY-193.

Data Availability Statement: The data supporting the results of this study was obtained from the NASA Prognostics Center of Excellence (PCoE). Available online: https://ti.arc.nasa.gov/tech/dash/groups/pcoe/prognostic-data-repository (accessed on 15 July 2022).

Conflicts of Interest: The authors declare no conflict of interest.

References

1. Ghorbanzadeh, M.; Astaneh, M.; Golzar, F. Long-term degradation based analysis for lithium-ion batteries in off-grid wind-battery renewable energy systems. *Energy* **2019**, *166*, 1194–1206. [CrossRef]
2. Xu, J.L.; Liu, B.L.; Zhang, G.Y.; Zhu, J.W. State-of-health estimation for lithium-ion batteries based on partial charging segment and stacking model fusion. *Energy Sci. Eng.* **2023**, *11*, 383–397. [CrossRef]
3. Li, J.; Li, Y.; Chen, G.; Lyu, C.; Wu, Y.; Xu, L.; Ma, S. Research on Feature Extraction and SOH Evaluation Methods for Retired Power Battery. *Proc. Chin. Soc. Electr. Eng.* **2022**, *42*, 1332–1346.
4. Pang, B.; Chen, L.; Dong, Z.M. Data-Driven Degradation Modeling and SOH Prediction of Li-Ion Batteries. *Energies* **2022**, *15*, 5580. [CrossRef]
5. Shen, S.Q.; Liu, B.C.; Zhang, K.; Ci, S. Toward Fast and Accurate SOH Prediction for Lithium-Ion Batteries. *IEEE Trans. Energy Convers.* **2021**, *36*, 2036–2046. [CrossRef]
6. Chen, D.; Meng, J.H.; Huang, H.Y.; Wu, J.; Liu, P.; Lu, J.W.; Liu, T.Q. An Empirical-Data Hybrid Driven Approach for Remaining Useful Life prediction of lithium-ion batteries considering capacity diving. *Energy* **2022**, *245*, 12. [CrossRef]
7. Iurilli, P.; Brivio, C.; Carrillo, R.E.; Wood, V. Physics-Based SoH Estimation for Li-Ion Cells. *Batteries* **2022**, *8*, 204. [CrossRef]
8. Wen, J.C.; Zou, Q.R.; Chen, C.G.; Wei, Y.J. Linear correlation between state-of-health and incremental state-of-charge in Li-ion batteries and its application to SoH evaluation. *Electrochim. Acta* **2022**, *434*, 10. [CrossRef]

9. Wu, T.Z.; Liu, S.Z.; Wang, Z.K.; Huang, Y.H. SOC and SOH Joint Estimation of Lithium-Ion Battery Based on Improved Particle Filter Algorithm. *J. Electr. Eng. Technol.* **2022**, *17*, 307–317. [CrossRef]
10. Sun, S.; Sun, J.Z.; Wang, Z.L.; Zhou, Z.Y.; Cai, W. Prediction of Battery SOH by CNN-BiLSTM Network Fused with Attention Mechanism. *Energies* **2022**, *15*, 4428. [CrossRef]
11. Li, Q.; Zhang, X.; Zhao, K.; Sun, J.; Wang, K. State of Health Estimation of Lithium-ion Battery Based on Ant Lion Optimization and Support Vector Regression. In Proceedings of the 2021 IEEE International Conference on Electrical Engineering and Mechatronics Technology (ICEEMT), Qingdao, China, 2–4 July 2021; pp. 334–337.
12. Wang, Y.; Ni, Y.; Zheng, Y.; Shi, X.; Wang, J. Remaining Useful Life Prediction of Lithium-ion Batteries Based on Support Vector Regression Optimized and Ant Lion Optimizations. *Proc. Chin. Soc. Electr. Eng.* **2021**, *41*, 1445–1457.
13. Xu, J.; Ni, Y.; Zhu, C. Remaining Useful Life Prediction for Lithium-Ion Batteries Based on Improved Support Vector Regression. *Trans. China Electrotech. Soc.* **2021**, *36*, 3693–3704.
14. Ye, J.Y.; Yang, Z.X.; Li, Z.L. Quadratic hyper-surface kernel-free least squares support vector regression. *Intell. Data Anal.* **2021**, *25*, 265–281. [CrossRef]
15. Cheng, Y.; Zheng, L.; Liu, J. Lithium battery health state estimation based on mode decomposition and time series. *J. Power Supply* **2023**. Available online: https://kns.cnki.net/kcms/detail/12.1420.TM.20230131.1102.001.html (accessed on 21 February 2023).
16. Khumprom, P.; Yodo, N. A Data-Driven Predictive Prognostic Model for Lithium-ion Batteries based on a Deep Learning Algorithm. *Energies* **2019**, *12*, 660. [CrossRef]
17. Xu, Z.Y.; Guo, Y.J.; Saleh, J.H. A physics-informed dynamic deep autoencoder for accurate state-of-health prediction of lithium-ion battery. *Neural Comput. Appl.* **2022**, *34*, 15997–16017. [CrossRef]
18. Hu, X.; Guo, Y.; Zhang, R. Review of State-of-health Estimation Methods for Lithium-ion Battery. *J. Power Supply* **2022**, *20*, 126–133.
19. Meng, J.H.; Cai, L.; Stroe, D.I.; Huang, X.R.; Peng, J.C.; Liu, T.Q.; Teodorescu, R. An Automatic Weak Learner Formulation for Lithium-Ion Battery State of Health Estimation. *IEEE Trans. Ind. Electron.* **2022**, *69*, 2659–2668. [CrossRef]
20. Zhang, C.; Zhao, S.; He, Y. State-of-health Estimate for Lithium-ion Battery Using Information Entropy and PSO-LSTM. *J. Mech. Eng.* **2022**, *58*, 180–190.
21. Zhang, Y.N.; Lian, Z.; Fu, W.L.; Chen, X. An ESR Quasi-Online Identification Method for the Fractional-Order Capacitor of Forward Converters Based on Variational Mode Decomposition. *IEEE Trans. Power Electron.* **2022**, *37*, 3685–3690. [CrossRef]
22. Ding, G.R.; Wang, W.B.; Zhu, T. Remaining Useful Life Prediction for Lithium-Ion Batteries Based on CS-VMD and GRU. *IEEE Access* **2022**, *10*, 89402–89413. [CrossRef]
23. Xue, J.K.; Shen, B. Dung beetle optimizer: A new meta-heuristic algorithm for global optimization. *J. Supercomput.* **2022**, *79*, 7305–7336. [CrossRef]
24. Wei, R.; Mao, T.; Gao, H.; Peng, J.; Yang, J. Health state estimation of lithium ion battery based on TWP-SVR. *Energy Storage Sci. Technol.* **2022**, *11*, 2585–2599.
25. Zhou, X.; Li, N.; Pan, Y.; Sun, L. Optimized SVR based on artificial bee colony algorithm for leaf area index inversion. *J. Remote Sens.* **2022**, *26*, 766–780.
26. Zhou, S.; Yang, C.C.; Su, Z.N.; Yu, P.; Jiao, J. An Aeromagnetic Compensation Algorithm Based on Radial Basis Function Artificial Neural Network. *Appl. Sci.* **2023**, *13*, 136. [CrossRef]
27. Huang, K.; Ding, H.; Guo, Y.; Tian, H. Prediction of Remaining Useful Life of Lithium-Ion Battery Based on Adaptive Data Preprocessing and Long Short-Term Memory Network. *Trans. China Electrotech. Soc.* **2022**, *37*, 3753–3766.

energies

MDPI

Article

Improved Battery Balancing Control Strategy for Reconfigurable Converter Systems

Guangwei Wan [1], Qiang Zhang [1,*], Menghan Li [2], Siyuan Li [1,*], Zehao Fu [1], Junjie Liu [1] and Gang Li [3]

1 School of Energy and Power Engineering, Shandong University, Jinan 250061, China; wgwedu@foxmail.com (G.W.); fuzehao2077@163.com (Z.F.); 18352869916@163.com (J.L.)
2 School of Energy and Environmental Engineering, Hebei University of Technology, Tianjin 300401, China; sdulmh@163.com
3 College of Mechanical and Electronic Engineering, Northwest A&F University, Xianyang 712100, China; cn_ligang@163.com
* Correspondence: sduzhangqiang@sdu.edu.cn (Q.Z.); lisiyuan@sdu.edu.cn (S.L.)

Abstract: In order to address the issue of battery cell disparity in lithium-ion battery systems, battery balancing techniques are required. This paper proposes an improved battery balancing strategy within a reconfigurable converter system. The strategy is based on the state of charge (SOC) of batteries, and utilizes the reconfigurable converter system to transfer energy from battery modules with high SOC to those with lower SOC. Additionally, it allows for battery module balancing while supplying power to loads. A MATLAB/Simulink simulation model with five batteries was built to validate the effectiveness of the proposed balancing strategy under unloaded and loaded conditions. The simulation results demonstrate that the proposed strategy achieves more efficient and accurate battery module balancing compared to the previous balancing modes.

Keywords: reconfigurable battery; balancing; integrated converter; state of charge (SOC)

Citation: Wan, G.; Zhang, Q.; Li, M.; Li, S.; Fu, Z.; Liu, J.; Li, G. Improved Battery Balancing Control Strategy for Reconfigurable Converter Systems. *Energies* **2023**, *16*, 5619. https://doi.org/10.3390/en16155619

Academic Editor: King Jet Tseng

Received: 3 July 2023
Revised: 15 July 2023
Accepted: 24 July 2023
Published: 26 July 2023

1. Introduction

Due to the cell-to-cell variation in lithium-ion battery systems, individual cells may become overcharged or over-discharged during charging and discharging processes. Failure to perform timely and effective balancing may result in decreased battery pack lifespan, reduced capacity, performance degradation, and even safety hazards such as fire [1,2]. Therefore, battery balancing plays an important role in improving overall battery pack lifespan, ensuring battery safety and reliability, and increasing energy utilization efficiency [3,4].

Battery balancing methods can be classified into passive balancing and active balancing, depending on whether energy dissipation is involved [5].

Passive balancing is mainly achieved by using resistors to discharge high-energy battery cells, consuming their excess energy in the form of heat to achieve energy consistency among all cells. The advantages of passive balancing include simple and feasible circuit structure, low cost, small circuit volume, and easy control. However, since energy is dissipated in the form of heat, the overall energy utilization efficiency of the battery module is relatively low [6,7].

Active balancing mainly utilizes energy storage components such as capacitors and inductors to transfer energy between cells, which can be achieved with different circuit structures. Active balancing has high energy utilization efficiency, high balancing efficiency, and fast balancing speed. However, it usually comes with complex circuit structures and control strategies, resulting in higher implementation costs [8].

Balancing circuits typically use battery voltage or state of charge (SOC) as the balancing indicators. When using battery voltage as the balancing indicator, the advantage lies in its simplicity and ease of implementation. However, the disparate internal resistance of batteries can lead to suboptimal balancing effects. On the other hand, using SOC as the balancing indicator offers the advantage of superior balancing effectiveness, but it entails complex calculations and greater implementation difficulties.

Many researchers have conducted research on battery balance using battery voltage as the balancing indicator. For example, K. Nishijima et al., 2000.proposed a PWM-controlled DC-DC converter technique that utilized battery voltage as the balancing indicator, which was comparatively simple and efficient [9]. Siqi Li et al.,2012. designed a balancing circuit composed of Metal-Oxide-Semiconductor Field-Effect Transistors (MOSFETs) and a multi-winding transformer. It achieved energy transfer from batteries with higher voltages to those with lower voltages by controlling the turning on and off of the MOSFETs [10].

As technology progresses, battery balancing methods based on SOC as the balancing indicator have gradually become mainstream. Cao et al., 2020 studied a hierarchical SOC balancing control method for battery energy storage systems, which achieved SOC balancing between battery cells and modules. This was accomplished through modulation of the duty cycle proposed for power converter switches [11]. Wang et al., 2022 proposed a layered Model Predictive Control (MPC)-based balancing approach utilizing adaptive estimation. Different balancing topologies were explored, and Hardware-in-the-Loop (HIL) testing was conducted to verify the real-time feasibility of the proposed MPC balancing strategy. Finally, the impact of series-connected battery cell count and the adopted topology on balancing performance were discussed [12]. The balancing scheme proposed by Yun et al. uses SOC as the balancing variable and a closed-loop flyback converter as the energy transfer circuit, which can effectively solve the inconsistency of battery energy [13]. Li et al. proposed a two-layer equalization method using SOC as the equalization index, combining the reconfigurable topology with the converter active equalization method. A simulation circuit of was built consisting of 12 battery cells in MATLAB/Simulink to verify its equalization effect [14]. The MATLAB version used in this paper is version 2022b.

Currently, reconfigurable battery energy storage systems have attracted increasing attention due to their ability to dynamically reconfigure the battery topology in real time to adapt to specific application requirements [15–19]. This can more effectively utilize battery resources, isolate corresponding batteries according to their current state of charge and health status without affecting the charge and discharge processes of other batteries, and extend the battery's service life while reducing the possibility of module failure [20,21].

The state of charge (SOC) is commonly used to characterize the amount of charge in a battery cell. A battery cell's SOC is defined by the ratio of the cell's present amount of charge to its rated charge capacity [22]. There is relatively little research on the combination of reconfigurable and integrated converters for battery systems. In [23], the authors proposed an integrated reconfigurable converter structure that can be used for high-voltage battery systems. In [24,25], the authors proposed load-sharing balancing strategies and distributed balancing control for battery modules based on the circuit structure of the integrated reconfigurable converter. In [26], the authors improved the structure of the integrated reconfigurable converter system and proposed a new balancing strategy. However, this balancing strategy failed to achieve free energy exchange between battery modules and only demonstrated effective balancing when the SOC of batteries followed a strict hierarchical order.

In this paper, we improve on the battery balancing strategy proposed in [26] and conduct a comparative analysis. Due to only discussing the balancing strategy, the structure of the reconfigurable converter is simplified. During the balancing process, this balancing strategy is capable of transferring energy from the battery module with the highest SOC to the one with the lowest SOC. Only the battery modules with the highest and lowest SOC are involved in the balancing process, while other battery modules remain unaffected. This approach effectively avoids repetitive charge–discharge cycles in certain batteries during balancing. As a result, it enhances system stability and prolongs battery lifespan.

This paper is organized as follows: Section 2 introduces the integrated system of reconfigurable battery and converter and illustrates its working principle. Section 3 describes the balancing strategy and design of the control system. Section 4 presents simulation experimental results and analysis. Section 5 discusses the experimental results and equilibrium strategies. Section 6 concludes this paper.

2. Structure and Working Principle of Integrated Reconfigurable Converter

The integrated reconfigurable converter is shown in Figure 1. In the figure, *S1–S10* and *Sb* are IGBT, and *D1* and *D2* are diodes. *L* is the inductance, C is the capacitance. *B1–B3* are batteries. For simplicity and convenience, three battery modules are used as examples for illustration. However, the same configuration can be applied to systems with a higher number of battery modules. The system consists of a battery module selector and boost converters. The boost converter's configuration is displayed in Figure 2. The battery modules can be dynamically reconfigured to select different input voltages. Table 1 shows the different battery modules that the battery module selector can choose from. When the input voltage changes, the range of output voltage that the boost converter can provide also changes accordingly. The relationship between the output voltage V_{out}, input voltage V_{in}, and the duty cycle D of the boost converter is expressed in Equation (1).

$$V_{\text{out}} = \frac{1}{1-D} V_{\text{in}} \tag{1}$$

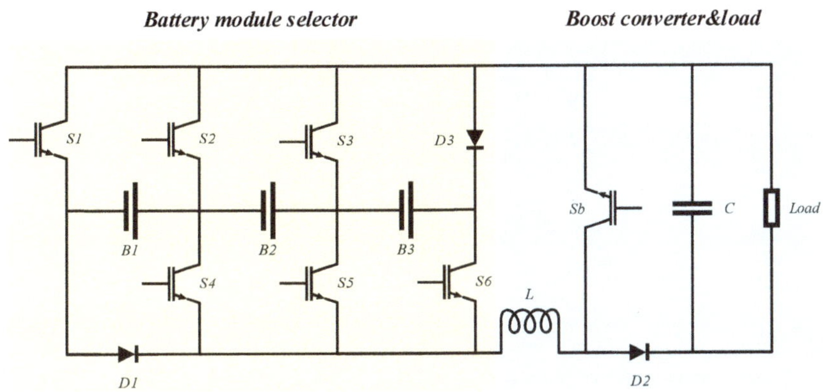

Figure 1. Integrated reconfigurable converter structure.

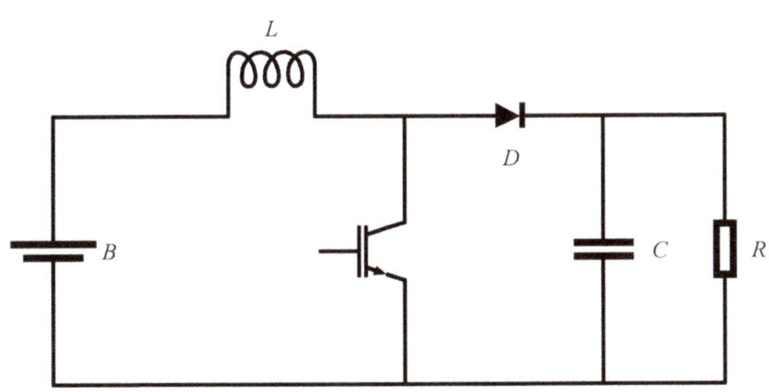

Figure 2. Boost converter.

Table 1. Different input modes of battery modules.

Selected Module	S1	S2	S3	S4	S5	S6
B1	1	0	0	1	0	0
B2	0	1	0	0	1	0
B3	0	0	1	0	0	1
B1, B2	1	0	0	0	1	0
B2, B3	0	1	0	0	0	1
B1, B2, B3	1	0	0	0	0	1

The Battery Management System (BMS) can identify battery modules with relatively high States of Charge (SOC), and the battery module selector prioritizes their discharging by switching corresponding switches. The BMS can dynamically reconfigure battery modules to have two different working modes: power supply mode and balancing mode. These two modes will be elaborated on in detail below.

2.1. Power Supply Mode

Figure 3a,b illustrate the configuration of the battery system when they are powering the load. In this mode, the converter operates in boost mode, and the control system can select different battery modules through the battery module selector to discharge at different input voltages. In the case of a failure among switches S1–S6, a flow path is required to release the energy from the inductor to protect the circuit from damage. Figure 4 shows the path along which the inductor current flows. This path can also be used to return the energy from the inductor back to all battery modules, which not only protects the circuit, but also avoids wasting energy.

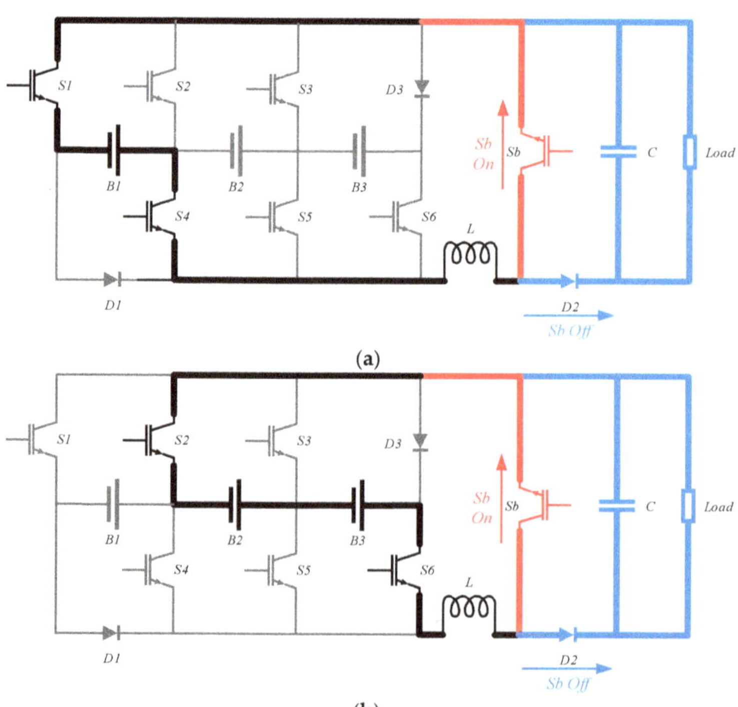

Figure 3. Battery system power supply mode: (**a**) B1 provides power to the load, (**b**) B2 and B3 provide power to the load.

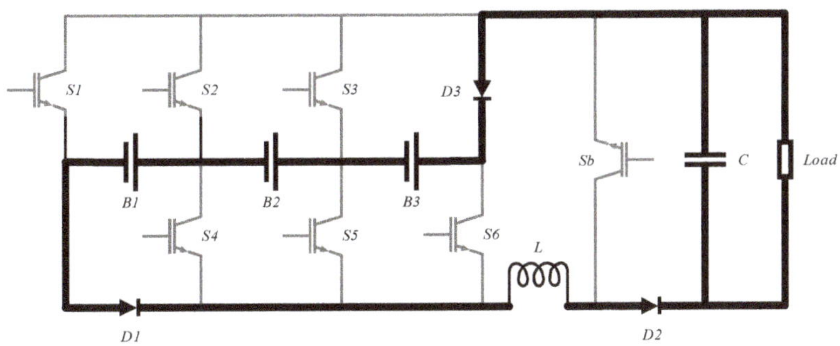

Figure 4. The freewheeling path for the inductor current.

2.2. Balance Mode

The balancing mode enables energy transfer from battery modules with high SOC to those with lower SOC, and this energy transfer is achieved by utilizing the inductor L. Figure 5a–c demonstrate the leftward energy transfer. Figure 5a shows the energy transfer between one module and another. In this figure, switches Sb and S3 are always on, and by turning on switch S6 and turning off switch S4, the energy in B3 shifts to inductor L. Then, by turning off switch S6 and turning on switch S4, the energy in inductor L is released and charges B2, achieving energy transfer from B3 to B2. The same switching cycle is repeated until the two modules reach equilibrium. Figure 5b illustrates the energy transfer from one module to two modules, and Figure 5c shows the energy transfer from two modules to one module.

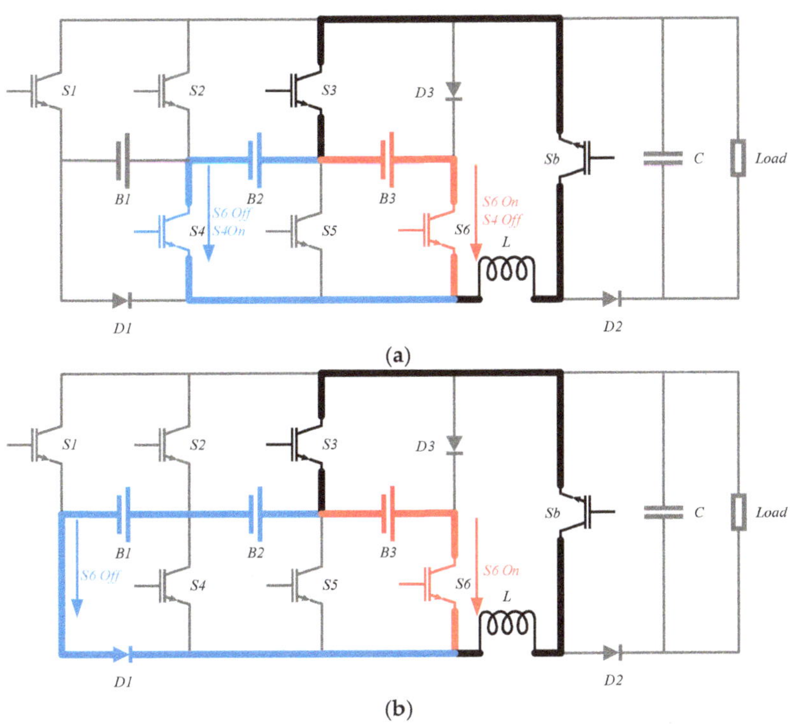

Figure 5. *Cont.*

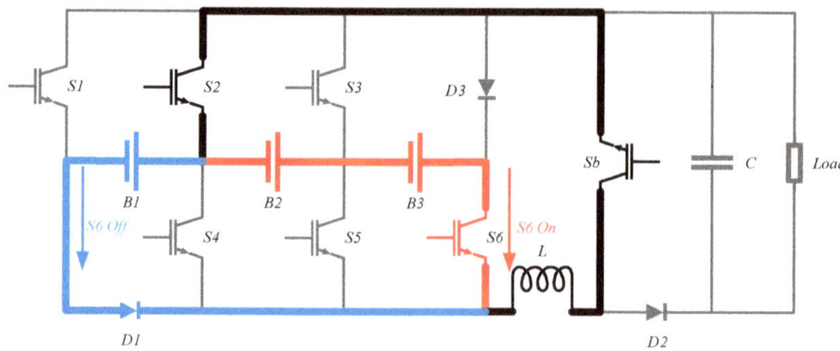

(c)

Figure 5. Illustration of balancing mode during one switching cycle. (**a**) Module to a module (B3 is discharged into B2). (**b**) Module to modules (B3 is discharged into B1 and B2). (**c**) Modules to a module (B2 and B3 are discharged into B1).

3. Balancing Strategy and Control System Design

This section describes the system balancing strategy and the design of the control system.

3.1. Balancing Strategy

The proposed new balancing strategy first identifies the battery module with the highest SOC and charges the inductor. Since the current flowing through the inductor cannot change immediately, the energy on the inductor is transferred to the battery module with the lowest SOC by changing the switch status. This balancing strategy can accelerate the balancing speed and can also be applied when batteries are powering the load. The specific balancing strategies for scenarios with and without load usage are explained below:

When there is no load usage, the balancing measure of the battery system is to have the battery module with the highest SOC store energy in the inductor, and then change the switch status to release the energy stored on the inductor to the battery module with the lowest SOC. As shown in Figure 6a,b, when battery module B3 has the highest charge, while B2 has the lowest, the balancing strategy adjusts such that battery module B3 first charges the inductor through switches S3 and S6, then maintains the closure of switch S3, and turns S4 On and S6 Off to transfer the energy from the inductor to B2. Figure 6b shows the energy transfer from battery module B3 to battery module B1 when B3 has the highest charge while B1 has the lowest.

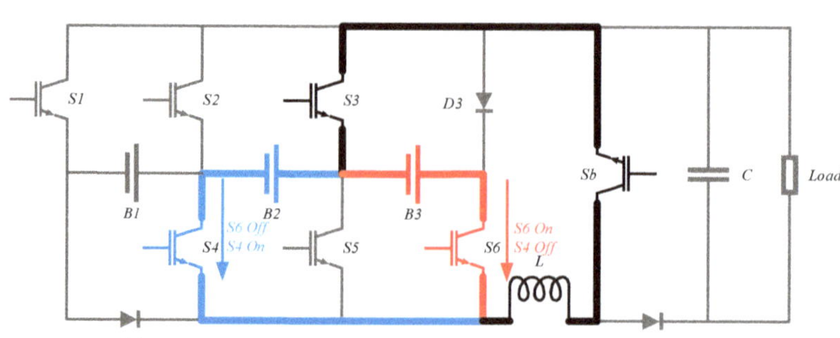

(a)

Figure 6. *Cont.*

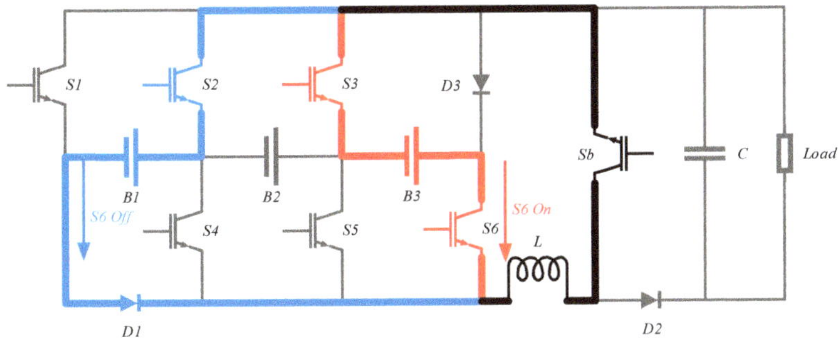

(b)

Figure 6. Balancing strategy when not using load: (**a**) B3 is discharged into B2. (**b**) B3 is discharged into B1.

When the load needs to be used, the converter is connected, and the balancing strategy of the battery system is adjusted so that the battery module with the highest charge level supplies power to both the load and the inductor, then, by changing the switch status, the energy stored on the inductor is released to the battery module with the lowest SOC. As shown in Figure 7a,b, when battery module B1 has the highest charge level while B2 has the lowest, with the load operating, the balancing strategy adjusts such that battery module B1 supplies power to the inductor and the load via switches S1 and S4, then maintains the closure of switch S4, and turns switch S3 On and S1 Off to transfer the energy from the inductor to the load and B2. Figure 7b shows the case when battery module B1 has the highest charge level while B3 has the lowest.

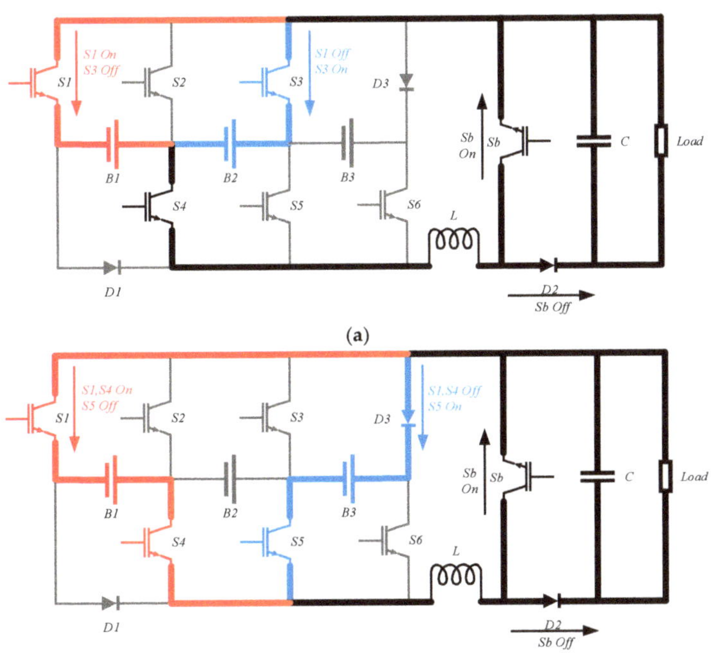

(b)

Figure 7. Balancing strategy when not using load: (**a**) B3 is discharged into B2. (**b**) B3 is discharged into B1.

Figure 8 illustrates the flow diagram of the balancing system: First, all SOC values of battery modules are obtained and sorted to find the module with the highest and lowest SOC levels. If the difference in SOC between two modules exceeds the threshold value Δe, the control system controls the corresponding switches of these two modules to transfer excess charge from the highest SOC module to the lowest SOC module. If there is no load during the balancing process, switch Sb remains closed. If the load is used, a PWM signal is used to control the On/Off state of switch Sb to connect the converter, ensuring that a single battery module can provide the required voltage. When the SOC difference between any two modules is lower than the threshold value Δe, the balancing process ends, and the required battery modules can be connected according to Table 1 to change the working voltage range of the boost converter. Once the SOC difference exceeds the threshold value Δe, the balancing procedure starts over. This iterative process ensures effective SOC balancing and prevents overcharging or over-discharging, thus improving the overall life and safety performance of the battery.

3.2. Control System Design

This section describes the design of a balancing system with and without load, and the balancing system controllers are both PI controlled.

3.2.1. Controller Design for the Balancing Operation When No Load Is Used

The system diagram of the PI controller without load balancing mode is shown in Figure 9. In order to design the PI controller, the small-signal modeling shown in Figure 10 is first derived.

The average state equation for balancing mode without load usage is as follows:

$$L\frac{di_L}{dt} = dv_i - (1-d)v_o \tag{2}$$

Introducing AC perturbations into the above equation yields

$$L\frac{d(I_L+\hat{i}_L)}{dt} = (D+\hat{d})(V_i+\hat{v}_i) - [1-(D+\hat{d})](V_o+\hat{v}_o) \tag{3}$$

The small-signal model can then be written as

$$L\frac{d\hat{i}_L}{dt} = \hat{d}V_i + D\hat{v}_i - (1-D)\hat{v}_o + \hat{d}V_o \tag{4}$$

The above equation can be obtained by applying Laplace transform as follows:

$$sL\hat{i}_L = D\hat{v}_i - (1-D)\hat{v}_o + \hat{d}(V_o+V_i) \tag{5}$$

The transfer function of the balancing mode without load usage can be obtained from Equation (5):

$$T_p = \frac{\hat{i}_L}{\hat{d}} = \frac{V_i+V_o}{Ls} \tag{6}$$

The transfer function of the PWM modulator can be modeled as follows:

$$T_m = \frac{1}{\hat{V}} \tag{7}$$

where $\hat{V} = 1$ is the peak value of the sawtooth carrier signal. Using the small-signal transfer function (6), the PI controller parameters K_p and K_i were calculated to regulate the current on the inductor, namely the balancing current, for achieving the desired open-loop phase margin at the required cutoff frequency. The Bode plot of the control loop in the charging mode is shown in Figure 11a. From the Bode plot, it can be inferred that the system is stable, as the open-loop phase margin (PM) at the cutoff frequency is greater than zero.

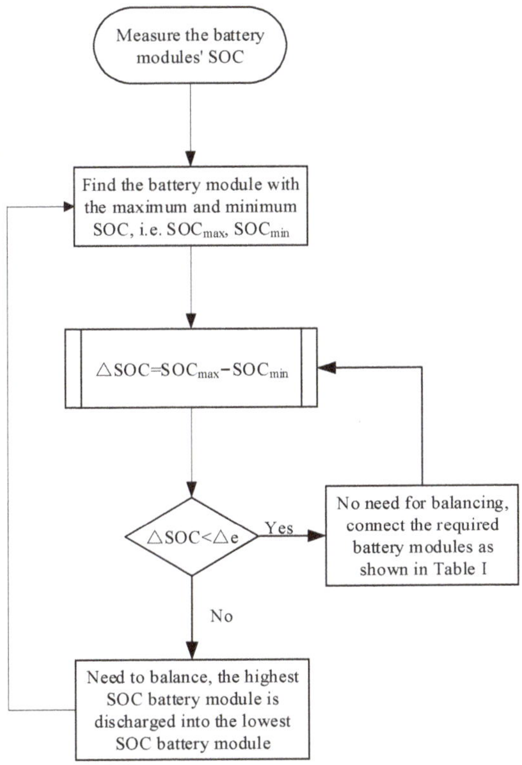

Figure 8. Flow diagram of the balancing management system.

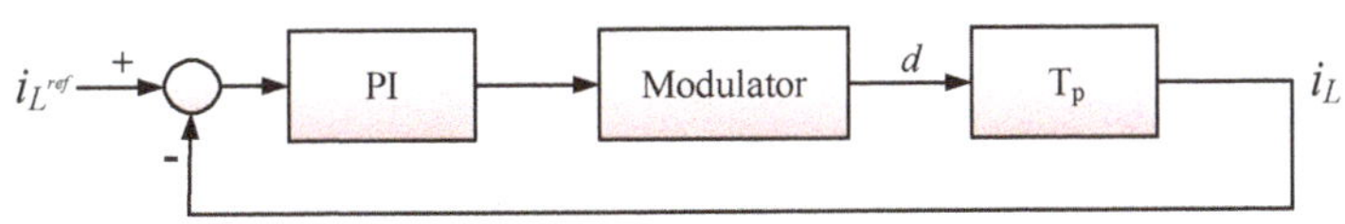

Figure 9. Controller design for the balancing operation when no load is used.

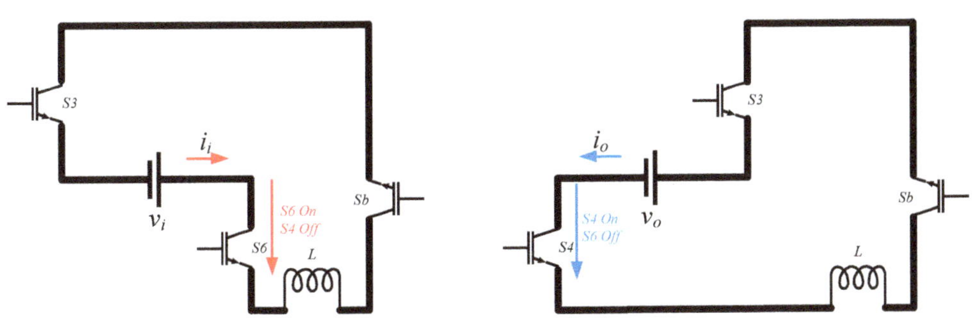

Figure 10. Simplified topology of the balancing mode without load usage in small-signal modeling.

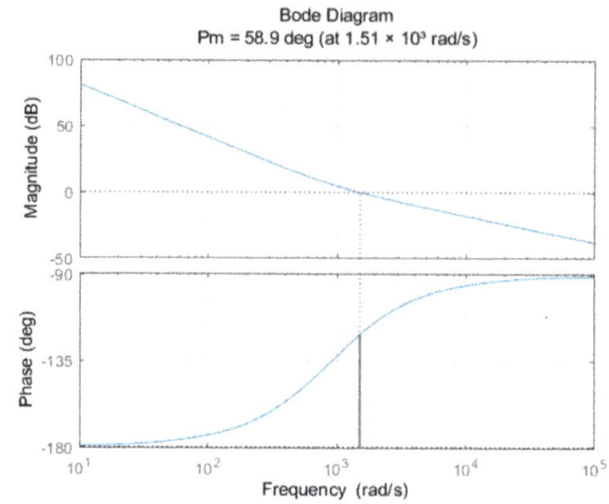

(a)

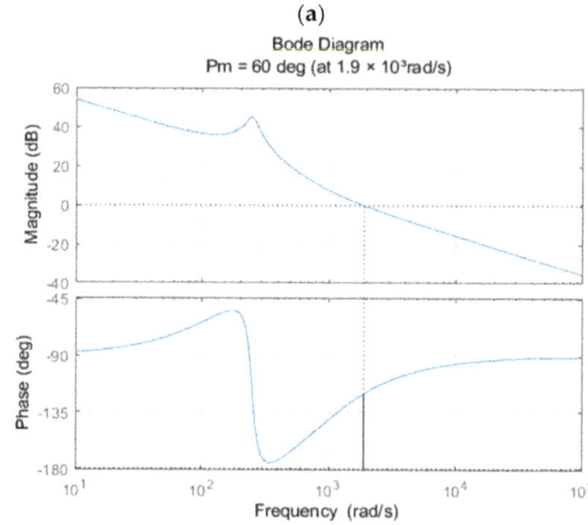

(b)

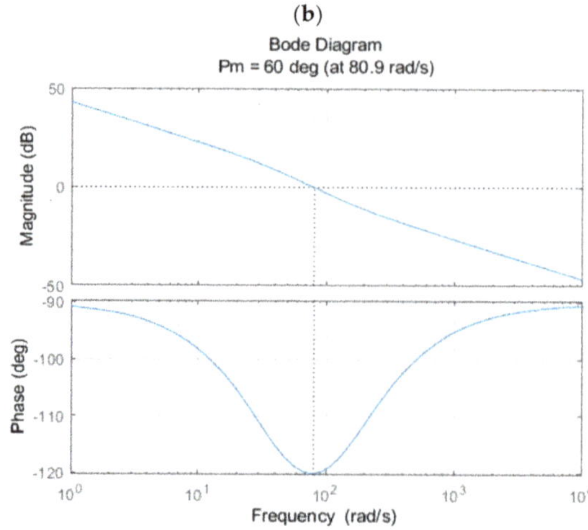

(c)

Figure 11. Bode diagrams of (**a**) balance mode control circuit when not using load, (**b**) internal current control circuit when using load, (**c**) external voltage control circuit when using load.

3.2.2. Controller Design for the Balancing Operation When Load Is Used

In the balancing mode when load is used, the double-loop control system is adopted for regulating the Buck–Boost circuit composed of the battery pack, inductor, and diode D2, as shown in Figure 12. The inner loop is a high-bandwidth current control loop, while the outer loop is a voltage control loop with lower bandwidth and slower response compared to the inner loop. The voltage of the outer loop adjusts the output voltage by providing a reference current signal to the current inner loop, which regulates the current on the inductor. Owing to the faster response of the inner loop, the outer loop can be treated separately in the circuit design process, in order to simplify the controller design.

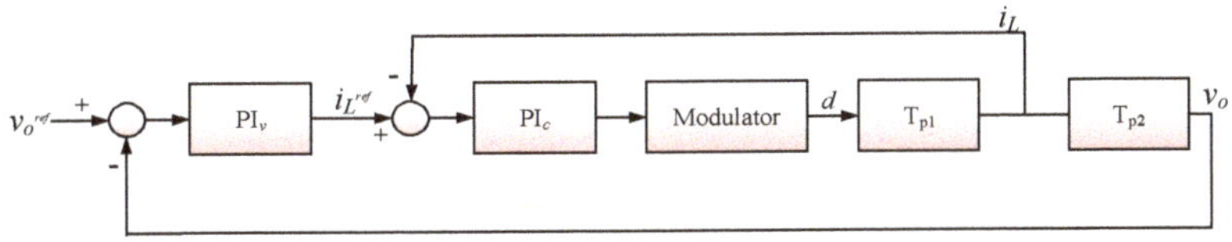

Figure 12. Controller design for the balancing operation when load is used.

1. Design of internal current control loop: In order to design the PI controller, the small-signal modeling as shown in Figure 13 is first derived.

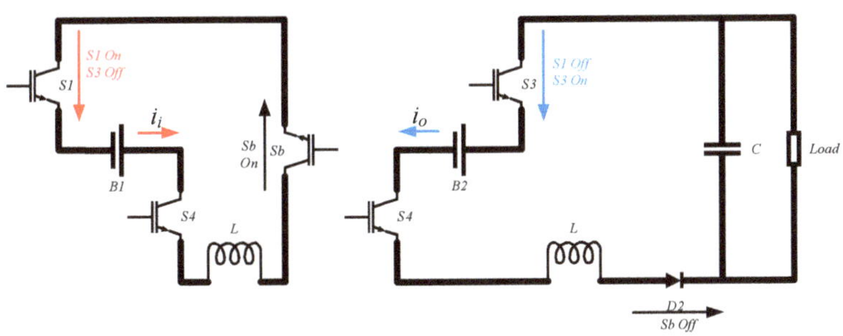

Figure 13. Simplified topology of the balancing mode when using load in small-signal modeling.

The average state equation for balancing mode when using load is as follows:

$$L\frac{di_L}{dt} = dv_i - (1-d)v_o \tag{8}$$

$$C\frac{dv_o}{dt} = (1-d)i_L - \frac{v_o}{R} \tag{9}$$

Introducing AC perturbations into the above equation yields

$$L\frac{d(I_L+\hat{i}_L)}{dt} = (D+\hat{d})(V_i+\hat{v}_i) - [1-(D+\hat{d})](V_o+\hat{v}_o) \tag{10}$$

$$C\frac{d(V_o+\hat{v}_o)}{dt} = [1-(D+\hat{d})](I_L+\hat{i}_L) - \frac{V_o+\hat{v}_o}{R} \tag{11}$$

The small-signal model can then be written as

$$L\frac{d\hat{i}_L}{dt} = D\hat{v}_i - (1-D)\hat{v}_o + \hat{d}(V_o+V_i) \tag{12}$$

$$C\frac{d\hat{v}_o}{dt} = (1-D)\hat{i}_L - \hat{d}I_L - \frac{\hat{v}_o}{R} \tag{13}$$

The above equation can be obtained by applying Laplace transform, as follows:

$$sL\hat{i}_L = D\hat{v}_i - (1-D)\hat{v}_o + \hat{d}(V_o + V_i) \tag{14}$$

$$sC\hat{v}_o = (1-D)\hat{i}_L - \hat{d}I_L - \frac{\hat{v}_o}{R} \tag{15}$$

The transfer function can be obtained from Equations (15) and (16):

$$T_{p1} = \frac{\hat{i}_L}{\hat{d}} = \frac{C(V_o + V_i)s + (2-D)I_L}{LCs^2 + \frac{L}{R}s + (1-D)^2} \tag{16}$$

Next, the PI controller parameters K_{pc} and K_{ic} are calculated to obtain the desired phase margin for the inner current control loop. The Bode plot of the inner loop is shown in Figure 11b, and indicates that the system is stable as the open-loop phase margin at the cutoff frequency is greater than zero.

2. Design of the outer voltage control loop: Due to the high bandwidth and fast current control characteristics of the inner loop, the transfer function of the inner current control loop can be neglected in the design of the voltage controller. Therefore, the duty cycle D can be assumed constant, and its transfer function is

$$T_{p2} = \frac{\hat{v}_o}{\hat{i}_L} = \frac{1-D}{Cs + \frac{1}{R}} \tag{17}$$

Then calculate the PI controller parameters K_{pv} and K_{iv} to obtain sufficient open-loop phase margin at the required cutoff frequency. Figure 11c shows the Bode diagram of the external voltage control circuit. The phase margin at the cut-off frequency is greater than zero, and the system is stable. Parameters of the control system designed as described in Section 3 are given in Table 2.

Table 2. Parameters of the control system.

Mode	PI Controller Parameters
Balancing mode when not using load	K_{pb} = 0.35, K_{ib} = 320
Balancing mode when using load	K_{pc} = 0.15, K_{ic} = 132 K_{pv} = 0.063, K_{iv} = 8.6

4. Simulation Results

To verify the effectiveness of the above SOC balancing strategy, a system model with five battery cells was built and simulated using Simulink for validation, and compared with the balancing strategy described in [26]. The balancing strategy described in [26] can be summarized as the discharge of the battery module with the highest level of charge to the entire battery module, but this will lead to the repeated charging and discharging of the battery module with the highest level of charge. It is worth noting that only three battery modules were used in [26], but in order to better illustrate the balancing strategy, this article uses five battery modules for simulation, and the same settings can be applied to more battery modules. When the control system detects that the difference between the maximum and minimum SOC of the battery module exceeds the set value, the balancing starts until the SOC of all battery modules reaches equilibrium, and the balancing process ends. The simulation model adopts a lithium-ion battery equivalent to 18,650 batteries, with a rated voltage of 3.7 V and capacity of 2000 mAh. The switching frequency is 20 kHz, the inductance is 2 mH, and the capacitance is 220 μF. SOC1~SOC5 are 59.5%, 59%, 60%,

58%, and 58.5%, respectively. The basic parameters of various devices are shown in Table 3. The overall block diagram of the system simulation is shown in Figure 14. Set the system step time simulation to 2.5×10^{-5} s. Note: the purpose of setting a smaller SOC difference between batteries is to accelerate simulation time to verify the effectiveness of the proposed balancing strategy.

Table 3. Parameters of simulation experiment.

Parameters	Size
V_{B1}~V_{B5}	3.7 V
C_{B1}~C_{B5}	2 Ah
$SOC1$	59.5%
$SOC2$	59%
$SOC3$	60%
$SOC4$	58%
$SOC5$	58.5%
L	2 mH
C	220 μF
R	100 Ω
f	20 kHz

4.1. Balanced Simulation When No Load Is Used

When no load was used, the balancing current was set to 1 A for the simulation experiment. To verify the effectiveness of the proposed new balancing control scheme, a comparison and analysis were conducted with the balancing strategy proposed in [26]. The SOC variation, balancing current, and output voltage of the battery module under the two balancing strategies are shown in Figures 15 and 16, respectively. The equilibrium time and integral of squared error criterion (ISE) of the two balancing strategies are shown in Table 4. Under the new balancing strategy, the battery module SOC reaches equilibrium in approximately 219 s, while the balancing strategy in [26] achieved equilibrium at around 415 s. The proposed balancing strategy improves the balancing speed by approximately 47.2%, with the balancing current remaining stable around 1 A, and its current ripple being relatively small, the ISE is approximately 0.267. Compared with the balance strategy in [26], the ISE is approximately 1.849, indicating that the new balancing strategy has better balancing performance.

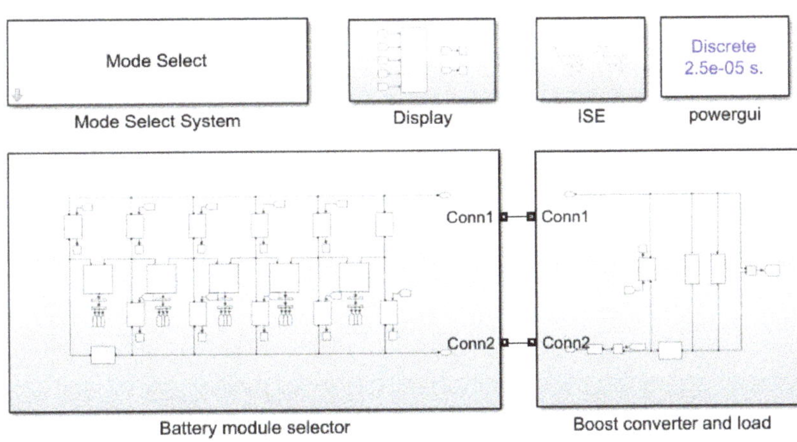

Figure 14. System simulation block diagram.

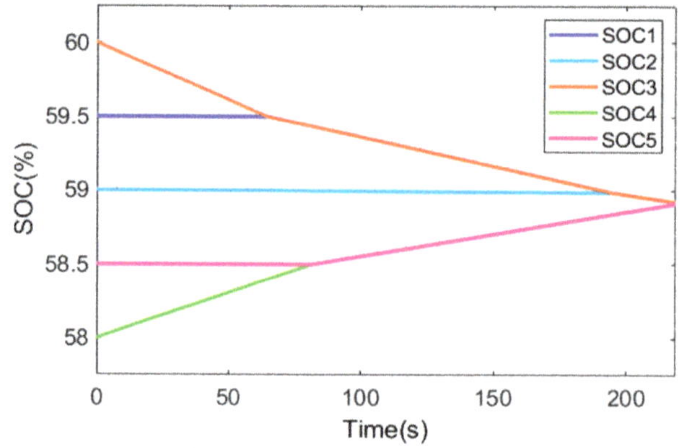

(a)

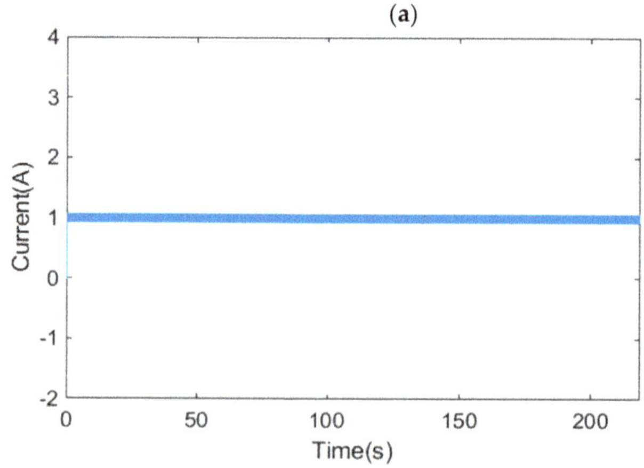

(b)

Figure 15. Simulation of the proposed balancing strategy when no load is used: (**a**) SOC variations of battery modules; (**b**) balanced current.

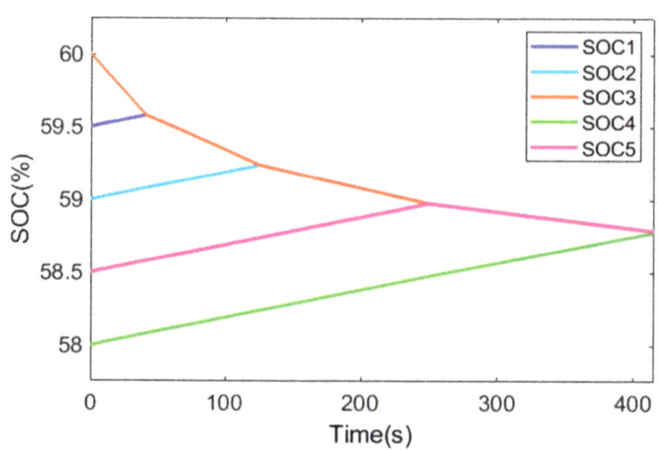

(a)

Figure 16. *Cont.*

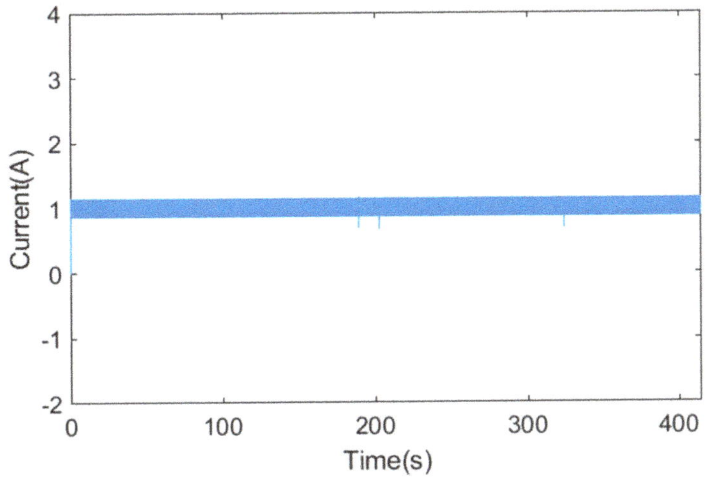

(**b**)

Figure 16. Balanced simulation when no load is used in [26]: (**a**) SOC variations of battery modules; (**b**) balanced current.

Table 4. Battery equalization time and the size of current ISE when no load is used.

Balancing Strategy	Battery Equalization Time	ISE
New	219 s	0.267
Old	415 s	1.849

4.2. Balanced Simulation When Using Load

When a load was used, the output voltage was set to 15 V for the simulation experiment. The SOC variation, balancing current, and output voltage of the battery module under the two balancing strategies are shown in Figures 17 and 18, respectively. The equilibrium time and ISE of the two balancing strategies are shown in Table 5. The battery module SOC reaches equilibrium at approximately 242 s with the proposed new balancing strategy, while the balancing strategy in [26] reached equilibrium at around 416 s. The proposed balancing strategy improves the balancing speed by approximately 41.8%. The balancing current in the proposed new balancing strategy remains stable at around 1 A, and the output voltage stays stable at 15 V with relatively small current ripple and voltage ripple, the ISE of the output voltage is approximately 1.686. In contrast, the balancing current in [26] was approximately 1 A, and the output voltage stayed stable at 15 V, and the ISE of the output voltage was approximately 2.517. Therefore, the output voltage ripple of the proposed new balancing strategy is relatively small.

To further validate the effectiveness of the proposed equilibrium strategy, the SOC value of the battery was modified to align with the approach presented in [26]. The SOC values of batteries B1 to B5 are set to 80%, 79.5%, 79%, 78.5%, and 78%, respectively. Subsequently, the simulation was conducted once again. This adjustment was made to ensure consistency and enable a comprehensive evaluation of the proposed strategy's performance.

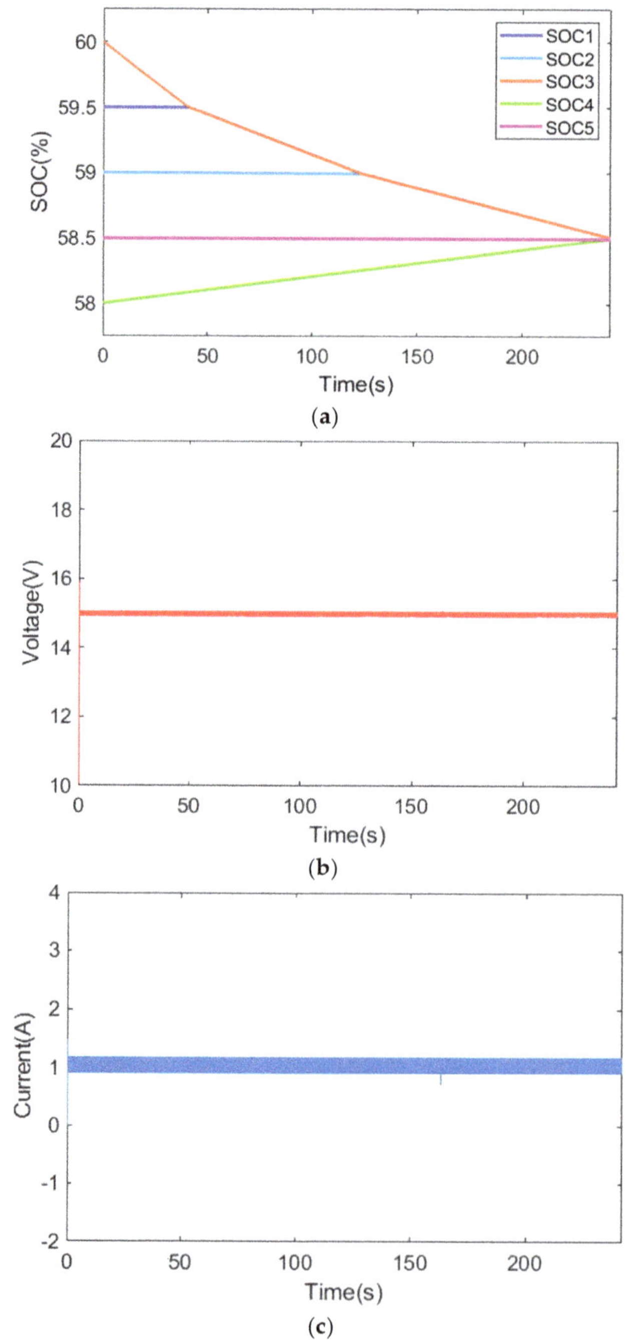

Figure 17. Simulation of the proposed balancing strategy when using load: (**a**) SOC variations of battery modules; (**b**) output voltage; (**c**) balanced current.

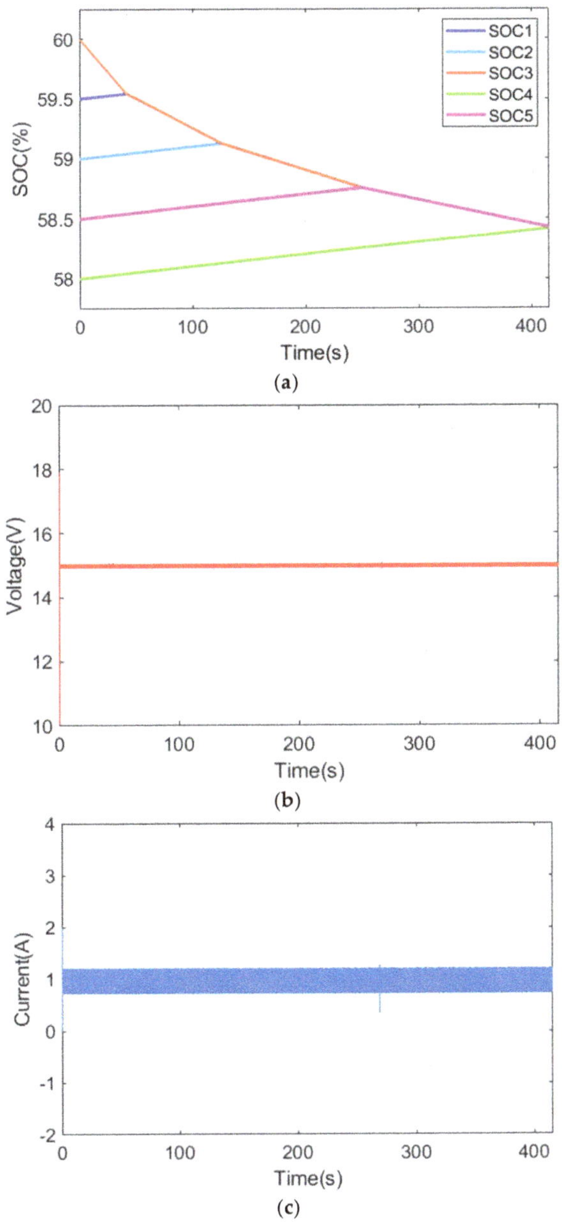

Figure 18. Simulation of the balancing strategy when using load in [26]: (**a**) SOC variations of battery modules; (**b**) output voltage; (**c**) balanced current.

Table 5. Battery equalization time and the size of voltage ISE when load is used.

Balancing Strategy	Battery Equalization Time	ISE
New	242 s	1.686
Old	416 s	2.517

In the simulation experiment, when there was no load, the balancing current was set to 1 A. The SOC variations and balancing current of the battery modules are illustrated in Figures 19 and 20, respectively. The equilibrium time and ISE of the two balancing strategies are shown in Table 6. With the new balancing strategy, the battery module's SOC reaches equilibrium in approximately 218 s, whereas the balancing strategy described in [26] achieved equilibrium at around 411 s. The proposed balancing strategy improves the balancing speed by approximately 47.0%. Additionally, the ISE for the balancing current in the proposed equilibrium strategy is approximately 0.268, whereas the ISE for the balancing current in [26] was approximately 1.845.

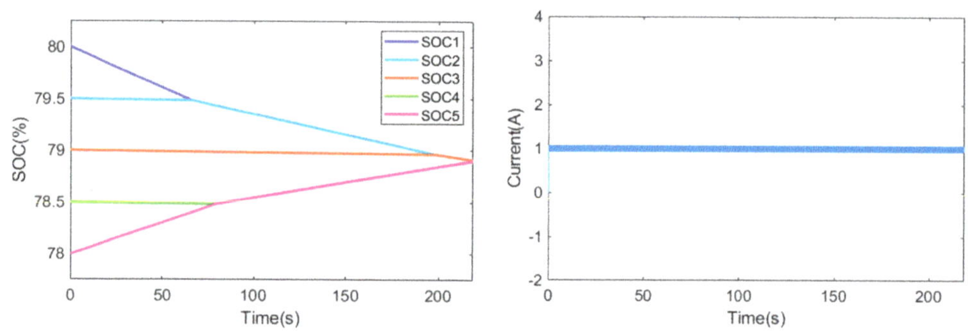

(a) (b)

Figure 19. Simulation of the proposed balancing strategy without using load: (**a**) SOC variations of battery modules; (**b**) balanced current.

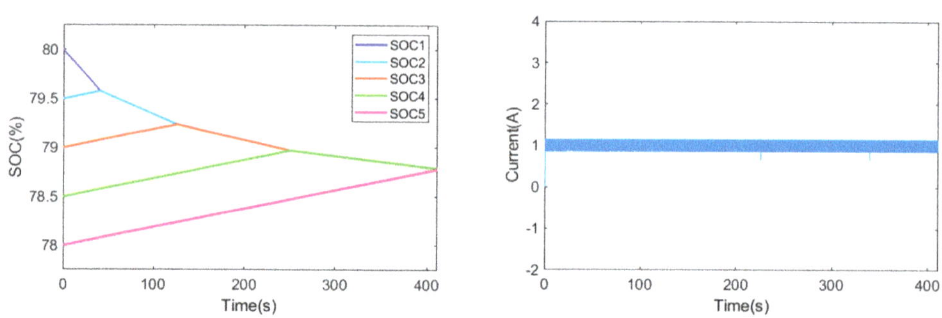

(a) (b)

Figure 20. Balanced simulation of unused load in [26]: (**a**) SOC variations of battery modules; (**b**) balanced current.

Table 6. The second working condition-Battery equalization time and the size of current ISE when no load is used.

Balancing Strategy	Battery Equalization Time	ISE
New	218 s	0.268
Old	411 s	1.845

In the simulation experiments using a load, the output voltage was set to 15 V. The SOC variations and output voltage of the battery module under the two balancing strategies are depicted in Figures 21 and 22, respectively. The equilibrium time and ISE of the two balancing strategies are shown in Table 7. With the proposed new balancing strategy, the battery module's SOC reaches equilibrium at approximately 244 s, while the balancing strategy mentioned in [26] achieved equilibrium at approximately 413 s. The proposed balancing strategy improves the balancing speed by approximately 40.9%. In terms of the output voltage, the ISE for the proposed strategy is approximately 1.683, whereas the ISE for the strategy in [26] was approximately 2.514. Hence, the proposed new balancing strategy exhibits relatively small output voltage ripple.

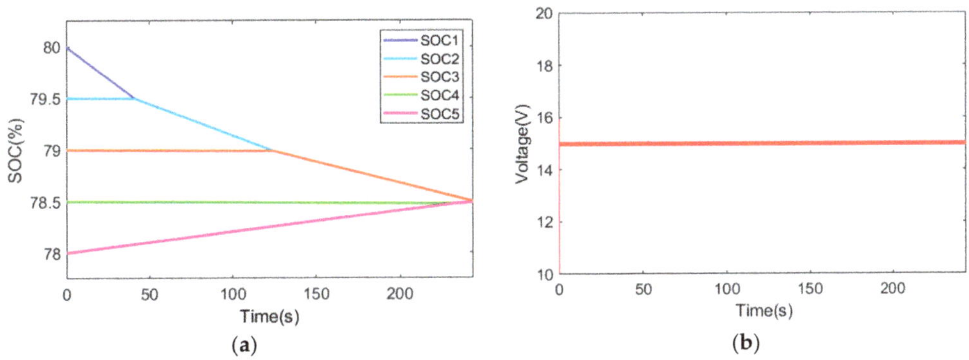

Figure 21. Simulation of the proposed balancing strategy when using load: (**a**) SOC variations of battery modules; (**b**) oOutput voltage.

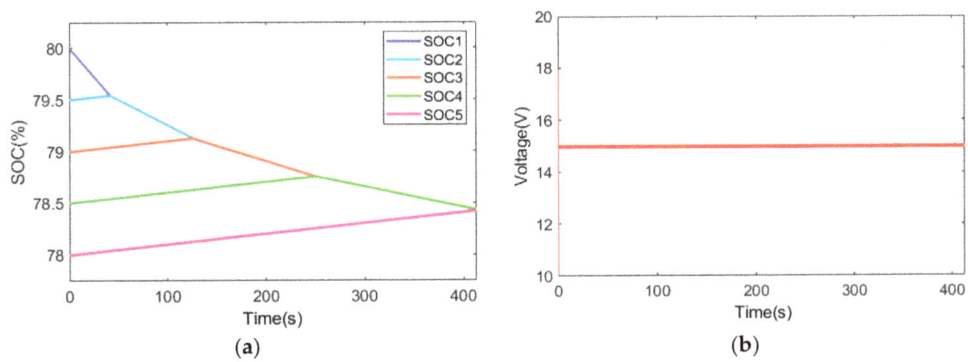

Figure 22. Simulation of the balancing strategy when using load in [26]: (**a**) SOC variations of battery modules; (**b**) output voltage.

Table 7. The second working condition-Battery equalization time and the size of voltage ISE when load is used.

Balancing Strategy	Battery Equalization Time	ISE
New	244 s	1.683
Old	413 s	2.514

Based on the above simulation results, the following conclusions can be drawn:

1. The newly proposed equalization strategy results in a significant enhancement of balancing speed, regardless of the presence or absence of load.
2. Without utilizing load balancing, the equalizing current remains at 1 A. With the implementation of the new strategy, the current ripple is reduced, indicating an improvement in system stability.

3. When using load, where the output voltage on the load is set to 15 V, the new balancing strategy demonstrates diminished voltage ripple, which implies an augmentation in overall system stability.

5. Discussion

The proposed new balancing strategy showed significant improvements in balancing speed compared to the balancing strategy described in [26]. Furthermore, there was a certain enhancement in system stability, indicated by the reduced current ripple and voltage ripple during balancing. This can be attributed to the fact that the balancing strategy operates on a single-cell-to-single-cell basis, transferring energy from the battery with the highest state of charge to that with the lowest state of charge, thereby avoiding repeated charge and discharge cycles in certain batteries. However, the limitation of this balancing strategy lies in the need for a larger duty cycle when load balancing is performed. This is due to the fact that only one battery is supplying power during the process, necessitating a higher duty cycle to meet the requirements of output voltage. However, it should also be noted that this integrated reconfigurable converter system only enables adjacent batteries to be connected in series in order to supply power together, and cannot bypass intermediate batteries. For instance, B1 and B2 can be connected in series to simultaneously supply power to the load, but B1 and B3 are unable to be connected in series with the load by bypassing B2. This limitation provides potential for improvement in future research.

6. Conclusions

In this paper, an improved battery balancing strategy was presented for application in integrated reconfigurable converter systems. The integrated reconfigurable converter system combines a reconfigurable battery system with a converter system, which can be configured into different operating modes based on the battery's state: supplying power to the load mode and balancing mode. When supplying power to the load without the need for battery balancing, the reconfigurable battery system combined with a boost converter can select the input voltage according to the demand. When there is a significant charge difference between the batteries and the system enters the balancing mode, the improved balancing strategy allows balancing to be performed both during no-load conditions and while supplying power to the load. The balancing mode adopts a PI controller for control, and a simulation model was established using MATLAB/Simulink to validate the effectiveness of the simulation strategy and compare it with previous strategies. The simulation results demonstrate that regardless of the presence or absence of a load during balancing, the balancing speed is improved, and the system's stability is also enhanced.

Author Contributions: Conceptualization, G.W. and Q.Z.; methodology, G.W. and M.L.; software, G.W.; validation, M.L., S.L., Z.F. and J.L.; formal analysis, Q.Z.; investigation, Z.F.; resources, J.L.; data curation, G.W.; writing—original draft preparation, G.W.; writing—review and editing, Q.Z.; visualization, M.L.; supervision, S.L. and G.L.; project administration, Q.Z.; funding acquisition, Q.Z. and G.L. All authors have read and agreed to the published version of the manuscript.

Funding: This research was funded by the Natural Science Foundation of Shandong Province, China (ZR2021ME163) and the Key Research and Development Program of Shaanxi Province (No. 2022NY-203).

Data Availability Statement: The data that support the findings of this study are available from the corresponding author upon reasonable request.

Conflicts of Interest: The authors declare no conflict of interest.

References

1. Daowd, M.; Antoine, M.; Omar, N.; Van den Bossche, P.; Van Mierlo, J. Single Switched Capacitor Battery Balancing System Enhancements. *Energies* **2013**, *6*, 2149–2174. [CrossRef]
2. Zun, C.-Y.; Park, S.-U.; Mok, H.-S. New Cell Balancing Charging System Research for Lithium-ion Batteries. *Energies* **2020**, *13*, 1393. [CrossRef]

3. Xu, J.; Li, S.; Mi, C.; Chen, Z.; Cao, B. SOC Based Battery Cell Balancing with a Novel Topology and Reduced Component Count. *Energies* **2013**, *6*, 2726–2740. [CrossRef]
4. Zhang, D.-H.; Zhu, G.-R.; He, S.-J.; Qiu, S.; Ma, Y.; Wu, Q.-M.; Chen, W. Balancing Control Strategy for Li-Ion Batteries String Based on Dynamic Balanced Point. *Energies* **2015**, *8*, 1830–1847. [CrossRef]
5. Zhang, Y.; Liu, Z.; Chen, Z. Smart-Leader-Based Distributed Charging Control of Battery Energy Storage Systems Considering SoC Balance. *Batteries* **2023**, *9*, 18. [CrossRef]
6. Daowd, M.; Antoine, M.; Omar, N.; Lataire, P.; Van Den Bossche, P.; Van Mierlo, J. Battery Management System—Balancing Modularization Based on a Single Switched Capacitor and Bi-Directional DC/DC Converter with the Auxiliary Battery. *Energies* **2014**, *7*, 2897–2937. [CrossRef]
7. Van, C.N.; Vinh, T.N.; Ngo, M.-D.; Ahn, S.-J. Optimal SoC Balancing Control for Lithium-Ion Battery Cells Connected in Series. *Energies* **2021**, *14*, 2875. [CrossRef]
8. Cui, X.; Shen, W.; Zhang, Y.; Hu, C. A Novel Active Online State of Charge Based Balancing Approach for Lithium-Ion Battery Packs during Fast Charging Process in Electric Vehicles. *Energies* **2017**, *10*, 1766. [CrossRef]
9. Nishijima, K.; Sakamoto, H.; Harada, K. A PWM controlled simple and high performance battery balancing system. In Proceedings of the 2000 IEEE 31st Annual Power Electronics Specialists Conference. Conference Proceedings (Cat. No. 00CH37018), Galway, Ireland, 23 June 2000; IEEE: Piscataway, NJ, USA, 2000; pp. 517–520.
10. Li, S.; Mi, C.C.; Zhang, M. A high-efficiency active battery-balancing circuit using multiwinding transformer. *IEEE Trans. Ind. Appl.* **2012**, *49*, 198–207. [CrossRef]
11. Cao, Y.; Qahouq, J.A.A. Hierarchical SOC balancing controller for battery energy storage system. *IEEE Trans. Ind. Electron.* **2020**, *68*, 9386–9397. [CrossRef]
12. Wang, Y.-X.; Zhong, H.; Li, J.; Zhang, W. Adaptive estimation-based hierarchical model predictive control methodology for battery active equalization topologies: Part I–Balancing strategy. *J. Energy Storage* **2022**, *45*, 103235. [CrossRef]
13. Yun, Z.; Qin, W.; Shi, W.; Wu, C. Research on active state of charge balance of battery pack based on two controllable flyback converters. *J. Energy Storage* **2023**, *57*, 106183. [CrossRef]
14. Li, Y.; Yin, P.; Chen, J. Active Equalization of Lithium-Ion Battery Based on Reconfigurable Topology. *Appl. Sci.* **2023**, *13*, 1154. [CrossRef]
15. Engelhardt, J.; Zepter, J.M.; Gabderakhmanova, T.; Rohde, G.; Marinelli, M. Double-String Battery System with Reconfigurable Cell Topology Operated as a Fast Charging Station for Electric Vehicles. *Energies* **2021**, *14*, 2414. [CrossRef]
16. Haller, S.; Alam, M.F.; Bertilsson, K. Reconfigurable Battery for Charging 48 V EVs in High-Voltage Infrastructure. *Electronics* **2022**, *11*, 353. [CrossRef]
17. Karunathilake, D.; Vilathgamuwa, M.; Mishra, Y.; Corry, P.; Farrell, T.; Choi, S.S. Degradation-Conscious Multiobjective Optimal Control of Reconfigurable Li-Ion Battery Energy Storage Systems. *Batteries* **2023**, *9*, 217. [CrossRef]
18. Tashakor, N.; Dusengimana, J.; Bayati, M.; Kersten, A.; Schotten, H.; Götz, S. General Decoupling and Sampling Technique for Reduced-Sensor Battery Management Systems in Modular Reconfigurable Batteries. *Batteries* **2023**, *9*, 99. [CrossRef]
19. Wang, S.; Liu, D.; Zhou, J.; Zhang, B.; Peng, Y. A Run-Time Dynamic Reconfigurable Computing System for Lithium-Ion Battery Prognosis. *Energies* **2016**, *9*, 572. [CrossRef]
20. Ci, S.; Lin, N.; Wu, D. Reconfigurable battery techniques and systems: A survey. *IEEE Access* **2016**, *4*, 1175–1189. [CrossRef]
21. Xu, J.; Cao, B.; Wang, J. A Novel Method to Balance and Reconfigure Series-Connected Battery Strings. *Energies* **2016**, *9*, 766. [CrossRef]
22. Han, W.; Zou, C.; Zhou, C.; Zhang, L. Estimation of cell SOC evolution and system performance in module-based battery charge equalization systems. *IEEE Trans. Smart Grid* **2018**, *10*, 4717–4728. [CrossRef]
23. Momayyezan, M.; Hredzak, B.; Agelidis, V.G. Integrated reconfigurable converter topology for high-voltage battery systems. *IEEE Trans. Power Electron.* **2015**, *31*, 1968–1979. [CrossRef]
24. Momayyezan, M.; Hredzak, B.; Agelidis, V.G. A load-sharing strategy for the state of charge balancing between the battery modules of integrated reconfigurable converter. *IEEE Trans. Power Electron.* **2016**, *32*, 4056–4063. [CrossRef]
25. Morstyn, T.; Momayyezan, M.; Hredzak, B.; Agelidis, V.G. Distributed control for state-of-charge balancing between the modules of a reconfigurable battery energy storage system. *IEEE Trans. Power Electron.* **2015**, *31*, 7986–7995. [CrossRef]
26. Zhao, C.; Liu, S.; Fan, B. State of Charge Balancing Control for Battery System Based on the Reconfigurable Converter. In Proceedings of the 2021 IEEE 5th Conference on Energy Internet and Energy System Integration (EI2), Taiyuan, China, 22–24 October 2021; IEEE: Piscataway, NJ, USA, 2021; pp. 4215–4220.

energies

MDPI

Article

Design of Zonal E/E Architectures in Vehicles Using a Coupled Approach of k-Means Clustering and Dijkstra's Algorithm

Jonas Maier * and Hans-Christian Reuss

University of Stuttgart, Faculty 7: Engineering Design, Production Engineering and Automotive Engineering (F07), Institute of Automotive Engineering (IFS), 70569 Stuttgart, Germany
* Correspondence: jonas.maier@ifs.uni-stuttgart.de

Abstract: Electromobility and autonomous driving has started a transformation in the automotive industry, resulting in new requirements for vehicle systems. Due to its functions, the electrical/electronic (E/E) architecture is one of the essential systems. Zonal E/E architecture is a promising approach to tackle this issue. The research presented in this paper describes a methodology for determining the optimal number of zones, the position of the zone control units (ZCU), and the assignment of electric components to these zones and ZCUs. Therefore, the design of the power supply and the wiring harness is essential. This approach aims to identify the most suitable system architecture for a given vehicle geometry and a set of electric components. For this purpose, the assignment of electric components is accomplished by k-means clustering, and Dijkstra's algorithm is used to optimize the cable routing. As ZCUs will be the hubs for the in-vehicle data and information transport in zonal architectures, their position and their number are crucial for the architecture and wiring harness development. Simulations show a suitable zonal architecture reduces wiring harness length as well as weight and brings functional benefits. However, the number of zones must be chosen with care, as there may also be functional limitations.

Keywords: E/E architecture; clustering; wiring harness; zonal architecture; zone control unit

1. Introduction

The automotive industry is undergoing a fundamental transformation, which began a few years ago with the rise of electric mobility [1,2]. The energy density and thus the range of battery electric vehicles (BEV) is continually increasing as well as safety, which is indicated by falling fire accident numbers [3]. Katis summarizes the advancement in autonomous driving by abilities like autopilot and assisted navigation, wherefore further sensor technologies like lidar and radar as well as data analysis approaches using artificial intelligence are essential [4]. The rising number of electric vehicle production and sales leads to reduction in component cost according to the International Council on Clean Transportation (ICCT) [5]. High research focus is put on new battery materials, which are supposed to bring a strong increase in energy density of batteries [3–5].

Autonomous driving, digitalization, and electric mobility place new demands on vehicles and their systems such as the introduction of voltages above 60 V, the integration of additional sensors (lidar, radar) and data processing units for autonomous driving (computer vision, artificial intelligence), cloud services, and better connectivity within the vehicle and with its environment [6]. With a rising number of functions, the complexity of the vehicle increases, thus the in-vehicle networks need to fulfill higher and completely new requirements, which makes a profound change indispensable. Looking at the vehicle's energy supply system, this means fundamental changes in the area of hardware and software. With the growing demand for BEV, platforms are being developed specifically to better meet their needs. However, they are still based on the E/E architecture of previous vehicle generations using internal combustion engines. One architectural approach supporting the advancing digitalization and connectivity is the zonal E/E architecture [7].

Citation: Maier, J.; Reuss, H.-C. Design of Zonal E/E Architectures in Vehicles Using a Coupled Approach of k-Means Clustering and Dijkstra's Algorithm. *Energies* **2023**, *16*, 6884. https://doi.org/10.3390/en16196884

Academic Editor: Chunhua Liu

Received: 29 August 2023
Revised: 8 September 2023
Accepted: 15 September 2023
Published: 29 September 2023

The E/E architecture of vehicles has grown rapidly in recent years. The first architectures included a few components and connections, including the 12 V battery, the lighting, and the ignition system. With the steady influx of electrical as well as electronic hard- and software, bus systems were necessary to reduce the number of interconnect lines. Over the years, these architectures have evolved and today almost all vehicles still use a decentralized architecture and domain controllers to bundle all functionalities of a domain. Continuous integration of additional functions, following the "one function per ECU"-approach, increased the number of ECUs to as many as 100. In addition, the wiring harnesses can be up to five kilometers long and weigh 60 to 80 kg, with individual cables having a length of up to eight meter [7,8]. Although it is common to arrange interrelated functions in one bus system, several ECUs, even from different busses, are usually required to realize a customer-related function. The advantage of this architecture is a relatively uncomplicated extension by single functions. The manufacturer integrates a new ECU with the corresponding function into the vehicle and connects it to the communication structure. However, decentralized architectures show some disadvantages regarding the changing requirements portfolio. The importance of software in vehicles has increased significantly in recent years and this trend will continue. Therefore, the extension of existing functions as well as the addition of new functions mainly involves new software. Due to the "one function per ECU" approach and its hardware parts developed for one specific function, it is also necessary to change parts of the hardware in current architectures. Furthermore, this approach leads to poor utilization of the available weight and volume. This is demonstrated by two ECUs, both of which perform separate functions. If combined, the weight and volume of the computing unit will increase, but the peripheral elements would increase relatively less since it is sufficient to have only one of them. The length and weight of the wiring harness in technical systems are proportional to the complexity of the architecture, i.e., the number of its components, connections, and interfaces. As the number of ECUs increases, so does the amount of cabling. This leads to weight increase, as well as restrictions in the composition of the systems, thus causing an increase in cost and time. Another challenge pushing the decentralized architecture to its limits is the high number of variants [9]. The required combinations of hard- and software determine the complexity of a variant. One variant consists of the basic functionalities in combination with the equipment- and market-specific functions [10]. A high number of variants, which is proportional to the quantity of ECUs, demands great effort for developing architectures and extending them. In addition, a high number of software variants and their specific requirements can increase the number of hardware variants [11].

Although vehicle manufacturers are aware of the problems of the decentralized architecture, current applications show little innovation. The large number of dependencies on other systems means that changes must be made over multiple product generations [12,13]. Since a first generation of zone controllers is already available, e.g., from Bosch or Vitesco, it can be expected that these will gradually replace existing ECUs in the coming years [14,15]. VW and Cariad follow a similar approach called E^3, combining multiple ECUs into one physical ECU with multiple virtual ones [16]. To take full advantage of these technologies, methods and tools are necessary to design zonal architectures. Therefore, the aim of this paper is to create a method that helps define the architecture and the ZCU. This will help developers to evaluate concepts in an early phase and provide further requirements for the design of the ZCUs.

The article is structured as follows. After introducing current architectures, Section 2 presents the state of the art through two aspects. First, development trends of E/E architecture and features of zonal E/E architectures are given. Based on this, a practical approach to implement a zonal architecture in a test vehicle is shown. Secondly, the already existing measures for the optimization of vehicle power supply systems are presented considering the packaging and wiring harness design. These are checked in regard to their suitability to achieve a system that is optimum even under changing conditions. Section 3 deals with the novel method to develop and evaluate the zonal E/E architectures based on a design of the power supply system and wiring harness. It consists of several steps for analysis, conceptual design, dimensioning, and evaluation, including k-means and Dijkstra algorithm. In Section 4, the autoSHUTTLE, one of the UNICARagil vehicles, is evaluated for zone numbers of one to eight. Section 5 identifies findings of this paper and potentials for further research.

2. Power Supply in Zonal E/E Architectures: Basic Concepts and Research Gap

2.1. Basic Concepts: Shift in E/E Architecture

One centralization approach is the division of the E/E architecture into domains, such as powertrain, infotainment, and body, enabling a more structured architecture. When it was still possible to implement one function per controller, the domain-centralized architecture made it easy to add and verify individual functions. However, this advantage is dwindling with the increase in cross-domain functions and greater connectivity within the vehicle and with its environment [17]. Moreover, the domain-driven structure comes with two further drawbacks: The first is the high communication effort between communication nodes, as ECUs of different domains need to interact for many customer-related functions. The other is the blurring of boundaries between domains by strongly cross-domain functions. This results in a fuzzy structure, especially for autonomous driving and more digitalized vehicles. One possible approach addressing these issues is a cross-domain centralized architecture. Meaning, the functionalities of several domains are bundled in one ECU. Vitesco Technologies combines all the functions that are needed for driving in one ECU [15]. Yet, there is one major disadvantage, namely the communication effort. As the functions in the cross-domain controller increase, so does the calculation effort. Thereby, the bandwidth necessary to exchange data between these ECUs increases, leading to a heavier and more complex wiring harness [18].

The zonal architecture is helpful in tackling this problem. In contrast to domain-based and cross-domain-based architectures, the zonal architecture does not bundle or center functions with respect to domains; instead, it structures them according to geometric aspects [19]. Therefore, the vehicle is divided into zones, each with its own ZCU [11]. All sensors and actuators located in the same zone are connected to the corresponding ZCU. There are approaches in which the ZCUs only transmit the data to a central processing unit. It is also possible to process zone-related functions directly on the ZCU in order to relieve the central processing unit and the communication volume [8,20]. An approach towards the ZCU are generic ECUs as proposed by Tomar [21]. Generic ECUs are general-purpose nodes, which can be used for different processing tasks within the vehicle. Bosch estimates a weight reduction in the wiring harness of 15 to 20% by using a zonal architecture, while EDAG expects potential savings of up to 30% [7,22]. The zonal approach also enables more point-to-point connections of safety-critical components, making it easier and more cost-effective to ensure functional safety. Such architectures are already under development. However, no manufacturer known to the authors is currently using such architecture in a series production vehicle. Most system suppliers typically divide the vehicle along the longitudinal axis and one or more times parallel to the transverse axis, thus obtaining four to eight zones [23–26].

As other applications, such as unmanned aerial vehicles (UAV) and robots, also have an E/E architecture, it may be possible to utilize the benefits of zonal architecture in them as well. Cui's research indicates a reduction in wiring harness length and weight by at least 16% [27]. Shrestha introduces a domain-based E/E architecture for UAVs and points out that the centralization of E/E architectures will likely introduce zonal architecture in UAVs in the future [28].

Figure 1 shows the geometrical and the functional view of the UNICARagil vehicle architecture. Figure 1a displays the geometrical view with the separation into four zones. The functional architecture with the components and their interfaces is shown in Figure 1b. In the UNICARagil project, the research consortium designed and built a platform for four autonomous vehicles [29–32]. This platform corresponds to a four-zone architecture, but without fully comprehensive zone controllers, as they were not available at the beginning of the project. Instead, the vehicle uses one sensor module per zone to process data from the optical sensors, one switch to connect components to the Ethernet network, and one battery including the power distribution unit (PDU) to supply the components in the zone. In addition to a reduction of the wiring harness weight, the zonal architecture has positive effects on safety, modularity, and offers new degrees of freedom in packaging [33]. To ensure the safety of the vehicle in the event of a driver absence, additional hardware is required, which is best provided in separate zones. Considering the increasing number of variants, modularity is an important factor. Interchangeable zones make the diversity of variants more manageable in terms of hardware and software. Efficient packaging leads to more cabin volume that is available for passengers. Furthermore, it offers additional degrees of freedom in the development of the overall vehicle, especially in the integration of other systems. In addition to handling the complexity of the system, freedom in packaging as well as a high degree of modularity reduces development time and lower costs.

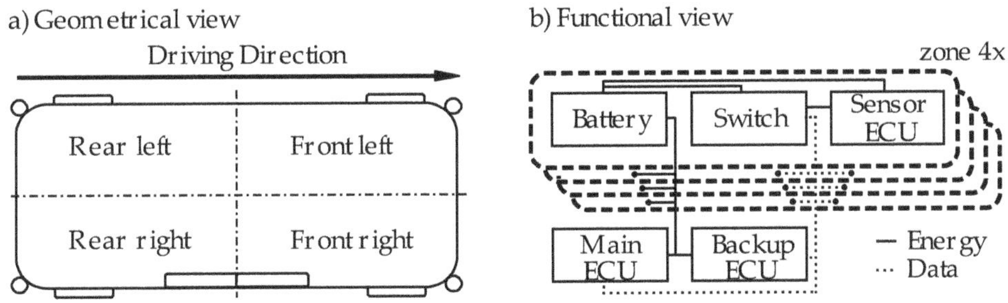

Figure 1. Four-zone architecture of the UNICARagil vehicles.

2.2. Related Work: Optimization of Vehicle Powernet in Conventional and Domain-Based Architectures

The E/E architecture of vehicles consists of two essential subsystems: the power supply system for storing and distributing electrical energy and the communication structure for transmitting information. When comparing these two, it becomes obvious that the power supply requires volume- and weight-intensive components such as batteries, DC/DC-converters, and power cables. The communication structure, on the other hand, is mostly defined by the connection logic, protocols, and the deployment of software functions on the ECUs. As the zones in zonal E/E architectures are defined by geometrical aspects, hardware has a greater influence on it. Therefore, the method focuses on the volume- and weight intensive hardware of the E/E architecture, as the power supply systems, the wiring harness and the ZCU. To ensure a suitable architecture for the communication structure, the required cables for the vehicle networks are also considered. Several approaches optimize the power supply systems in conventional and domain-based architectures, considering the electrical components with varying level of detail. Since the E/E architecture design was dominated by the functional criteria, the methods used for previous or current architectures have a different perspective. As it is necessary to combine package, power supply, and wiring harness design to realize a suitable zonal E/E architecture, recent methods focusing on these aspects will be discussed.

Most approaches consider the battery as an essential component for BEVs, while further components such as the wiring harness and voltage converter are only covered by few. Package design usually takes the mass of the vehicle's electrical system into account as it influences the energy consumption. It focuses on the allocation of vehicle space to components, the suitable arrangement of the vehicle components, and thus the realization of the required space in the passenger cabin. Fuchs and Sethuraman focus on a holistic vehicle design, thus considering the components of the E/E architecture in a simplified way [34,35]. Their approaches to assess package compatibility are not suitable for zonal E/E architectures, as they do not consider the wiring harness. Berthold and Fay developed approaches modeling the major components of the powertrain, as well as the battery, by scaling the cells while neglecting other components of the E/E architecture [36,37]. This can be useful for a technological or economic investigation of vehicles, but is not suitable for the design of the E/E architecture in an early concept phase. Lübke developed one of the first systematic approaches for optimizing the vehicle wiring systems [38]. He simplifies the E/E architectures to nodes and their connecting elements, varying the node position with a genetic algorithm. The costs are minimized under various constraints, but it neglects the electrical design and packaging aspects. Diebig develops a methodology for the simulation-based design of automotive wiring systems with a focus on the current carrying capacity of the cables as well as the voltage stability [39]. The thermal modeling of the cables is detailed, but the process simplifies the package situation and does not include any battery design. Consequently, this method is optimal for designing the power supply system at a later stage of development. Wang assesses common power system topologies with respect to their voltage stability using high-power loads [40]. Even though it is not possible to apply his method to power systems above 12 V, and it does not consider cost aspects, there are meaningful findings. However, transferring these findings to practical applications, especially in the context of zonal architectures, proves challenging. Zhu develops a framework for optimizing wiring harnesses in aircraft, which begins with the overall wiring harness design and meticulously optimizes the branch structure in detail [41]. While path planning is optimized, and packaging limitations are taken into account, additional electrical components are not considered. Braun presents a methodology for designing conventional and high-voltage vehicle power systems [42]. She investigates the influence of degrees of freedom in the development process based on energy consumption and total cost of ownership (TCO), taking voltage stability into consideration. For this purpose, she employs models to simulate the electrical behavior of the power network components. Braun also integrates cable routing into the methodology to determine cable lengths. The procedure is suitable for vehicles with conventional and domain-based architectures, but the lack of packaging consideration renders it inapplicable to zonal architectures. Becker and Frank delve into detailed modeling of the high-voltage energy storage system [43,44]. Both focus on the battery design investigating the effects on the energy consumption and the driving performance. Therefore, the battery performance is verified, but compatibility with the vehicle package is not ensured. In addition to automotive applications, there are some methods from other disciplines. In addition to the overall vehicle and communication structure, the power system, as part of the E/E architecture, is subject to requirements from almost all vehicle systems. The vehicle systems have demands regarding the power necessary and the position at which it should be available. The communication network can be understood as one of these vehicle systems or as an electrical load. In safety-critical cases, it may also be necessary for a component to require two power supplies. The requirements of the overall vehicle are directly linked to customer needs such as maximum speed, range, or possible load. Further requirements may be functional and technical, such as maximum weight and volume or failure behavior.

In Table 1, we summarize the modeling characteristics of the described literature. In conclusion, it is neither necessary nor possible to consider a detailed performance of electric components or the powertrain, as these are not fully defined in early development phases. Furthermore, a TCO calculation is possible, when environmental data are used as an input, but since the method brings a small mass reduction in relation to the vehicles mass, there might not be a significant efficiency benefit. The consideration of the E/E architecture hardware components as well as the wiring harness with respect to the packaging is essential for designing the most suitable zonal E/E architecture.

Table 1. Literature review on electric system design with respect to packaging.

Reference	Traction Battery			Wiring Harness				DC/DC-Converter			LV-Battery			Electric Loads			Packaging
	E	M	V	E	M	R	V	E	M	V	E	M	V	E	M	V	
[34]	x	x	x						x	x		x	x		x	x	x
[35]	x	x	x											x	x	x	x
[36]	x	x												x	x		
[37]	x	x												x	x		
[38]				x		x											
[39]				x													
[40]				x				x			x			x			
[41]					x	x											x
[42]	x	x			x	x		x	x		x	x		x	x		
[43]	x	x	x														
[44]	x	x															

Modelling aspects: E = Electric; M = Mass; R = Routing; V = Volume.

2.3. Research Gap and Contributions

In addition to the overall vehicle and communication structure, the power system, as part of the E/E architecture, is subject to requirements from almost all vehicle systems. The vehicle systems have demands regarding the power necessary and the position at which it should be available. The communication network can be understood as one of these vehicle systems or as an electrical load. In safety-critical cases, it may also be necessary for a component to require two power supplies. The requirements of the overall vehicle are directly linked to customer needs such as maximum speed, range, or possible load. Further requirements may be functional and technical, such as maximum weight and volume or failure behavior.

The shift to a zonal E/E architecture adds more requirements for the vehicle power system. Instead of structuring the vehicle by domains according to functional affiliation, the position of the component in the vehicle is decisive. This leads not only to limitations in the software but also to physical boundaries in the vehicle structure. To achieve the highest possible modularity, zones should be self-contained units. This allows different vehicle variants and models to reuse zones with little effort. It follows that both components and cabling should remain within their defined zone. The number of components that do not belong to a zone must remain as small as possible in favor of modularity. A fully self-sufficient zone would also require one independent energy storage device per zone, which is actually not practical. Therefore, only a logical separation of the battery cells is accomplished. Since a high number of energy storage units can be associated with higher costs, engineers must weigh up which concepts is most suitable. Modularity and practicability of the solution constantly have to be assessed.

Electric and autonomous driving define additional requirements and constraints for tomorrow's architecture, particularly affecting the wiring harness. The introduction of HV components into vehicles has increased the number and the weight of the wiring harness because a second voltage level is necessary and the drives are supplied electrically. The zonal architecture solves this problem for the HV cables but only under certain conditions. Most concepts currently show the use of only one HV battery [7,8,20,45,46]. However, since the driver is no longer available as a fallback level due to autonomous driving functions, at least one energy storage device must be available in the event of a traction battery failure, providing the energy for driving an emergency trajectory. The introduction of new vehicle concepts, such as autonomous shuttles, increases the vehicle height and leads to further restrictions in regards of the packaging [30,47,48]. The high space requirement for the passenger cabin reduces the available volume for components. On the other hand, the number of antennas, sensors, and other electric loads rises at ceiling height, which also increases the number of cables in those areas. Hence, zone architectures are necessary to address these challenges [49]. In [7], an eight-zone architecture is assumed, while in [8], six to eleven, in [46] and [13] six, in [45] four, and in [50] two to four zones are defined. This already reveals a lack of a structured procedure for determining the number of zones. The criteria used to determine the number of zones are unknown. Likewise, the definition of zone controllers is imprecise, and a clear delimitation of the zone areas does not take place.

The authors are not aware of any holistic method for determining the optimal zonal E/E architecture or the optimal number of zones for same. In a holistic approach, it is necessary to consider packaging, the design of the electrical system, the cable layout, and their aggregation into an optimal architecture. Furthermore, one has to consider the requirements of the overall vehicle level and other vehicle systems. This paper presents a method supporting the comparison and selection of the optimal zone architecture already in early phases of product development with a focus on system specification. Since the selection of the architecture has extensive influence on the concept and costs of the vehicle, it is necessary to make the correct decision in early phases. For this purpose, the method carries out a zone optimization and designs the electrical system with few input data, such as vehicle dimensions, as well as the electrical data and the positions of the electrical components. The method combines optimization and design methods to map extensive wiring system architectures and compare them in early development phases. This reduces development and production effort of the cables, which results in an increase of redundancy and modularity of the electrical system architecture and the entire vehicle.

3. Coupled Approach of k-Means Clustering and Dijkstra's Algorithm for Zonal E/E Architecture Optimization

To realize a holistic approach considering zone, battery, and wiring harness design, several optimization and design steps are necessary. Figure 2 shows the entire methodology to design a zonal E/E architecture with consideration of the power supply and the wiring harness. The methodology starts with the architecture definition to determine boundary conditions and discretization. In the next step, clustering techniques identify the optimal cluster position for the ZCU. The cluster position is corrected, considering package boundary conditions. Then, a genetic algorithm (GA) optimizes the HV traction battery, and a further procedure allocates the cells to each zone by their power share in relation to the vehicle's entire power. Hereafter, the remaining components of the power supply system are designed, comprising the 12 V battery and the 48 V supercapacitors (SCAP). The algorithm designs and positions these components in the vehicle to ensure package and concept compatibility. The routing of the cables is optimized utilizing the Dijkstra algorithm and a map-based edge weighting. This allows determining the shortest cable paths while considering requirements from the overall vehicle level and the packaging.

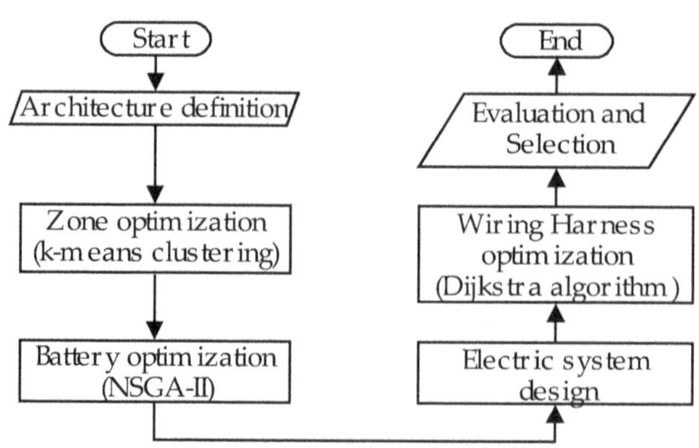

Figure 2. Design method for zonal E/E architectures.

3.1. Architecture Definition

For generating a three-dimensional grid of nodes and edges, the outer shell and dimensions of the passenger cabin are needed. This forms the basis of the geometric modeling. The nodes correspond to discrete points in the vehicle architecture, and the edges connect them according to the graph theory. After the removal of unused nodes or edges, the remaining ones are available to the algorithm. The electrical components are mapped to the node closest to their coordinates. The mapping of the other systems is performed according to the position and the energy demand of their assigned electronic components. Both sets of information are available, albeit with limited accuracy, prior to the elaboration of the systems. In the following, the set of available edges, nodes, and the power and position of the electrical loads is called the structure. In addition to the geometric representation, an electrical one is necessary, defining the energy and information flows in the architecture. Since it is not clear yet how the interface design of ZCUs will look, the architecture specified in Figure 3 is assumed. It shows a block definition diagram using the system modeling language (SysML). The E/E architecture is based on one or more zones. Each zone requires a ZCU, which incorporates a DC/DC converter and a switch, at least one HV battery cell, and electric components. The number of HV battery cells is proportional to the electric power consumption in a zone and thus also proportional to the number of electrical loads.

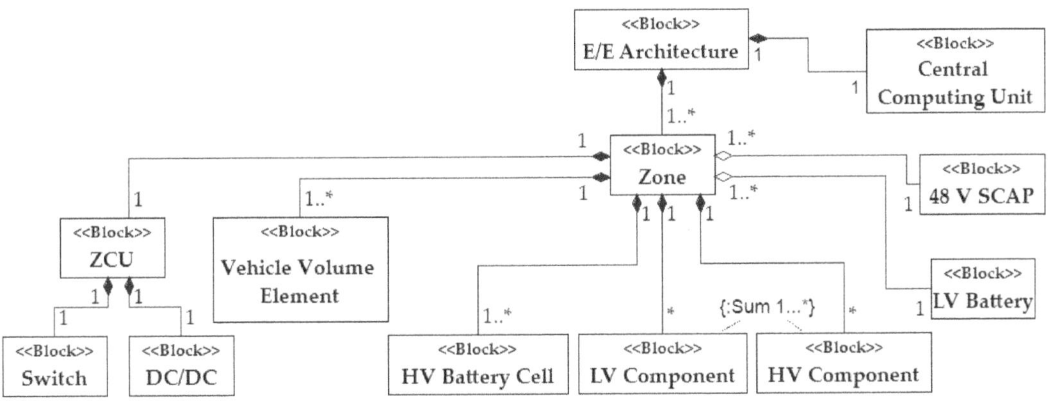

Figure 3. Block definition diagram of the hardware relations for the zonal E/E architecture in vehicles.

The 48 V SCAP and the LV battery are shared between several zones. Besides the components, a zone is defined by the assigned space of the vehicle. Therefore, vehicle volume elements are assigned to a zone, defining spatial affiliation and delimitation. In addition to the structure, there are some functional dependencies. The ZCU controls the energy flow on the LV level within a zone. An HV line, which is connected to the HV energy storage via a PDU, supplies the ZCU. HV cables connect all other HV consumers in the zone to the HV energy storage via the PDU. The information flow within a zone corresponds to a star topology with the zone controller as the star point [51]. A ring line connects the zone controllers, and one of them may operate as the central computing unit to reduce wiring harness length. If a ring line is not applicable due to network specifications, the central computing unit is connected to all ZCUs as in a star topology [52].

3.2. Zone Optimization

In order to utilize the potential for reducing cable length and weight, it is necessary to minimize the distance between ZCUs and their dedicated loads. This is achieved by clustering the position of the electrical loads and positioning the ZCU at the cluster centers. Therefore, the method described by Maier is used and extended [53]. Since the possible number of clusters is defined based on the literature review as $k = \{1, 2, \ldots, 8\}$, k-means clustering is most suitable. The aim of the k-means algorithm is to minimize the sum of the quadratic Euclidean distances D between data points and cluster centers.

$$\min D = \sum_{i=1}^{k} \sum_{x_j \in S_i} \left\| x_j - \mu_i \right\|_2^2, \tag{1}$$

wherein $\|x_j - \mu_i\|_2$ indicates the Euclidean distance (l_2 norm) between the centroid coordinates $\mu = \{\mu_1, \ldots, \mu_k\}$, and the data points $X = \{x_1, \ldots, x_n\}$, $1 \leq j \leq n$, which are categorized into clusters $S = \{S_1, \ldots, S_k\}$, $1 \leq i \leq k$; $k, n \in N^{>0}$. Clusters are groups of data that are similar regarding specific features. In this analysis, the features are the x-, y-, and z-coordinates of the electric components. k-means algorithm comprises three steps: initialization, assignment, and update [54]. In the initialization, the k cluster centers are positioned randomly in the search space. In the second step, the data points X are assigned to the clusters S, which has the nearest mean to the data point according to

$$S_i = \left\{x_j : \|x_j - \mu_i\|_2^{\ 2} \leq \|x_j - \mu_{i^*}\|_2^{\ 2}\ \forall i^*, 1 \leq i^* \leq k\right\}. \quad (2)$$

For the update step, the new cluster centers are calculated based on the features of the assigned data points. Steps two and three are repeated until the algorithm converges, meaning a further iteration causes no change in the data point-to-cluster assignment. The k-means ++ -algorithm uses the same procedure as k-means, except for a different initialization [55]. The first cluster is initialized randomly. Based on the distance between each data point and the nearest cluster point, the next cluster point is selected. The probability that a datapoint is selected as a new cluster center is proportional to the squared distance to its nearest cluster center. The benefit of the k-means++ initialization step is a shorter convergence time with similar results, as the initial cluster centers are already closer to the final ones. The sum of the Euclidean distances is proportional to the wiring harness cost and enables the identification of suitable ZCU positions. Manhattan distance describes the cable routes better, but as there are limitations like the passenger cabin, they are optimized in a later step using the Dijkstra algorithm.

Nevertheless, there are two issues that k-means algorithm cannot handle: The approximation of the vehicle by a grid and a uniform cluster size. As the vehicle is approximated by a three-dimensional grid of nodes and edges, components may not be positioned exactly on a node and therefore need to be assigned to a node. Starting from the optimal centroid position, the next available node is searched for and determined as the ZCU position, and the Euclidean distances are recalculated based on this node. The consistent cluster size is essential to realize uniform ZCUs, which will increase the quantity of similar produced units. This will lead to increased flexibility for hardware and software and a reduction of the cost. Based on the number of data points n and the number of clusters k, the maximum number of components per cluster, respectively, ZCU x_{max} is determined using

$$\forall S_i \in S : |x_{max}| \leq \left\lceil \frac{n}{k} \right\rceil. \quad (3)$$

To ensure that the clusters comprise a similar number of data points, the clusters c_i are categorized into three groups by the following condition, wherein $|x_i|$ is the number of data points assigned to a cluster:

- I $|x_i| < |x_{max}|$, the cluster is in the receiver group,
- if $|x_i| > |x_{max}|$, the cluster is in the deliverer group,
- if $|x_i| = |x_{max}|$, the cluster is in the neutral group, which is not considered in the balancing process.

The balancing process sorts data points from the group of delivering clusters to the nearest cluster of the receiving group. After each resorting step, the receiving and the delivering clusters are evaluated by the conditions above and regrouped. The balancing process continues until all clusters are in the receiver or in the neutral group. While the balancing process increases the sum of the Euclidean distance, it simultaneously reduces costs for the ZCUs. ZCU costs C_{ZCU} are calculated using a third-degree polynomial that is assumed to account for the effect of cost degression and is expressed as follows:

$$C_{ZCU} = f_{ZCU} \cdot \left(p_1 \cdot k^3 + p_2 \cdot k^2 + p_3 \cdot k + p_4\right), \quad (4)$$

where f_{ZCU} is the cost factor of a ZCU, k is the number of zones, respectively, the number of ZCUs in an architecture. The coefficients of the polynomial, p_1, p_2, p_3, and p_4, are listed in Appendix A.

3.3. Battery Optimization

The HV battery is a decisive component for the vehicle powernet, as it is one of the components with the highest weight and volume. Therefore, an optimal design of the battery offers potential for increasing vehicle performance and improving the packaging efficiency considering other requirements. Becker describes an approach for designing the optimal traction battery by mixing different battery cells using a genetic algorithm, namely the Covariance Matrix Adaptation Evolution Strategy (CMA-ES) [33]. As the disadvantages of different cell types outperform the benefits, most vehicle manufacturers and suppliers do not pursue this approach further. The traction battery in this article is thus also generated using one cell type. The optimal cell number for one cell type is determined using the Non-Dominated Sorting Genetic Algorithm II (NSGA-II), which shows good results for various optimization problems [56]. The optimization problem concerns the optimal design of the energy storage system considering the criteria acceleration, component cost, range, volume, and weight. It is assumed that there is an exponential relationship between the number of cells and acceleration, while a linear relationship is supposed for the other criteria. A genetic algorithm is employed to solve the optimization problem efficiently and identify a sufficiently robust solution. The number of cells is used as the optimization variable via binary encoding. Beginning with an initial population, the genetic algorithm generates new populations that are either equivalent or superior. Three mechanisms are available for this purpose: mutation, recombination, and a specific selection mechanism described later on. Recombination combines several existing individuals, while mutation modifies an individual at one or more positions in the binary encoding. Recombination is essential for improving the solution, while mutation helps prevent constraints imposed by local minima. The choice of the recombination and mutation rate decisively influences the quality of the solution as well as the computation time. The selection mechanism consists of a binary tournament pairing selection as well as sorting by displacement distance. For tournament pairing selection, it is necessary to evaluate the performance of each individual concerning the target functions. The sum of the equally considered five objectives cost f_C, mass f_m, volume f_V, acceleration f_a, and range f_R is the fitness value g according to

$$\min g = f_C + f_m + f_V + f_a + f_R \tag{5}$$

with the lower boundary for the number of cells n_C defined as

$$n_C = \frac{E_{Veh}}{E_C}, \tag{6}$$

wherein E_{Veh} is the energy requirement for the demanded vehicle range and E_C is the energy of a single cell. The energy demand of the vehicle is determined using a one-dimensional driving model. Based on acceleration and rolling resistance as well as drag, the energy consumption of the vehicle drives is determined for an autonomous shuttle bus using Worldwide Harmonized Light Vehicles Test Procedure (WLTP). The consumption of the electric components is simplified by their average power consumption, which is based on the power consumption and the duration of usage for each device. The necessary electric loads were determined based on the components of the UNICARagil platform. As the vehicles based on this platform are able to fulfill functions specified in SAE automated driving level 4, the dataset is considered as representative for shuttles of level 4 and below [57]. The electric data of the components do not match those of the UNICARagil vehicles as these are prototypes. To achieve better alignment with the values of production vehicles, literature and datasheets were consulted. The data for the electric components are listed in Appendix A.

The algorithm stops when the termination criteria are met. Using the NSGAII algorithm, a Pareto front is generated, which consists of a set of optimal solutions. When no weighting is applied to the cost factors, the variant with the lowest sum of cost factors is chosen. Depending on requirements, development preferences, and vehicle classification, a weighting factor can be obtained to identify the most suitable solution from the Pareto front. To achieve high modularity, it is necessary to allocate the optimized number of battery cells for the entire vehicle to its various zones. This allocation is based on the power requirements of each zone, as energy can be shifted between zones during operation. The shifting of power between zones would be necessary in short time, which comes with higher cost, as all electric components need to withstand higher currents. The number of battery cells $n_{B,i}$ per zone is determined using

$$n_{B,i} = \sum_{i=1}^{k} \frac{P_i}{P_{Veh} \cdot n_{B,tot}}, \tag{7}$$

where P_i is the electric power consumption in zone i, P_{Veh} is the electric power consumption of the vehicle, and $n_{B,tot}$ is the total number of cells.

In addition to optimizing the system design and placement of the power supply system, the aim is to achieve overall vehicle optimization. In terms of the vehicle's performance, this is considered in the evaluation criteria of the genetic algorithm. Requirements regarding the position of the battery cell are considered too. The battery's volume is constrained by the space below the cabin and between the wheels. The battery's height is determined by the number of cells necessary to fulfill the range requirements. This ensures that the battery can still be manufactured and mounted as a single component and a logical separation of the battery cells $B = \{B_1, \ldots, B_m\}$, $1 \leq i \leq m$, to the zones brings modularity and reliability. To avoid compromising driving dynamics, the battery cells are positioned on the chassis floor symmetrically to the longitudinal axis. The placement in the transverse direction is determined by the position and power distribution of the drive units. The UNICARagil platform uses four drive units with equal power share, each in one corner resulting in a symmetrical placement in reference to the transverse axis. Many passenger cars have one drive unit in the front, causing a shift towards the front of the vehicle. Once the entire battery is placed, the cells are logically separated based on their nearest zone, considering the assigned number of battery cells. Therefore, there are no battery cells B_{i^*} with a shorter Euclidean distance to the cluster center μ_i than its assigned cells B_i, so the following applies:

$$B_i : \|B_i - \mu_i\|_2 \leq \|B_{i^*} - \mu_i\|_2, \forall i^* = 1 \leq i^* \leq k. \tag{8}$$

As a result, all battery cells are positioned at the vehicle bottom and closest to their respective zone. Another approach is the separation of cells from the main battery and therefore realizing decentralized energy storages. This is not investigated in detail here, as separate auxiliary components like the battery management system (BMS) and the housing would lead to disproportionately higher additional costs. With respect to housing, there is research about approaches to tackle this issue, using batteries that are integrated into the chassis and do not have their own housing [58].

3.4. Electric System Design

Besides the HV battery, it is necessary to design and position LV components. For this purpose, the architecture presented in Section 3.1 is assumed. Since the ZCUs serve as the central power distributors, the DC/DC converters will be integrated into them. This means that the 12 V battery and an optional 48 V SCAP remain for design and positioning. In order to guarantee the function of the LV components in case of a failure in the HV voltage level, two 12 V batteries and 48 V SCAPs are used. Two of each are used to avoid large and heavy components. The components with safety-critical functions of the dedicated zones are connected to the battery or the SCAP depending on their power level. The design is performed based on a worst-case scenario, assuming that the zone with the most powerful components dedicated to the battery or SCAP, stops operating. The time for an emergency halt maneuver t_{EH} and the power of the safety-critical components in this zone on the corresponding voltage level, $\sum P_{C12V,i}$ for 12 V and $\sum P_{C48V,i}$ for 48 V, are used to determine the necessary energy of the batteries $E_{Bat,i}$ or SCAP $E_{SC,i}$ according to

$$E_{Bat,i} = \sum_{i=1}^{2} P_{C12V,i} \cdot t_{EH} \tag{9}$$

and

$$E_{SC,i} = \begin{cases} \sum_{i=1}^{2} P_{Z,48V,i} \cdot t_{EH}, & P_{C,max} \leq \sum P_{Z,48V} \\ \sum_{i=1}^{2} P_{R,max} \cdot t_{EH}, & P_{C,max} > \sum P_{Z,48V} \end{cases}. \tag{10}$$

Recuperation is considered in the design of the 48 V SCAP. Therefore, the power of the 48 V loads $P_{Z,48V}$ and the maximum charging power due to the recuperation $P_{R,max}$ of the drive units are compared, and the higher one is selected. The position of the 12 V batteries and the 48 V SCAPs is determined by minimizing the distance to the connected components.

3.5. Cable Routing Optimization:

The components of an E/E architecture are connected by cables for energy and information transmission. Cables for energy transmission are denoted by $E = \{E_1, \ldots, E_m\}$, the cables for information transmission by $I = \{I_1, \ldots, I_n\}$. Since there is currently no standard for energy and information transmission on the same cable in the automotive sector (cf. USB in the consumer sector), a distinction between cables for energy and information transmission is made. In the following, the term wiring harness indicates all cables considering power or data transmission. As the vehicle structure is approximated by a three-dimensional graph of nodes and edges, the cable routing is treated as a single-source shortest path (SSSP) problem. Dijkstra algorithm is a well-established algorithm to solve SSSP problems as it considers the edge weight, which represents requirements of other vehicle systems, e.g., temperature or EMC. Using Dijkstra's algorithm, the length of each cable is minimized regarding edge cost. Dijkstra's algorithm aims to minimize the distance of a path; for orthogonal three-dimensional grids, it is similar to the Manhattan distance, which is defined as the sum of the absolute differences of all vector components [59]. The nodes are defined by the properties distance and predecessor. Furthermore, they are classified into two sets of nodes: the unvisited and the visited ones. In the initialization step, all nodes are put into the set of unvisited ones. The algorithm starts by inspecting the starting node, while the predecessor is empty for all nodes. Therefore, the distance for the starting node is set to zero, for all other nodes it is infinite. Beginning with the starting node, all nodes are visited. When a node is investigated, the distances of all its neighbor nodes are checked and updated if the shortest path to the visited node plus the distance to the neighboring node is less than the current distance. After completing this procedure for all neighboring nodes, the current node is put into the set of visited nodes and never visited again. The next inspected node is the unvisited node having the shortest distance to the starting node. The stopping criteria depends on the aim of the algorithm. The variant used here, plans the shortest path between start and end node. Therefore, it stops as soon as the target node is put in the set of visited nodes. The other variant identifies the shortest path between a starting node and all other nodes and terminates when there are only unvisited nodes with distance infinite left. This means that there is no path to the start node.

Based on the system definition described in Figure 3, start and end point are specified as follows: when the voltage of the start component is below 60 V, the target component is the ZCU, otherwise it is the HV PDU. For information processing, the ZCUs are always the targets, except for the cables connecting the ZCUs. The cost of the cables is determined using

$$C_{C,\,EI} = \sum_{x_j \epsilon X} \|x - \gamma\|_1 \cdot (c_I + c_E \cdot f_E \cdot d(x_j)), \tag{11}$$

wherein $\|x - \gamma\|_1$ is the Manhattan distance (l_1 norm) of the starting node x, which are the electric components, and the target node γ, which is a ZCU or the HV PDU. c_I is the cost factor for information transmission cables in €/m, c_E is the cost factor for energy transmission cable in €/m, which is depending on the diameter of the cable, and f_E is the factor for the ground line, which is one for components that operate on a voltage below 48 V. Components that operate on 48 V or above need a separate ground line, therefore the factor is two for them. Besides these cables, there is a ring line for data transmission between all ZCUs, respectively, the centroids μ. Its length corresponds to the Manhattan distance between the ZCUs, which is determined according to

$$C_{C,ZCU} = \begin{cases} \sum_{i=1}^{k} \|\mu_i - \mu_{i+1}\|_1 \cdot c_I, \ i < k \\ \sum_{i=1}^{k} \|\mu_i - \mu_1\|_1 \cdot c_I, \ i = k \end{cases}. \tag{12}$$

The effort required for cable installation can vary at local level. Influencing factors may be preferred cable routing, temperature, distance to ZCUs as well as occupancy of existing paths [49]. To reduce manufacturing time and cost, car manufacturers define main cable routes. It is requested to use these when designing the cable route between components. There are three common variants for designing the main cable paths: the E-variant, the H-variant, and the double-H-variant [60]. According to Neckenich, especially for long vehicles, the double H-variant is dominant. As the UNICARagil vehicles are shuttle busses, the double-H-variant appears to be most suitable. This routing variant consists of two paths along the longitudinal axis, one on each side, and two paths along the transverse axis, approximately at one third and two thirds of the vehicle length. The routing is extended to include four paths along the vertical axis of the vehicle, which are located in the four corners, and a second double-H structure in the vehicle roof. In order to consider these influences, a characteristic map for the graph is required. In Figure 4, the cost of all usable edges for the Dijkstra algorithm of the autoSHUTTLE are visualized. The characteristic map represents the requirements of the overall vehicle and the packaging to ensure that the Dijkstra algorithm respects them in the optimization of the wiring harness. The extended double-H-paths are indicated by lower edge weight in the vehicle floor and roof.

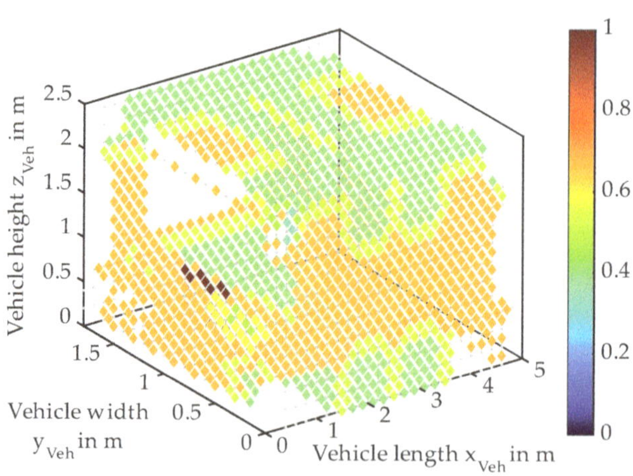

Figure 4. Characteristic edge cost map for cable routing optimization based on vehicle requirements.

3.6. Evaluation

Based on the previous optimization steps, the architectures are compared and evaluated by cost, weight, system complexity, state of function (SOF), and zone balance. Component cost and weight are estimated based on the electrical power or capacity utilizing cost functions. The definition of complexity for E/E architectures $\Lambda_{E/E}$ is based on the number of components, their dynamic, and interfaces according to

$$\Lambda_{E/E} = 1 - \frac{\lambda_{p,i} + \lambda_{p,m} + \lambda_{p,d} + \lambda_{v,i} + \lambda_{v,m} + \lambda_{v,d}}{6} \tag{13}$$

wherein $\lambda \in \mathbb{Q}$ [0, 1] indicates the components of complexity [53]. While the first index represents the dimension (p for physical, v for virtual), the second index names the aspect (i for interdependence, m for multiplicity, d for dynamic) [61].

Besides the development aspects like cost, weight, and complexity it is important to consider the functional differences caused by topologies. By separating the vehicle in zones, it is possible to realize self-sufficient functional units. They are supposed to operate independently, and therefore a measure that determines the remaining functions, whenever one zone has an error, is necessary. As the available energy $E_{el,SOF}$ and power $P_{el,SOF}$ are proportional to the functions, they are used to define the E/E architecture state of function SOF_{EE}. For the determination of the SOF_{EE}, the worst-case scenario is considered, and therefore the most powerful zone quits operating. The SOF_{EE} is determined based on

$$SOF_{EE} = 1 - \frac{\left(\frac{E_{el,SOF}}{E_{el,tot}} + \frac{P_{el,SOF}}{P_{el,tot}}\right)}{2}, \tag{14}$$

with $E_{el,tot}$ as total energy and $P_{el,tot}$ as total power of the architecture. The zone balance is the sum of the relative deviations of all zone powers of an architecture from their average power. The algorithm prioritizes these criteria according to the vehicle type and the user requirements.

Software has a higher share in vehicle development today, which will further increase in the future. As product development for software components is faster and more agile than for hardware components, it will be important for these architectures to be updateable and more flexible for changes. As the assignment of components defines the architectures, it is representative to evaluate the quality of the clustering as a measurement for flexibility of an architecture. There are different internal clustering validation measures as the Calinski–Harabasz index (CHI), the Davies–Bouldin index (DBI), or silhouette coefficient (SC), which are compared by Liu et al. [62]. The CHI is based on the similarity of a data point to its cluster (separation) compared with the similarity to other clusters (cohesion). The separation is defined by the distances between the data points in a cluster to their centroid, and cohesion is determined using the distance of the cluster centroids to the global one. The Davies–Bouldin index measures the quality of the clusters by comparing the distance between clusters with the size of the clusters themselves. The silhouette coefficient uses the average distance of an object to other objects in the same cluster and subtracts it from the average distance of the object to the objects of the nearest other cluster. As electric components in vehicles are distributed inhomogeneous, the distance between components can be higher than the size of a cluster. Nevertheless, it is reasonable to put them in the same cluster. Therefore, DBI and SC have a disadvantage against CHI, since they use a global cluster centroid for comparison. As the clusters of electric components are classified by geometric features reducing the total distance, the delimitation of clusters might not be as clear as with a set of problem-specific features. Therefore, the results of DBI and SC indicate a low to medium quality in cluster results, whereas CHI calculates an absolute value. As the target is to identify the most suitable architecture, it is more reasonable to use an index that puts the focus on the comparison of architectures instead of an index evaluating the quality of the architectures. CHI is determined according to

$$CHI = \frac{SS_B}{SS_W} \times \frac{(n-k)}{(k-1)}, \tag{15}$$

using SS_B as between-cluster variance, SS_W as within-cluster variance, k as the number of clusters, and n the number of data points in X [63]. The between-cluster variance SS_B and the within-cluster SS_W variance are determined using

$$SS_B = \sum_{i=1}^{k} |x_i| \cdot \|\mu_i - x_m\|^2, \tag{16}$$

and

$$SS_W = \sum_{i=1}^{k} \sum_{x \in s_i} \|x_j - \mu_i\|^2. \tag{17}$$

$|x_i|$ is the number of data points of the dedicated cluster S_i, x_m is the mean of all data points x_j, and $\|\mu_i - x_m\|$ and $\|x_j - \mu_i\|$ are the Euclidean distance between two vectors.

4. Discussion

The methodology described is verified by means of simulations. For this purpose, the structure of the prototype vehicle autoSHUTTLE and the electric components based on literature are used. The main dimensions of the vehicle are shown in Appendix A while the component data are listed in Appendix A. The vehicle has four-wheel hub drives and four sensor modules with cameras, lidars, and radars in each of the four corners of the vehicle. The other loads are distributed unevenly throughout the vehicle. Table 2 shows the simulation results for the designed zonal architectures. The criteria are SOF_{EE}, balance, flexibility, the mass of the E/E architecture m_{EE}, and the length of the wiring harness L_{WH}. The SOF_{EE} and balance are normalized values reaching from zero to one, where one represents the best possible fulfillment of the criterion. Rising zone numbers go along with an increase in the SOF_{EE}, as the power per zone and therefore the number of supplied functions decreases. Higher $\mathrm{SOF_{EE}}$ values indicate a lower complexity of the system, which is meaningful as the wiring harness is highly simplified by zone separation. The breakdown in several zones also reduces the complexity of the electronic components as the number of states, the variants, and their dynamic behavior simplifies. The balance decreases as zone numbers rise, caused by an increasing divergence in the power of the individual zones. The higher the balance, the more similar are the zones and their ZCUs, leading to higher modularity and less variants. Therefore, balance has a high influence on the variant cost, but no influence on functional aspects. The flexibility, which is characterized by the CHI, decreases as well with rising zone numbers.

Table 2. Functional criteria evaluation based on simulation of zonal E/E architectures from zone numbers of one to eight.

Zone Number k	1	2	3	4	5	6	7	8
Λ_{EE}	0.029	0.392	0.518	0.527	0.555	0.595	0.631	0.629
SOF_{EE}	-	0.437	0.558	0.624	0.669	0.717	0.701	0.767
Balance	-	0.937	0.82	0.871	0.847	0.866	0.76	0.773
Flexibility	-	100.5	62.01	32.7	45.2	29.0	28.9	24.6

An increased number of zones will come with both advantages and disadvantages. While the availability of function improves, there is a slight decrease of the balance, which is related to modularity. Flexibility lowers with a rising zone number, beside the five-zone architecture, which stands out with an increased value in opposition to the trend. Surjekat et al. discuss five different zonal architectures, having 3, 5 or 6 ZCUs [52]. In these architectures, the ZCUs are interconnected using a star topology, ring topology, or a combination of both. However, it is important to note that while they assess these architectures in terms of flexibility and safety, comparing their results to the findings outlined in this paper may not be meaningful. This is because the evaluation criteria are not clearly defined nor is the wiring harness considered in detail in their assessment.

Table 3 lists the mass m_{WH} and length L_{WH} of the entire wiring harness as well as the separate values for the information transmission ($m_{WH,I}$, $L_{WH,I}$) and energy transmission ($m_{WH,E}$, $L_{WH,E}$) cables. The wiring harness mass is reduced between 6.8 and 8.8%. While all architectures show a reduced weight in comparison to the reference architecture with one zone, the lowest weight is obtained for seven zones. Regarding the wiring harness mass of the one zone architecture, one needs to consider that four drive units in each corner of the vehicles were used, which is a major drawback for architectures with few zone numbers, while the mass of almost 60 kg is still meaningful [7,8]. The mass for the information transmission cables is reduced by up to one kilogram in accordance with their length decrease. The length of the cables for energy transmission rises while their mass decreases sharply. This is caused by the fact that more power cables are necessary as all ZCUs need to be connected to the HV battery. The weight of the power cables is reduced, as the cable diameter from the ZCU to the component is reduced. Evaluating the length of the entire wiring harness for the different zone numbers, a reduction of 5.6 to 13.6% is accomplished. The overall decrease in mass and length shows that there is potential to improve the E/E architecture by centralization, using ZCUs to optimize the wiring harness. Cui reduces the length and mass of its wiring harness by 16% and more considering only cables for information transmission [27]. With this method, their length is reduced by 6.7 to 13.6% and their mass by 5.7 to 13.6%. The difference is caused by the different application. In the following, the volumes of humanoid robots and vehicles will be compared, so the terms height, width, and length will be specified for the robot as follows: Height is the dimension from the back of the body to the front, that is, from back to chest. Width is the dimension between the outsides of both hands, and length is the dimension between the bottom of the feet and the top of the head. Comparing the volumes of the humanoid robot and the vehicle, it can be seen that the longitudinal extent of both is significantly greater than the extent in the other two axes. The ratio of length to height and width is significantly smaller for the humanoid robot. A further difference is the limited cable paths in robots, since no cables can be laid between the extremities. Thus, most cable paths are longitudinal connections; elevated and transverse connections are only useful for a few routes. In vehicles, on the other hand, connections are possible in all spatial directions; only the vehicle interior prevents connections inside the vehicle. The restriction to one spatial direction in humanoid robots leads to a higher percentage of weight and length savings, since practically only the connections to the furthest away components and short branches to the closer components are determined. Park reports reduction in length of 24.7% and in mass of 24.6% [6]. The pure values of Park seem superior, while one needs to consider that he focuses on end-to-end timing of messages and does not provide information regarding the design procedure of the wiring harness as well as the distribution of the ECUs in the vehicle. The end-to-end timing of messages is influenced by the wiring harness design, while the results are still meaningful for the architecture comparison. Regarding the wiring harness length and weight, the meaningfulness is reduced as power distribution as well as wiring harness routing is not considered. Besides the functional and package criteria, the cost is crucial when it comes to architecture design. The cost for the entire E/E architecture C_{EE} is decreasing with rising zone number caused by decreasing cost for the information transmission $C_{WH,I}$ and energy transmission $C_{WH,P}$ cables, whereby one should consider that $C_{WH,E}$ for one zone might be too high, as in a one-zone architecture are power distribution devices to reduce cable mass. Comparing the values listed in Table 4 with a recent study of the ICCT shows that the predicted costs are reasonable [62]. The summed cost projection by the ICCT for the electric components in BEVs are up to approx. USD 2000 (EUR 11,182.86, 5 September 2023) for 2019, while they are decreasing to approx. USD 5000 in 2030.

Table 3. Packaging criteria evaluation based on simulation of zonal E/E architectures from zone numbers of one to eight.

Zone Number k	1	2	3	4	5	6	7	8
m_{WH} in kg	58.48	23.6	22.92	16.93	16.4	15.74	14.49	14.77
$m_{WH,I}$ in kg	7.08	6.12	6.45	6.68	6.23	6.24	6.38	6.60
$m_{WH,E}$ in kg	51.40	17.48	16.47	10.25	10.17	9.50	8.11	8.17
L_{WH} in m	529.0	467.8	499.0	514.8	492.2	503.0	523.2	546.6
$L_{WH,I}$ in m	236.0	204.0	215.2	222.8	207.8	208.2	212.8	220.2
$L_{WH,E}$ in m	293.0	263.8	283.8	292.0	284.4	294.8	310.4	326.4

Table 4. Cost evaluation based on simulation of zonal E/E architectures from zone numbers of one to eight.

Zone Number k	1	2	3	4	5	6	7	8
C_{EE} in EUR	13,160	11,535	11,520	11,242	11,265	11,285	11,271	11,308
$C_{WH,I}$ in EUR	177.0	153.0	161.4	167.1	155.85	156.15	159.6	165.15
$C_{WH,E}$ in EUR	2570.0	874.3	823.6	512.7	508.5	475.3	405.8	408.5
C_{ZCU}	150.0	225.0	265.7	300.1	343.0	396.7	450.4	480.6

5. Conclusions

In early development phases, technical details of most electrical components are unclear. Since the power supply system interfaces with almost all systems, its design is highly dependent on them. Late consideration in the design process leads to problems in packaging, especially in regards to the wiring harness and the traction battery. This paper introduces a method to tackle this issue. Based on a minimum of information about, respectively, position, power consumption, and voltage level, a zonal architecture is designed in early development phases. As there is an intense interaction between the E/E architecture and its electronic components, developing one without the other will cause significant problems. Having a closer lock on the components with the highest problem potential, the traction battery and the wiring harness are the most critical of the power supply system. The wiring harness spans the entire vehicle structure, and the battery has a high weight and volume share of the vehicle. To tackle these problems, a coupled approach of k-means clustering and Dijkstra algorithm is introduced. The k-means algorithm identifies the most likely zone numbers for the components, the zone classification, and the position of its central power and data device, the ZCU. This is accomplished by clustering the position of the electric components and determining the best position for the centroid with respect to the vehicle packaging. To ensure that the proposed architecture is package-compatible, the design of the traction battery and the wiring harness is integrated. Therefore, the battery for a given range is determined and integrated in the vehicle packaging. The wiring harness is routed based on the definition of the system using SysML v1.6 and with respect to requirements such as temperature or EMC. This enables a synthesis of concepts from different disciplines and systems already in an early phase. This design method enables the determination of the optimal number of zones according to application-specific criteria. In this paper, functional, packaging, and cost criteria are used to evaluate the variant. The selection of the best variant is dependent on the development aim. For compact vehicles, the version with the lowest cost might be the best one, and for high-class vehicles, an architecture with the best fulfillment of the functional criteria is more suitable. With the presented method, it is possible to identify the optimal zone architecture and design a vehicle power supply system, even when facing uncertainties. Early identification of the optimal zone and power supply design increases the quality of the solution and reduces iteration loops, thereby cutting development time and costs.

The research identified that it is possible to design a zonal architecture with limited information and to utilize the potential of this new architecture approach. Even with such rudimentary data as position, power consumption, and voltage of the electric loads, the optimization procedure is able to structure the electric loads and reduce wiring harness length and weight. As the approach uses k-means clustering and Dijkstra's algorithm, a model compatible with both was identified and adapted. Furthermore, an adaption for k-means clustering was developed, which realizes an equal cluster size with minimal length increase. By combining electrical, package, and wiring harness design, this method realizes a guideline for the development of zonal E/E architectures in early design phases.

The results of the simulation show that there is a dependency of the clusters on the distribution of the components, but their electric power is not considered in the clustering process. The influence and the meaningfulness of its consideration in the clustering procedure needs to be investigated in the future. Furthermore, the sensitivity of the clusters to position data changes has to be examined to identify further potential for improvement in the clustering procedure. The relation of the optimization result to the consumer position, distribution, and power is essential in this case. A detailed sensitivity investigation is necessary to ensure the extensibility of the optimized architectures in both electrical and geometrical terms. For a more accurate determination of the power consumption, a traffic simulation should be added. To ensure that the cost projection for the architecture is reliable in the future, the declining costs for batteries and electronic components must be considered.

Author Contributions: Conceptualization, J.M.; methodology, J.M.; software, J.M.; validation, J.M.; formal analysis, J.M.; investigation, J.M.; resources, J.M.; data curation, J.M.; writing—original draft preparation, J.M.; writing—review and editing, H.-C.R.; visualization, J.M.; supervision, H.-C.R.; project administration, H.-C.R.; funding acquisition, H.-C.R. All authors have read and agreed to the published version of the manuscript.

Funding: This research was funded by the Federal Ministry of Education and Research of Germany (BMBF), AUTOtechagil (FKZ 01IS22088x). This publication was funded by the German Research Foundation (DFG) grant "Open Access Publication Funding/2023–2024/University of Stuttgart" (512689491).

Data Availability Statement: No new data were created or analyzed in this study. Data sharing is not applicable to this article.

Conflicts of Interest: The authors declare no conflict of interest.

Nomenclature

B	Battery cell
$C_{C,El}$	Cable cost, €
c_E	Cost factor for energy tra. cables
c_I	Cost factor for information tra. cables
$C_{C,ZCU}$	Cable cost ring line ZCU, €
C_{ZCU}	Zone control unit cost, €/Unit
D	Total Euclidean distance, m
$d(x)$	Diameter of cable x
E	Cable for energy transmission
E_{Bat}	Energy of the 12 V battery, J
E_C	Energy of a battery cell, Wh
E_{SC}	Energy of the 48 V SCAP, J
$E_{el,SOF}$	Remaining electric energy, J
$E_{el,tot}$	Total electric energy, J
E_{Veh}	Necessary energy for the vehicle range, J
f_a	Cost factor acceleration
f_c	Cost factor material cost
f_E	Cost factor for ground line
f_m	Cost factor mass
$n_{B,tot}$	Total number of battery cells
p_1–p_4	Parameter for ZCU cost function
P_i	Power of a zone, W
P_{C12V}	Safety-critical power of a zone on 12 V, W
P_{C12V}	Safety-critical power of a zone on 48 V, W
$P_{el,SOF}$	Remaining electric power, W
$P_{el,tot}$	Total electric power, W
P_{Veh}	Power consumption of the vehicle, W
S_i	Cluster
SOF_{EE}	State of Function of E/E architecture
SS_B	Between-cluster variance
SS_W	Within-cluster variance
t_{EH}	Emergency halt maneuver time, s
x_j	Data points of electric components
x_m	Mean of all data points x
x_{max}	Max. number of data points per cluster
$\lvert x \rvert$	Number of data points in a cluster
x	Vehicle length, m
y	Vehicle width, m

f_R	Cost factor Range	z	Vehicle height, m
f_V	Cost factor volume	γ	Target node
f_{ZCU}	Cost factor ECU, €	$\lambda_{p,i}$	Physical interdependence
g	Fitness value	$\lambda_{p,d}$	Physical dynamic
I	Cable for information transmission	$\lambda_{p,m}$	Physical multiplicity
k	Zone number	$\lambda_{v,i}$	Virtual interdependence
n	Number of data points	$\lambda_{v,d}$	Virtual dynamic
n_C	Lower boundary cell number	$\lambda_{v,m}$	Virtual multiplicity
$n_{B,i}$	Number of battery cells per zone	μ_i	Centroid

Appendix A

Table A1. Parameters.

Symbol	Description	Value	Units
E_C	Energy of a battery cell	11.5	Wh
F_{ZCU}	Cost ZCU	150	€
p_1	p_1 of ZCU Cost Function	−0.003571	-
p_2	p_2 of ZCU Cost Function	0.06667	-
p_3	p_3 of ZCU Cost Function	−0.425	-
p_4	p_4 of ZCU Cost Function	1.362	-
t_{EH}	Emergency halt maneuver time	30	s
x_{Veh}	Vehicle length	4.97	m
y_{Veh}	Vehicle width	2.05	m
z_{Veh}	Vehicle height	2.44	m

Table A2. Assumed electric loads for the E/E architecture of SAE level 4 vehicles.

Component	Maximum el. Power P_{max} [W]	Usage Time t_U [%]	Average el. Power P_{avg} [W]	Number [-]	Reference
Camera	2	100	2	7	[64]
CPU for	96	100	96	2	[64]
DSRC (dedicated short-range communication)	2	100	2	1	[64]
Lidar sensor	60	100	60	4	[64]
Radar sensor	8	100	8	8	[64,65]
Sonar/Ultrasound sensor	0.15	100	0.15	2	[64]
Air conditioner front	-	100	600	1	[66]
Anti-fog light	35	0	0	3	[67]
Blind zone radar	5	100	5	1	[67]
Braking light	21	50	10.5	3	[67]
Cabin lights	20	100	20	1	[67]
Door module	-	100	360	2	[66]
Electric roof	300	5	15	1	[67]
Front window heater	1500	-	120	1	[68]
Headlamps	60	100	60	2	[67]
Rear window heating	-	-	120	1	[69]
Reversing light	21	10	2.1	2	[67]
Body Control Modul	360	100	360	1	[70]
Seats electronics	300	5	15	2	[67]
Turning light	21	20	4.2	6	[67]
5G Router	68.4	100	68.4	1	[71]
GNSS	0.55	100	0.55	1	[72]
Audio system	25	100	25	1	[67]
Multimedia screens	30	100	30	2	[67]
Navigation	15	100	15	1	[67]
ABS	-	100	600	1	[73]
Brake-by-wire	1300	-	150	1	[74]
Cooling pump	1200	100	500	1	[68]
Cooling radiator	500	50	100	1	[69]
Engine ECU	-	100	700	1	[74]
ESP	7800	-	3000	1	[66]
Steer-by-wire	1400	-	28	1	[74]

Table A2. *Cont.*

Component	Maximum el. Power P_{max} [W]	Usage Time t_U [%]	Average el. Power P_{avg} [W]	Number [-]	Reference
Suspension pump	1000	20	200	1	[67]
Vehicle control unit (Control and Fusion)	1000	-	750	1	[74]
Drive Modul	13,000	25	3250	4	[75]
BMS	24	100	24	1	*
Backup ADAS ECU	500	100	500	1	[76]

* Measurement.

References

1. Bernhart, W.; Riederle, S.; Hotz, T.; Olschewski, I.; Busse, A. E-Mobility Index 2021. Available online: https://www.rolandberger.com/publications/publication_pdf/roland_berger_e_mobility_index_2021_en.pdf (accessed on 28 August 2023).
2. Kampker, A.; Offermanns, C.; Heimes, H.; Bi, P. Meta-analysis on the Market Development of Electrified Vehicles. *ATZ Worldw.* **2021**, *123*, 58–63. [CrossRef]
3. He, H.; Sun, F.; Wang, Z.; Lin, C.; Zhang, C.; Xiong, R.; Deng, J.; Zhu, X.; Xie, P.; Zhang, S.; et al. China's battery electric vehicles lead the world: Achievements in technology system architecture and technological breakthroughs. *Green Energy Intell. Transp.* **2022**, *1*, 100020. [CrossRef]
4. Katis, C.; Karlis, A. Evolution of Equipment in Electromobility and Autonomous Driving Regarding Safety Issues. *Energies* **2023**, *16*, 1271. [CrossRef]
5. Lutsey, N.; Cui, H.; Rujie, Y. *Evaluating Electric Vehicle Costs and Benefits in China in the 2020–2035 Time Frame*; White Paper; International Council for Clean Transportation: Washington, DC, USA, 2021.
6. Park, C.; Park, S. Performance Evaluation of Zone-Based In-Vehicle Network Architecture for Autonomous Vehicles. *Sensors* **2023**, *23*, 669. [CrossRef] [PubMed]
7. Maul, M.; Becker, G.; Bernhard, U. Service-oriented EE zone architecture key elements for new market segments. *ATZ Elektron. Worldw.* **2018**, *13*, 36–41. [CrossRef]
8. Klaus-Wagenbrenner, J. Zonal EE Architecture: Towards a Fully Automotive Ethernet–Based Vehicle Infrastructure. In Proceedings of the Automotive E/E Architecture Technology Innovation Conference, Shanghai, China, 7–8 November 2019.
9. Otten, S.; Glock, T.; Hohl, C.P.; Sax, E. Model-based Variant Management in Automotive Systems Engineering. In Proceedings of the 5th IEEE International Symposium on Systems Engineering, Edinburgh, UK, 1–3 October 2019. [CrossRef]
10. Mischo, S.; Kornhaas, R.; Krauter, I.; Kersken, U.; Schöttle, R. E/E architectures at the crossroads. *ATZ Elektron. Worldw.* **2008**, *3*, 10–13. [CrossRef]
11. Bandur, V.; Selim, G.; Pantelic, V.; Lawford, M. Making the Case for Centralized Automotive E/E Architectures. *IEEE Trans. Veh. Technol.* **2021**, *70*, 1230–1245. [CrossRef]
12. Buttle, D.; Gold, S. MCUs and Virtualization in Zone E/E Architectures. *ATZ Elektron. Worldw.* **2022**, *17*, 18–24. [CrossRef]
13. Schäfer, C.; Denkelmann, R. Sustainable E/E Architecture Power Supply and Data Transmission for Autonomous Driving. *ATZ Elektron. Worldw.* **2018**, *13*, 16–21. [CrossRef]
14. Robert Bosch GmbH. Vehicle Control Unit: Die Vehicle Control Unit als zentrale E/E-Architekturkomponente für Alle Powertrain Topologien. Available online: https://www.bosch-mobility.com/de/loesungen/steuergeraete/vehicle-control-unit/ (accessed on 28 August 2023).
15. Vitesco Technologies GmbH. Vitesco Technologies Entwickelt Master Controller für Eine Neue Ära in der Antriebssteuerung. Available online: https://www.vitesco-technologies.com/de-de/press-events/press/tech-day-master-controller (accessed on 28 August 2023).
16. Stroh, C.A. E/E-Architekturen Frischzellenkur. Available online: https://www.automotiveit.eu/exklusiv/frischzellenkur-210.html (accessed on 28 August 2023).
17. Askaripoor, H.; Hashemi Farzaneh, M.; Knoll, A. E/E Architecture Synthesis: Challenges and Technologies. *Electronics* **2022**, *11*, 518. [CrossRef]
18. Brunner, S.; Roder, J.; Kucera, M.; Waas, T. Automotive E/E-architecture enhancements by usage of ethernet TSN. In Proceedings of the 2017 13th Workshop on Intelligent Solutions in Embedded Systems (WISES), Hamburg, Germany, 12–13 June 2017; IEEE: Piscataway, NJ, USA, 2017; pp. 9–13, ISBN 9781538611579.
19. Zhu, H.; Zhou, W.; Li, Z.; Li, L.; Huang, T. Requirements-Driven Automotive Electrical/Electronic Architecture: A Survey and Prospective Trends. *IEEE Access* **2021**, *9*, 100096–100112. [CrossRef]
20. Frigerio, A.; Vermeulen, B.; Goossens, K.G.W. Automotive Architecture Topologies: Analysis for Safety-Critical Autonomous Vehicle Applications. *IEEE Access* **2021**, *9*, 62837–62846. [CrossRef]
21. Tomar, A.S. Modern Electrical/Electronic Infrastructure for Commercial Trucks: Generic Input/Output nodes for sensors and actuators in Commercial Trucks. Master's Thesis, KTH Royal Institute of Technology, Stockholm, Sweden, 2017.
22. Robert Bosch GmbH. *E/E-Architecture in a Connected World*; Robert Bosch GmbH: Gerlingen, Germany, 8 March 2017.

23. Niklas-Höret, M. Impact of the Zone Architecture on the in Vehicle SW Distribution. Available online: https://ahkfinnland.de/filehub/deliverFile/e9a3077f-b2ce-425f-8835-a61b2e62513b/1229026/Continental_Keynote_1229026.pdf (accessed on 28 August 2023).
24. Avery, G.; Dawson, J.; Dornbusch, G. *Zonal Architecture: Making the Car of the future Possible*; White paper; Molex: Lisle, IL, USA, 2023.
25. Infineon Technologies AG. *Dependable Solutions for Future Mobility and Zonal E/E Architectures*; Infineon Technologies AG: Neubiberg, Germany, 2021.
26. Aberl, P.; Haas, S.; Vemuri, A. How a Zone Architecture Paves the Way to a Fully Software-Defined Vehicle; White Paper. Available online: https://www.ti.com/lit/wp/spry345a/spry345a.pdf?ts=1681378174830&ref_url=https%253A%252F%252Fwww.google.com%252F (accessed on 28 August 2023).
27. Cui, C.; Park, C.; Park, S. Physical Length and Weight Reduction of Humanoid In-Robot Network with Zonal Architecture. *Sensors* **2023**, *23*, 2627. [CrossRef] [PubMed]
28. Shrestha, R.; Kim, D.; Choi, J.; Kim, S. A Novel E/E Architecture for Low Altitude UAVs. In Proceedings of the 2022 IEEE International Symposium on Circuits and Systems (ISCAS), Austin, TX, USA, 27 May–1 June 2022; IEEE: Piscataway, NJ, USA, 2022; pp. 346–350, ISBN 9781665484855.
29. Zaheri, D.; Niedballa, D.; Leuffen, M.; Bilkei-Gorzo, G. Practical Implementation of a Zonal E/E Architecture for Autonomous Vehicles in UNICARagil. *ATZ Electron. Worldw.* **2023**, *18*, 8–12. [CrossRef]
30. Woopen, T.; Lampe, B.; Böddeker, T.; Lutz, E.; Kampmann, A.; Alrifaee, B.; Kowalewski, S.; Moormann, D.; Torben, S.; Jatzkowski, I.; et al. UNICARagil—Disruptive Modular Architectures for Agile, Automated Vehicle Concepts. In Proceedings of the 27th Aachen Colloquium Automobile and Engine Technology, Aachen, Germany, 8–10 October 2018.
31. Goth, M.; Keilhoff, D.; Reuss, H.-C. Fault tolerant electric energy supply system design for automated electric shuttle bus: Band 2. In *20. Internationales Stuttgarter Symposium*; Springer: Berlin/Heidelberg, Germany, 2020; pp. 441–455. [CrossRef]
32. Gehringer, D.; Kuthada, T.; Wagner, A. Thermal Management System of the UNICARagil Vehicles—A Comprehensive Overview. *World Electr. Veh. J.* **2023**, *14*, 6. [CrossRef]
33. Moran, R.; Sindaco, F.; Birnie, A.; Vermeulen, B. Der Plattform-Ansatz als Revolution der Fahrzeugarchitektur: Von Domänen zu Zonen: Architektur für Autonome Autos Neu Gedacht. Available online: https://www.all-electronics.de/automotive-transportation/von-domaenen-zu-zonen-architektur-fuer-autonome-autos-neu-gedacht-197.html (accessed on 28 September 2023).
34. Fuchs, S. Verfahren zur Parameterbasierten Gewichtsabschätzung neuer Fahrzeugkonzepte. Ph.D. Thesis, Technische Universität München, Munich, Germany, 2014.
35. Sethuraman, G. Vehicle Component Configuration Design and Packaging in Virtual Environment for Autonomous Electric Buses. Ph.D. Thesis, Tokyo University of Agriculture and Technology, Tokyo, Japan, 2019.
36. Fay, T.-A. Methodischer Technologieverleich von Systemkonzepten für die Elektrifizierung von Stadtbussen. Ph.D. Thesis, Technical University of Berlin, Berlin, Germany, 2019.
37. Berthold, K. *Techno-Ökonomische Auslegungsmethodik für die Elektrifizierung Urbaner Busnetze*; Karlsruhe Institute of Technology: Karlsruhe, Germany, 2019.
38. Lübke, A. *Systematischer Bordnetzentwurf: Optimierung der Bordnetzarchitektur mit Hilfe von Genetischen Algorithmen*; Technische Universität Dresden: Dresden, Germany, 1999; ISBN 3896535609.
39. Diebig, M. Entwicklung einer Methodik zur Simulationsbasierten Dimensionierung von Kfz-Bordnetzen. Ph.D. Thesis, Technical University of Dortmund, Dortmund, Germany, 2016.
40. Wang, J. Simulationsumgebung zur Bewertung von Bornetz-Architekturen mit Hochleistungsverbrauchern. Ph.D. Thesis, University of Kassel, Kassel, Germany, 2016.
41. Zhu, Z.; La Rocca, G.; van Tooren, M.J.L. A methodology to enable automatic 3D routing of aircraft Electrical Wiring Interconnection System. *CEAS Aeronaut. J.* **2017**, *8*, 287–302. [CrossRef]
42. Braun, L. Modellbasierte Design-Space-Exploration Nicht-Funktionaler Auslegungskriterien des Fahrzeugenergiebordnetzes. Ph.D. Thesis, Karlsruhe Institute of Technology, Karlsruhe, Germany, 2018.
43. Becker, J.N. *Dimensioning and Optimization of Hybrid Li-Ion Battery Systems with different Optimization Goals*; RWTH Aachen University: Aachen, Germany, 2017.
44. Frank, F. Optimierter Hochvoltbatterieentwurf Hinsichtlich Mechanischer, Thermischer und Elektrischer Randbedingungen unter Berücksichtigung der Produktionskosten. Ph.D. Thesis, University of Duisburg-Essen, Duisburg, Germany, 2019.
45. Meyer, G. Schlüsselrolle der EE-Architektur und der Bordnetze für das Automobil der Zukunft; White Paper. Available online: https://www.strategiekreis-automobile-zukunft.de/files/enova-position-bordnetze-und-fahrzeugarchitektur.pdf (accessed on 28 September 2023).
46. Müller, M. *Zonen-Basierte E/E-Architektur im Fokus: Automatisiertes und Vernetztes Fahren*; HANSER Automotive: Ulm, Germany, 2021; pp. 26–28.
47. Zoox Inc. Introducing Zoox: Built for Riders, Not Drivers. Available online: https://zoox.com/vehicle/ (accessed on 28 August 2023).
48. LOHR Group. Le Cristal: Mobilitätslösung 100% Elektrisch, Flexibel und Vernetzt. Available online: https://www.lohr.fr/uploads/2023/05/d18647c-2023-03-plaquette-cristal-de.pdf (accessed on 28 August 2023).

49. Maier, J.; Reuss, H.-C. A Framework for Optimizing the Wiring Harness of Automated Electric Vehicles. In Proceedings of the 6th Shanghai-Stuttgart-Symposium, Shanghai, China, 1–2 December 2022.
50. Bornemann, M. Zone Controllers Build Bridge to Tomorrow's Technology; White Paper. Available online: https://www.aptiv.com/docs/default-source/white-papers/2021_aptiv_whitepaper_zonecontroller.pdf?sfvrsn=c1f9163d_23 (accessed on 28 August 2023).
51. Liebetrau, T. E/E Architecture Transformation How it impacts value chain and networking technologies. In Proceedings of the AmE 2022: Automotive Meets Electronics: 13th GMM-Symposium, Dortmund, Germany, 29–30 September 2022; Wahl, M., Stockem, A., Eds.; VDE Verlag Gmbh: Offenbach, Germany, 2022; ISBN 9783800759552.
52. Surjekar, N.N.; Patwardhan, Y.; Konduju, V. A Case Study on Migrating towards Functionally Safe Zonal Architecture using MBSE. In Proceedings of the 33rd Annual INCOSE International Symposium, Honolulu, HI, USA, 15–20 July 2023; Wiley: Hoboken, NJ, USA, 2023; Volume 33, pp. 1403–1417. [CrossRef]
53. Maier, J.; Reuss, H.-C. Handling System Complexity in Zonal E/E Architectures. *Transp. Eng.* **2023**, *13*, 100195. [CrossRef]
54. Lloyd, S. Least squares quantization in PCM. *IEEE Trans. Inform. Theory* **1982**, *28*, 129–137. [CrossRef]
55. Arthur, D.; Vassilvitskii, S. k-means++: The advantages of careful seeding. In Proceedings of the 18th Annual ACM-SIAM Symposium on Discrete Algorithms, New Orleans, LA, USA, 7–9 January 2007; pp. 1027–1035.
56. Deb, K.; Pratap, A.; Agarwal, S.; Meyarivan, T. A fast and elitist multiobjective genetic algorithm: NSGA-II. *IEEE Trans. Evol. Computat.* **2002**, *6*, 182–197. [CrossRef]
57. SAE International. *Taxonomy and Definition for Terms Related to Driving Automation Systems for On-Road Motor Vehicles*; Revision April 2021 (J3016); SAE International: Warrendale, PA, USA, 2021.
58. Asp, L.E.; Bouton, K.; Carlstedt, D.; Duan, S.; Harnden, R.; Johannisson, W.; Johansen, M.; Johansson, M.K.G.; Lindbergh, G.; Liu, F.; et al. A Structural Battery and its Multifunctional Performance. *Adv. Energy Sustain. Res.* **2021**, *2*, 2000093. [CrossRef]
59. Dijkstra, E.W. A Note on Two Problems in Connexion with Graphs. In *Edsger Wybe Dijkstra: His Life, Work, and Legacy*; ACM Books: New York, NY, USA, 2022; pp. 287–290. [CrossRef]
60. Neckenich, J. 3D-Master-Leitungssatz—Konzept zur Entwicklung von Leitungssätzen als 3D-Masterin Einem Realistischen, Vollständigen DMU-Modell. Ph.D. Thesis, Saarland University, Saarbrücken, Germany, 2017.
61. Riesener, M.; Dölle, C.; Keuper, A.; Fruntke, M.; Schuh, G. Quantification of complexity in cyber-physical systems based on key figures. *Procedia CIRP* **2021**, *100*, 445–450. [CrossRef]
62. Liu, Y.; Li, Z.; Xiong, H.; Gao, X.; Wu, J. Understanding of Internal Clustering Validation Measures. In Proceedings of the 2010 IEEE International Conference on Data Mining, Sydney, Australia, 13–17 December 2010; IEEE: Piscataway, NJ, USA, 2010; ISBN 9781424491315.
63. Calinski, T.; Harabasz, J. A dendrite method for cluster analysis. *Comm. Stats.-Theory Methods* **1974**, *3*, 1–27. [CrossRef]
64. Gawron, J.H.; Keoleian, G.A.; de Kleine, R.D.; Wallington, T.J.; Kim, H.C. Life Cycle Assessment of Connected and Automated Vehicles: Sensing and Computing Subsystem and Vehicle Level Effects. *Environ. Sci. Technol.* **2018**, *52*, 3249–3256. [CrossRef] [PubMed]
65. Kutila, M.; Virtanen, A.; Tarkiainen, M.; Peussa, P. Sensor System Power Adaptation for Automated Vehicles. In Proceedings of the 2019 IEEE 15th International Conference on Intelligent Computer Communication and Processing (ICCP), Cluj-Napoca, Romania, 5–7 September 2019; Nedevschi, S., Potolea, R., Slavescu, R.R., Eds.; IEEE: Piscataway, NJ, USA, 2019; ISBN 9781728149141.
66. Kohler, T.P. *Prädiktives Leistungsmanagement in Fahrzeugbordnetzen*; Springer Fachmedien Wiesbaden: Wiesbaden, Germany, 2014; ISBN 9783658050115.
67. Kavalchuk, I.; Arisoy, H.; Stojcevski, A.; Maun Than Oo, A. Advanced Simulation of Power Consumption of Electric Vehicles. *Int. J. Electr. Comput. Electron. Commun. Eng.* **2015**, *9*, 53–59.
68. Wallentowitz, H.; Reif, K. *Handbuch Kraftfahrzeugelektronik: Grundlagen, Komponenten, Systeme, Anwendungen*; Wallentowitz, H., Ed.; Vieweg: Wiesbaden, Germany, 2006; ISBN 352803971X.
69. *Bosch Automotive Electrics and Automotive Electronics*; Springer Fachmedien Wiesbaden: Wiesbaden, Germany, 2014; ISBN 9783-658017835.
70. Bosch Engineering. Body Computer Module (BCM); Data Sheet. Available online: https://www.bosch-engineering.cn/media/de/pdfs/einsatzgebiete_1/produktdatenblaetter/110621_BEG-BCM_EN.pdf (accessed on 5 September 2023).
71. Sierra Wireless, S.A. AirLink®XR90—Router Specifications; Data Sheet. Available online: https://www.sierrawireless.com/router-solutions/xr90#specs (accessed on 5 September 2023).
72. TerrisGPS. Harxon HX-AULT006 High-Precision GNSS Vehicle Antenna; Data Sheet. Available online: https://www.terrisgps.com/product/harxon-hx-ault006/ (accessed on 5 September 2023).
73. Heinemann, D. Strukturen von Batterie- und Energiemanagementsystemen mit Bleibatterien und Ultracaps. Ph.D. Thesis, Technical University of Berlin, Berlin, Germany, 2007.
74. Schumi, S.; Graf, A. Energy and Supply Concepts for Automated Drving. In Proceedings of the AmE 2018—Automotive Meets Electronics, 9th GMM-Symposium, Dortmund, Germany, 7–8 March 2018; VDE: Berlin, Germany, 2018; pp. 95–99, ISBN 3800745240.

75. Harkort, C.; Kesselgruber, D.; Kraus, M.; Moseberg, J.; Wuebbolt-Gorbatenko, B. Mobil in der stadt von morgen: Die Verschmelzung von Antrieb und Chassis. In *Mobility for Tomorrow*; Schaeffler Technologies AG & Co., KG: Herzogenaurach, Germany, 2018; p. 397.
76. Grubmüller, S.; Stettinger, G.; Nešić, D.; Watzenig, D. Concepts for improved availability and computational power in automated driving. *Elektrotech. Inftech.* **2018**, *135*, 316–321. [CrossRef]

energies

Article

Non-Iterative Coordinated Optimisation of Power–Traffic Networks Based on Equivalent Projection

Wei Dai [1,*], Zhihong Zeng [1], Cheng Wang [2], Zhijie Zhang [1], Yang Gao [1] and Jun Xu [1]

[1] School of Electrical Engineering, Guangxi University, Nanning 530004, China; 2112392006@st.gxu.edu.cn (Z.Z.); zhijiezh@163.com (Z.Z.); 2112301084@st.gxu.edu.cn (J.X.)
[2] Gui'an Power Supply Bureau of Guizhou Power Grid Co., Ltd., Gui'an 550025, China; 18965167687@163.com
* Correspondence: weidai2019@163.com or weidai@gxu.edu.cn

Abstract: The exchange of sensitive information between power distribution networks (PDNs) and urban transport networks (UTNs) presents a difficulty in ensuring privacy protection. This research proposes a new collaborative operation method for a coupled system. The scheme takes into account the schedulable capacity of electric vehicle charging stations (EVCSs) and locational marginal prices (LMPs) to handle the difficulty at hand. The EVCS hosting capacity model is built and expressed as the feasible area of charging power, based on AC power flow. This model is then used to offer information on the real schedulable capacity. By incorporating the charging loads into the coupling nodes between PDNs and UTNs, the issue of coordinated operation is separated and becomes equal to the optimal problem involving charging loads. Based on this premise, the most efficient operational cost of PDNs is transformed into a comparable representation of cost information in PDNs. This representation incorporates LMP information that guides charging decisions in UTNs. The suggested collaborative scheduling methodology in UTNs utilises the collected projection information from the static traffic assignment (STA) to ensure data privacy protection and achieve non-iterative calculation. Numerical experiments are conducted to illustrate that the proposed method, which uses a smaller amount of data, achieves the same level of optimality as the coordinated optimisation.

Keywords: electric vehicles (EV); coordinated optimisation; equivalent model; non-iterative; power and traffic system

Citation: Dai, W.; Zeng, Z.; Wang, C.; Zhang, Z.; Gao, Y.; Xu, J. Non-Iterative Coordinated Optimisation of Power–Traffic Networks Based on Equivalent Projection. *Energies* **2024**, *17*, 1899. https://doi.org/10.3390/en17081899

Academic Editors: Stefano Bracco, Ruixin Yang, Jinhao Meng, Qian Xiao, Zeyu Chen, Yanan Wang, Dongdong Zhao, Daniel Stroe and Ji Wu

Received: 2 February 2024
Revised: 7 April 2024
Accepted: 13 April 2024
Published: 16 April 2024

1. Introduction

EVs have gained considerable international recognition due to concerns such as the oil crisis and carbon emissions. They have emerged as a prominent alternative to replace cars that run on petrol [1]. Global EV forecast research [2] predicts that the number of EVs in use globally will increase to 270 million by 2030, representing almost 14% of the total number of vehicles on the road. However, the extensive integration of electric vehicles (EVs) is expected to establish a mutually dependent relationship between PDNs and UTNs [3,4]. The travel patterns of EVs will be impacted by different road conditions, leading to changes in the spatial and temporal distribution of traffic flow. However, the charging price and queueing time at EVCSs are expected to influence the preferences of EV drivers for charging stations and, as a result, change the distribution of the electrical demand. Therefore, it is crucial to include coupling parameters in the coordination and scheduling of PTNs [5–9].

In this regard, researchers have recently focused on studying the interactions in PTNs to effectively accommodate the widespread use of EVs. Therefore, due to concerns regarding computational efficiency and data security, a substantial amount of study is focused on examining the coordination of joint flow. One specific form of research focuses on centralised optimisation by using joint modelling of PTNs and applying various acceleration strategies for computation. In the given example, a stochastic optimisation framework is developed to analyse the interconnections of PTNs in [10,11]. These interconnections include relationships between EVs/fuel-based automobiles, charging costs, and charging power. In [12], a coordination model is proposed that is based on the generalised user equilibrium in power–traffic coupled networks. This model effectively reduces the pressure in the power distribution networks with EVs. Furthermore, a model that combines dynamic user equilibrium and that is based on [8,13,14] is introduced to accurately represent the flows in PTNs. The paper [15] presents a scheduling technique that combines the assignment of charging stations and the allocation of charging power to ensure an appropriate charging plan for each EV. In their study, the authors of [16] develop a two-stage architecture that integrates optimal pathways and active and reactive power regulation for EVs in order to minimise the cost of charging. A comprehensive model is developed in [17–19] to address the intricate relationship between EVs, the power grid, and photovoltaics. This model incorporates multiple stages and takes into account the routing and scheduling of EVs to effectively handle complex traffic scenarios. In order to address congestion in UTNs, a pricing model that incorporates both LMPs and congestion charges is proposed. This model is built on a variational inequality framework, as described in [20].

The previously described research on centralised optimisation enhances computational efficiency by integrating electricity and transportation modelling. However, due to the fact that the electricity and traffic systems are controlled by various entities with distinct information security needs, it is not feasible to implement a centralised method that requires sharing of information [21].

In addition to centralised optimisation methods, decentralised procedures that rely on limited information iteration are also crucial for coordinating scheduling in networked systems. A bi-level coordinate operation framework is built using the alternative direction method of multipliers (ADMM) in [14,22,23], considering both systemic and individual views. In [24], a decentralised collaborative pricing method is proposed which uses variational inequalities. Based on this premise, a decentralised and decoupling architecture is constructed to effectively address the issue. In order to achieve the best outcome and maintain the anonymity of information, a scheme has been developed that combines the numerous individual decisions of EVs in PTNs. This strategy involves two separate network operators and uses small data, as described in reference [25]. A decentralised approach is utilised in [26] to address the collaborative pricing model, which encompasses road tolls and charging costs.

Several academics have examined several variables that impact the process of charging EVs. As an illustration, the researchers in [27] conducted a research study to analyse the effect of charging station placements on PDNs. The failure of a charging station does not impact the charging behaviour of electric buses, as seen in [28]. In a previous study, scientists employed genetic algorithms to forecast forthcoming charging requirements for EVs and strategise the most advantageous sites for charging stations [29]. A comparison analysis was conducted in [30] to compare wireless charging with traditional charging models. The writers of [31] concentrate on long-distance transportation for EVs and devise the most efficient sites for charging stations to guarantee rapid charging capabilities. In their study, the authors of [32] used an optimisation model to determine the optimal placement of charging stations and the appropriate size of electric cars. They took into account factors such as time-of-use energy price and the behaviour of electric buses to ensure the safe and efficient operation of PTNs. In order to address the charging requirements and enhance the charging effectiveness for electric vehicle users, a dispatch model for electric vehicles is suggested in [33]. This approach employs price advice to reduce the burden on charging stations. The authors of [34] consider the unpredictability of wind power generation and optimise the charging behaviour of electric buses to efficiently utilise renewable energy resources and decrease the use of non-clean energy sources. In order to synchronise the economic dispatch in PDNs with the traffic assignment in transportation networks (TNs), a decentralised architecture is suggested to develop the most efficient charge price in [35]. In their study, the authors of [36] investigate social optimal welfare by examining the charging fees and the interactions between cooperative Charging Network Operators, mobile EVs, and bulk power infrastructures. In addition, the authors of [37] suggest a framework that integrates pricing for charging and scheduling of power.

Using a decentralised strategy to synchronise power–traffic flows is in perfect harmony with the practical reality of their operation by two separate entities. This strategy successfully addresses the difficulties presented by restricted data sharing while guaranteeing the highest level of data privacy. However, due to the dependence on information iteration, there are several disadvantages to traditional decentralised approaches. (1) Within the domain of conventional distributed algorithms, such as Lagrangian relaxation [38], Benders decomposition [39], and generalised Benders decomposition [40], these methods may face challenges related to slow convergence or even the potential for convergence failure. (2) In order to maintain equilibrium in the coupled system, the increased frequency of information exchange will place a greater strain on communication resources. (3) The increasing number of distribution networks and transportation systems will result in a significant increase in the iteration count [41].

This study focuses on implementing the analogous projection approach to address the previously mentioned limitations in PTNs. The strategy, initially presented in [42], aims to achieve system reduction and has been proven to provide the same level of optimality as the primal model. Moreover, the technique is utilised in [43] to efficiently synchronise the optimisation process between the transmission and distribution of electrical power networks. In [44], the authors successfully characterised the charging power area of EVCS using this method. However, they have not yet included the projection of optimal cost information in PDNs.

The aforementioned approaches are specifically employed to address the interconnections of power systems. However, the interactions of traffic systems are considerably more complex, involving factors such as the unpredictability of individual behaviour of EVs and the multitude of road pathways. In order to achieve this goal, this study aims to create a thorough modelling framework using network equivalent projection that includes the schedulable capacity of EVCSs and LMPs. This framework will enable effective interaction and allow for reaching the optimal operating point with minimal information in PTNs. This paper presents the following primary contributions:

1. A novel non-iterative coordinated optimisation method for PTNs is created using network equivalent projection. The coupled networks incorporate the schedulable capacity of EVCSs and dynamic LMPs at charging stations, and map this information into the feasible region of boundary information for PDNs. Implementing this suggested methodology can safeguard data confidentiality without the need to share sensitive information.

2. A technique is introduced to map the optimal cost function in the PDN. The segmented cost function of the PDNs is produced by specifically addressing the economic dispatch model and the Karush–Kuhn–Tucker (KKT) conditions. The original linked model of the PTN is substituted with integrated equivalent restrictions and segmented cost functions incorporating the traffic model. This proposed approach will guarantee manageable computation in the PTN.

The subsequent sections of this work are structured in the following manner. The coordinated optimisation model of the PTNs is established in Section 2. Section 3 introduces a decoupled model of the PTNs based on the feasible region of the PDNs. This model is created using network equivalent projection and takes into account the schedulable capacity of EVCS and LMPs. A unique optimal cost mapping methodology for the PDN is devised in Section 4. A case study is conducted to validate the effectiveness and advantages of the suggested approach outlined in Section 5. The conclusions are presented in Section 6.

2. Power–Traffic Coupled Model Equation

2.1. Interaction of the Coupled PTNs

The interconnection between PDNs and UTNs is strongly interconnected as a result of EVs. The EVCS recharge facilities facilitate the interplay between the flow of power and the flow of traffic. Figure 1 illustrates the operational framework of the interconnected power–traffic system. Pricing methods employed by electricity and transportation system organisations have an impact on the driving and charging decisions made by vehicle users. EV drivers choose their driving routes and charging stations based on price signals that are limited by the PDN and road conditions, with the goal of minimising their travel expenses.

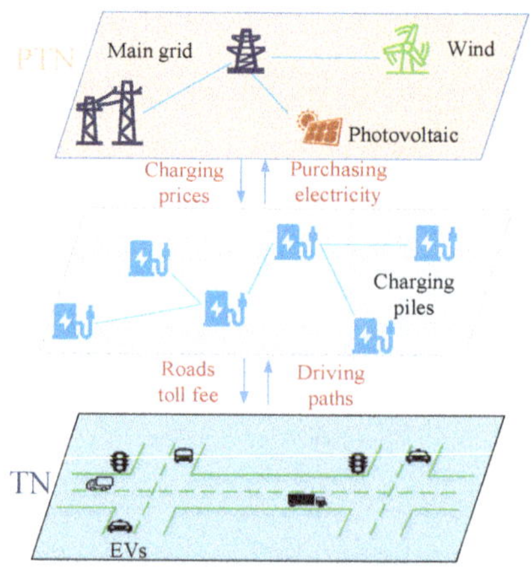

Figure 1. Operation framework of coupled power-traffic system.

2.2. *A Traffic Model for the Travel Characteristics of Vehicles*

From a graph theory standpoint, the fundamental components of the UTN are examined, with crossings being considered as nodes and road segments as connections. Moreover, the structure of the UTN is represented as (V, A), where V and A are collections of sequentially numbered intersections and road segments, respectively. An O-D pair represents the route used by electric vehicles (EVs) from a starting point to a destination, which indicates the traffic demand. Based on this premise, the traffic assignment problem is transformed into the calculation of traffic flow on various paths in a UTN, with each origin–destination pair specified.

An investigation is conducted to examine the travel characteristics of various EVs in order to meet the charging needs of EVs.

2.2.1. Road Congestion Analysis Based on Various Types of EVs

The structure of the transportation network consists of interconnected links and nodes. The links symbolise several paths or routes, while the nodes indicate the starting points, ending points, and points where different paths overlap. In order to model the travel patterns of EVs, the links in the UTN will be split into three equal segments: charging links, ordinary links, and bypass links [6].

1. Charging links with EVCSs

The travel time of EVs on a charging link is determined by the combined factors of charging time and queueing time. Hence, the improved Davidson function considering queueing theory is employed to quantify the travel time $t_a^c(x_a)$ of charging EVs, i.e.,

$$t_a^c(x_a) = t_a^{FCS}\left[1 + J\left(\frac{x_a}{c_a^{FCS} - x_a}\right)\right], \ \forall a \in T^C(A) \tag{1}$$

where t_a^{FCS} represents the free traveling time in the charging oracle, i.e., the charging time. x_a is the traffic flow of the a-th link. c_a^{FCS} is the traffic capacity of the a-th charging segment. J represents the parameter controlling the shape of the congestion function in the charging segment. $T^C(A)$ denotes the set of all charging links in UTNs.

2. Regular links without EVCSs

The widely used Bureau of Public Road (BPR) function is adopted in this work to represent the time spent on a regular link, i.e.,

$$t_a^R(x_a) = t_a^0\left[1 + 0.15\left(\frac{x_a}{c_a}\right)^4\right], \ \forall a \in T^R(A) \tag{2}$$

where t_a^0 is the free traveling time of the a-th link. c_a is the traffic capacity of the a-th link. $T^R(A)$ denotes the set of all regular links in UTNs.

3. Bypass links

A bypass link denotes a road segment with an EVCS where EVs skip and continue to drive without interruption. Due to the short length of the bypass oracle, it can be assumed that the travel time through this link is negligible and approximated as zero, i.e.,

$$t_a^b(x_a) = 0, a \in T^B(A) \tag{3}$$

where $T^B(A)$ is the set of all bypass links. $T(A)$ is the set of all links in UTNs: $T^C(A) \cup T^R(A) \cup T^B(A) = T(A)$.

2.2.2. Modelling the Costs of Vehicles Based on Different Driving Behaviours

Based on the above different types of congestion in UTNs, the travel cost functions of different paths are established for various transportation participants. For EVs with charging, a feasible path must include at least a charging station. The set of feasible paths K_{od} is expressed as

$$K_{od} = K_{od}^C \cup K_{od}^R \tag{4}$$

where K_{od}^C, K_{od}^R denote feasible path sets for EVs with recharging and regular vehicles, respectively.

1. The travel costs of EVs with charging

To estimate the travel cost of EVs with recharging, it is essential to consider three key components: the monetary value of time spent driving on regular links, the queueing time in EVCSs, and the charging cost. Accordingly, the travel time t_k^{od} and cost c_k^{od} of an EV with charging on the feasible path k-th are, respectively, represented by

$$t_k^{od} = \sum_{a \in T^C(A)} t_a^c(x_a)\delta_{a,k}^{od} + \sum_{a \in T^R(A)} t_a^c(x_a)\delta_{a,k}^{od}, \ \forall k \in K_{od}^C, \forall(o,d) \tag{5}$$

$$c_k^{od} = \omega t_k^{od} + \sum_{a \in T^C(A)} \left(\lambda_a^j P_{FCS} t_a^{FCS}\right)\delta_{a,k}^{od}, \ \forall k \in K_{od}^C, \forall(o,d) \tag{6}$$

where t_k^{od} and c_k^{od} denote the travel time and cost of a feasible path k between the O-D pair, respectively. ω is the unit travel cost coefficient. λ_a^j represents the charging price in EVCS a supplied by node j in a PDN. P_{FCS} is the charging power of EVs.

2. The travel costs of regular EVs

The total cost of regular vehicles is calculated by factoring in only the monetary value of time spent driving on regular links, i.e.,

$$t_k^{od} = \sum_{a \in T^R(A)} t_a^R(x_a)\delta_{a,k}^{od}, \ \forall k \in K_{od}^R, \forall(o,d) \tag{7}$$

$$c_k^{od} = \omega t_k^{od}, \ \forall k \in K_{od}^R, \forall(o,d) \tag{8}$$

2.2.3. A Traffic Model Based on User Equilibrium

As the charging behaviours of EV users affect the feasible path set and cause divergence from the path choices of non-charging users, it is vital to express the user equilibrium (UE) condition more explicitly. On this basis, an EV charging equilibrium model originating from [6] is utilised to characterise the equilibrium state in the UTN, i.e.,

$$\min \sum_{t \in T(T)} \left(\sum_{a \in T^R(A)} \int_0^{x_a} \omega t_a^R(\theta) d\theta + \sum_{a \in T^C(A)} \int_0^{x_a} \omega t_a^c(\theta) d\theta \right) \tag{9}$$

$$f_{k,t}^{od} \geq 0, \forall t \in T, \forall k \in K^{od}, \forall od \in (O, D) \tag{10}$$

$$r_t^{od} = \sum_{k \in K_{od}} \frac{f_{k,t}^{od} t_{k,t}^{od}}{\tau^{\mathrm{RF}}}, \forall t \in T, \forall k \in K \tag{11}$$

$$q_{od,t}^{mod} = q_{od,t} + \frac{1}{2} r_{t-1}^{od} - \frac{1}{2} r_t^{od}, \forall t \in T, \forall od \in (O, D) \tag{12}$$

$$\sum_{k \in K_{od}^C} f_{k,t}^{od} = \mu q_{od,t}^{mod}, \forall t \in T, \forall od \in (O, D) \tag{13}$$

$$\sum_{k \in K_{od}^R} f_{k,t}^{od} = (1 - \mu) q_{od,t}^{mod}, \forall t \in T, \forall od \in (O, D) \tag{14}$$

$$x_{a,t} = \sum_{o \in O} \sum_{d \in D} \sum_{k \in K_{od}} f_{k,t}^{od} \delta_{a,k}^{od}, \forall t \in T, \forall a \in T(A) \tag{15}$$

where f_k^{od} is the traffic flow on the k-th path connecting the O-D pair. q_{od} represents the total travel demand between the O-D pair. $\delta_{a,k}^{od}$ is a binary variable that represents the relationship between link a and path k connecting the O-D pair when $\delta_{a,k}^{od} = 0$ if link a is included in path k, and $\delta_{a,k}^{od} = 1$ otherwise. μ denotes the ratio of the number of EVs with charging to the total transportation demand.

In the given model, Equation (9) represents the objective of minimising the total cost of travel for traffic users. Equation (10) ensures that the traffic flow remains non-negative. Equations (11) and (12) describe the temporal relationship between traffic flow by incorporating the remaining flow from path k in the previous period into the traffic demand in the next period. Equations (13) and (14) represent the balance between traffic demand and path flow for traffic users on the feasible path k. Equation (15) states that the traffic flow in link a is equal to the sum of the traffic flows on all paths passing through this oracle.

2.3. Modelling of the Optimal Power Flow in the PDN

A radial PDN adopted in this part is represented by a directed graph (N, L), where N denotes the set of nodes and L means the set of branches. The initial node designated as $\{1\}$ is connected to the transmission power grid and purchases electricity directly from the main grid. The other nodes can be numbered sequentially as $N^+ = \{2, \ldots, n\}$, and so on. $(i, j) \in L$ corresponds to a branch from node i to node j. The sets N_{FCS} denote the collection of nodes that are connected to the EVCS. $\Phi(i)$ refers to the set of sub-nodes that are connected to node i, while $\Pi(j)$ means the set of nodes that stem from node j. The optimal power flow is regarded as Equations (16)–(24), in which the subscript t represents the time interval.

$$\min \sum_{t \in T} \left(\underbrace{\sum_{j \in N^+} \left(a_j (P_{DG,jt})^2 + b_j P_{DG,jt} \right)}_{\mathrm{i}} + \underbrace{\lambda_{\mathrm{MAIN},t} \sum_{j \in \Phi(1)} P_{1jt}}_{\mathrm{ii}} \right) \tag{16}$$

s.t.

$$\sum_{i=1}^{Ng} P_{Gi,t} - \sum_{i=1}^{Ne} P_{EVCSi,t} = P_{loss,t} + P_{L,t}, \forall t \in T \tag{17}$$

$$P_{i,t} - V_{i,t} \sum_{j=1} V_{j,t} \left(\frac{\cos \delta_{ij,t}}{r_{ij}} + \frac{\sin \delta_{ij,t}}{x_{ij}} \right) = 0, \forall t \in T \tag{18}$$

$$Q_{i,t} - V_{i,t} \sum_{j=1} V_{j,t} \left(\frac{\sin \delta_{ij,t}}{r_{ij}} - \frac{\cos \delta_{ij,t}}{x_{ij}} \right) = 0, \forall t \in T \tag{19}$$

$$P_{i,t}^2 + Q_{i,t}^2 \leq \left(S_{ij,t}^{MAX} \right)^2, \forall t \in T \tag{20}$$

$$0 \leq P_{DG,jt} \leq P_{DG,j}^{\max}, \ \forall j \in N^+ \quad \forall t \in T \tag{21}$$

$$0 \leq Q_{DG,jt} \leq Q_{DG,j}^{\max}, \ \forall j \in N^+ \quad \forall t \in T \tag{22}$$

$$V_{\min} \leq V_{j,t} \leq V_{\max}, \ \forall j \in N \quad \forall t \in T \tag{23}$$

$$P_{EVCS}^{\min} \leq P_{EVCSi,t} \leq P_{EVCS}^{\max} \ \ \forall t \in T \tag{24}$$

where $P_{1,jt}$ is the power purchased from the main grid. T denotes the entire scheduling period. a_j and b_j represent the cost coefficient of the controllable generations connected to node j in the PDN. $P_{DG,j}$ and $Q_{DG,j}$ refer to the active and reactive power output of the controllable generation at node j. λ_{MAIN} represents the purchasing electricity price from the main grid. P_i and Q_i denote the injective active and reactive power at node i. P_{ij} and Q_{ij} refer to the active and reactive power transmitted in the branch (*i,j*), respectively. r_{ij} and x_{ij} represent the equivalent resistance and reactance of the branch (I, j). V_j denotes the voltage magnitude at node j. δ_j is the voltage phase angle at node j. $P_{L,j}$ and $P_{loss,t}$ are conventional active loads and loss power at node j, respectively. $P_{EVCS,j}$ represent the charging loads of EVCSs located at node j. $P_{DG,j}^{\max}$ and $Q_{DG,j}^{\max}$ represent, respectively, the upper bound of the distributed units' active and reactive power output. V_{max} and V_{min} are the upper and lower bounds of the nodal voltage magnitudes. $P_{EVCS}^{\max}$ and $P_{EVCS}^{\min}$ represent the active power limits in the EVCSs, respectively.

The objective function of Equation (16) represents the minimum operational cost of the PDN, including (i) the generation cost of controllable distributed generation units and (ii) the purchase cost of electricity from the main grid. Equation (17) denotes the active power balance of the PDN. Equations (18) and (19) denote the power flow balance of the PDN. Equation (20) represents the active and reactive power transmission limits of branches. Equations (21) and (22) denote controllable generation capacity in the PDN. Equation (23) represents the nodal voltage magnitude bounds. Equation (24) is the capacity limit of EVCSs.

The power flow balance constraints (18) and (19) can be expressed using the linearization derivation from [45] in the following manner:

$$P_{i,t} \approx \sum_{j=1}^{n} \frac{V_{j,t}}{r_{ij}} - \sum_{j=1}^{n} \frac{\delta_{j,t}}{x_{ij}}, \forall t \in T \tag{25}$$

$$Q_{i,t} \approx -\sum_{j=1}^{n} \frac{V_{j,t}}{x_{ij}} - \sum_{j=1}^{n} \frac{\delta_{j,t}}{r_{ij}}, \forall t \in T \tag{26}$$

We employ the linearization method based on polygon approximation in [46] to address the branch transmission limitations. According to Equation (20), it is evident that there are many power circles being considered. Thus, it is capable of approximating the aforementioned power circles by using polygons that have a limited number of edges. Thus, Equation (20) can be converted into:

$$\overline{P}_{ij,t}P_{ij,t} + \overline{Q}_{ij,t}Q_{ij,t} \leq \left(S_{ij,t}^{\max}\right)^2, \forall t \in T \tag{27}$$

$$\overline{P}_{ij,t} = \cos\left[\left(p - \frac{E+4}{4}\right)\frac{2\pi}{E}\right], \forall p \in \left[1, 2, \ldots, \frac{E}{2}\right], \forall t \in T \tag{28}$$

$$\overline{Q}_{ij,t} = \sin\left[\left(p - \frac{E+4}{4}\right)\frac{2\pi}{E}\right], \forall p \in \left[1, 2, \ldots, \frac{E}{2}\right], \forall t \in T \tag{29}$$

where E represents the number of edges of the polygon approximating the power circles. The value of E can be chosen to be between 8 and 20, considering the trade-off between computational efficiency and accuracy. After evaluating the relationship between efficiency and precision, an inscribed regular dodecagon (i.e., a polygon with twelve edges) is opted to replace the power circles.

2.4. Modelling of the Coupled PTN

This section examines the interconnected boundary between the power and transportation systems, specifically focusing on the independent operation models for the UTN and the PDN mentioned earlier. The goal is to understand the relationship between the boundary information of these two systems. To achieve this, a coordinated scheduling model for the power–traffic coupling system is developed. More precisely, the interaction in a PTN occurs when EVs transfer energy by charging over the links connecting to the PDN for additional electrical power. Thus, in the PTN, the boundary information of the UTN pertains to the traffic flow of EVs entering the charging stations, while the boundary information on the PDN corresponds to the charging load of the EVCS. The correlation between the charging load linked to the PTN node j and the traffic flow in the UTN is represented as:

$$P_{FCS,j} = P_{FCS} \sum_{o\in O} \sum_{d\in D} \sum_{k\in K_{od}^{C}} f_k^{od} \delta_{a,k}^{od}, \ \forall a \in T^C(A), j \in N_{FCS} \tag{30}$$

The coordinated scheduling in the PTN is aimed at the minimum social cost, i.e.,

$$\min \sum_{a\in T^R(A)} \int_0^{x_a} \omega t_a^R(\theta)d\theta + \sum_{a\in T^C(A)} \int_0^{x_a} \omega t_a^c(\theta)d\theta + \sum_{t\in T}\left(\sum_{i\in N^+}\left(a_j(P_{DG,jt})^2 + b_j P_{DG,jt}\right) + \lambda_{\mathrm{MAIN}} \sum_{j\in\Phi(1)} P_{1j,t}\right) \tag{31}$$

s.t.

$$\begin{cases} (10)-(15) \\ (17),(21)-(29) \end{cases} \tag{32}$$

3. Decoupled Model of the PTN Based on the Feasible Region of the PDN

Definition 1 (schedulable capacity (SC)): *During the EV charging process, the PDN establishes a zone that includes all the possible operating locations of the charging loads. This region ensures stability and security by meeting the restrictions specified in (35).*

To facilitate the understanding of the model derivation, a concise representation of the PTN, comprising Equations (31) and (32), is provided.

$$\min \sum_t \left(C_{PDN,t}\left(y_{PDN,t}\right) + C_{UTN,t}\left(y_{UTN,t}^a\right)\right) \tag{33}$$

$$\text{s.t.} \begin{cases} h_{PDN,t}^{eq}\left(y_{PDN,t}, u_t, P_t^{EVC}\right) = 0 \\ h_{PDN,t}^{ineq}\left(y_{PDN,t}, u_t, P_t^{EVC}\right) \leq 0 \end{cases} \tag{34}$$

$$g_{UTN,t}\left(y_{UTN,t}^{a}, f_{k,t}^{od}\right) \leq 0 \tag{35}$$

$$\boldsymbol{A}_{CPT} P_t^{EVC} + \boldsymbol{B}_{CPT} \boldsymbol{y}_{UTN,t}^{a} = \boldsymbol{c}_{CPT,t} \tag{36}$$

The constraints of Equation (34) denote the sets formed by the equality Equations (17), (25) and (26), and inequality constraints Equations (21)–(24) and (27)–(29), respectively. The constraints of Equation (35) denote the constraint sets formed by the UTN in Equations (10)–(15). The $C_{PDN,t}$ denotes the operational cost function of the PDN. $y_{PDN,t}$ is the power vector injected from controllable generators and the main grid. u represents the vector of the state variable in the PDN. P_t^{EVC} is the EV maximum charging power from the PDN, which comprises the charging load and the available charging power. $C_{UTN,t}$ is the cost function of the UTN. $\boldsymbol{y}_{UTN,t}^{a}$ is a vector composed of the column vectors traffic flow x_a and time t_a, i.e., $\boldsymbol{y}_{UTN,t}^{a} = [x_a, t_a]$. A_{CPT}, B_{CPT}, and c_{CPT} represent coefficient vectors corresponding to the PDN and the UTN in Equation (30), respectively. P_{EVC} is the charging power of EVCS.

The polyhedral space Ω_{SC}^{pri} formed by the constraint of Equation (34) is denoted as follows, where the subscript t is neglected for simplicity:

$$\Omega_{SC}^{pri} = \left\{ \left(y_{PDN}, P^{EVC}\right) \in R^m \times R^n \middle| \begin{array}{l} h_{PDN}^{eq}\left(y_{PDN}, u\,, P^{EVC}\right) = 0 \\ h_{PDN}^{ineq}\left(y_{PDN}, u\,, P^{EVC}\right) \leq 0 \end{array} \right\} \tag{37}$$

where m and n denote the dimensions.

According to definition 1, it is shown that the schedulable capacity is interpreted as a projection from the constraints of space (y_{PDN}, u, P^{EVC}) to P^{EVC} in the PDN. The projection area Ω_{SC}^{map} is stated as

$$\Omega_{SC}^{map} = \left\{ \left(P^{EVC}\right) \in R^n \middle| \exists y_{PDN},\ \left(y_{PDN}, u\,, P^{EVC}\right) \in \Omega_{SC}^{pri} \right\} \tag{38}$$

where Ω_{SC}^{map} describes the range of EV charging loads accommodated by the PDN at any period t without violating safety operation constraints, as shown in Equation (34).

This polyhedron, i.e., the boundary information feasibility region, is expressed as:

$$\Omega_{EVC}^{\mathrm{ap}} = \left\{ \boldsymbol{A}_{EV} P_k^{EVC} \leq \boldsymbol{\beta}_{EV} \right\} \tag{39}$$

where A_{EV} is the coefficient matrix for different stress directions; and $\boldsymbol{\beta}_{EV}$ is a coefficient vector that describes the boundary of the EVSC. The dimensions of $\boldsymbol{A}_{EV}$ and $\boldsymbol{\beta}_{EV}$ correspond to the number of boundary points characterising the EVSC. Each determined boundary point leads to a set of constraints.

Figure 2 presents the coordinated operation framework of the coupled power–traffic system. The equivalent projection model of the PTN is expressed as:

$$\min\left(C_{PDN}\left(P_k^{EVC}\right) + C_{UTN}\left(\boldsymbol{y}_{UTN}^{a}\right)\right) \tag{40}$$

$$\boldsymbol{A}_{EV} P_k^{EVC} \leq \boldsymbol{\beta}_{EV} \tag{41}$$

$$g_{UTN}\left(\boldsymbol{y}_{UTN}^{a}, f_k^{od}\right) \leq 0 \tag{42}$$

$$\boldsymbol{A}_{CPT} P_k^{EVC} + \boldsymbol{B}_{CPT} \boldsymbol{y}_{UTN}^{a} = \boldsymbol{c}_{CPT} \tag{43}$$

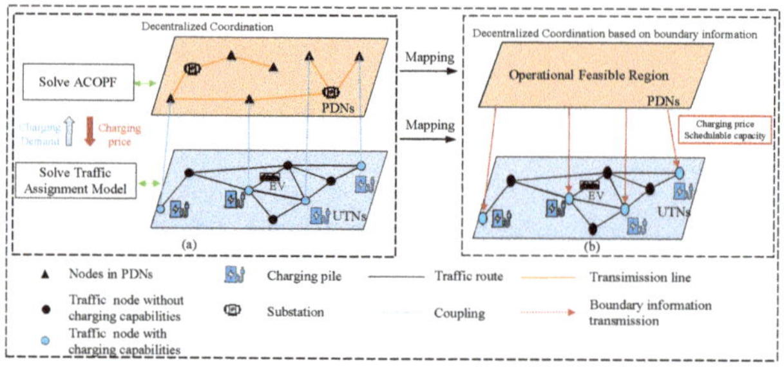

Figure 2. Coordinated operation framework of the coupled power–traffic system.

4. Mapping Optimal Costs of the PDNs

When it comes to connections including boundary EVCSs, the PTN aims to optimise the power and traffic flows. From the perspective of the PDN, charging stations are classified as variable loads since they consume electrical energy from the grid. Alternatively, when considering the transportation network, these stations can be seen as virtual sources that provide electric power to cars. The fundamental coupled model in the PTN is reconstructed and expressed as a result of boundary charging power.

$$
\begin{array}{c}
\min C_{PDN}(y_{PDN}) + C_{UTN}(y^{a}_{UTN}) \\
\text{s.t.} \quad \left\{ \begin{array}{l} h^{eq}_{PDN}(y_{PDN}, u, P^{EVC}) = 0 \\ h^{ineq}_{PDN}(y_{PDN}, u, P^{EVC}) \leq 0 \\ g_{UTN}\left(y^{a}_{UTN}, f^{od}_{k}\right) \leq 0 \end{array} \right. \\
\boldsymbol{A}_{PDN} \boldsymbol{y}_{PDN} + \boldsymbol{D}_{PDN} P^{EVC}_{k} = \boldsymbol{c}_{PDN} \\
\boldsymbol{B}_{UTN} \boldsymbol{y}^{a}_{UTN} + \boldsymbol{D}_{UTN} P^{EVC}_{k} = \boldsymbol{c}_{UTN}
\end{array} \tag{44}
$$

where vector D denotes the charging power at a charging station.

The equivalent model consisting of Equations (33)–(36) implies that each PDN and the UTN perform optimal social welfare individually with limited data exchange. Hence, the model of the PDN at the time t is reformulated and expressed as

$$
\begin{array}{c}
f_{PDN}(P^{EVC}_{k}) = \min C_{PDN}(y_{PDN}) \\
\text{s.t.} \quad \left\{ \begin{array}{l} h^{eq}_{PDN}(y_{PDN}, u, P^{EVC}) = 0 \\ h^{ineq}_{PDN}(y_{PDN}, u, P^{EVC}) \leq 0 \\ \boldsymbol{A}_{PDN} \boldsymbol{y}_{PDN} + \boldsymbol{D}_{PDN} P^{EVC}_{k} = \boldsymbol{c}_{PDN} \end{array} \right.
\end{array} \tag{45}
$$

where the objective function of $f_{PDN}(P^{EVC}_{k})$ means the optimal social cost of the PDN related to the boundary charging power P^{EVC}_{k}.

It has been demonstrated that there is an equivalence in optimality between the joint optimisation Equations (45) and (46), i.e.,

$$
\begin{array}{c}
\min\left(f_{PDN}(P^{EVC}_{k}) + C_{UTN}(y^{a}_{UTN})\right) \\
\text{s.t.} \quad \left\{ \begin{array}{l} g_{UTN}\left(y^{a}_{UTN}, f^{od}_{k}\right) \leq 0 \\ \boldsymbol{B}_{UTN} \boldsymbol{y}^{a}_{UTN} + \boldsymbol{D}_{UTN} P^{EVC}_{k} = \boldsymbol{c}_{UTN} \end{array} \right.
\end{array} \tag{46}
$$

From the modified model (see Equation (46)), it is crucial to derive the specific form of the function $C_{PDN}(D_{PDN})$. According to Equation (16), the form of the objective function in PDN is quadratic, which is converted and stated as:

$$
F_{PDN}(P_{DG}) = \frac{1}{2} P_{DG} A^{T}_{\text{Re}} P_{DG} + B^{T}_{\text{Re}} P_{DG} + \lambda_{MAIN} P_{1} \tag{47}
$$

where vector P_{DG} and P_1 denote all $P_{DG,j}$ and P_{1j}, respectively. The syntax $A_{\mathrm{Re}}^T = (a_1, \cdots, a_i)^T$ and $B_{\mathrm{Re}}^T = (b_1, \cdots, b_i)^T$ are utilised to catenate column vectors with the cost coefficient of the controllable generators.

With the charging power given, the optimal objective in Equation (47) is computed, and the solution corresponds to an optimal value, where the constraints are categorised into active and inactive ones. These constraints divided are rewritten, i.e.,

$$(\boldsymbol{A}_{PDN})_a P_{DG} + (\boldsymbol{D}_{PDN})_a P^{EVC} = (\boldsymbol{c}_{PDN})_a \tag{48}$$

$$(\boldsymbol{A}_{PDN})_{ina} P_{DG} + (\boldsymbol{D}_{PDN})_{ina} P^{EVC} \leq (\boldsymbol{c}_{PDN})_{ina} \tag{49}$$

where Equation (48) means the active constraints, denoted by subscript a. Additionally, Equation (49) is the inactive constraints with subscript ina.

From the perspective of optimisation theory, the optimal solution of the model is not affected by inactive constraints. Hence, Equation (46) is modified equivalently, i.e.,

$$\begin{gathered} \min F_{PDN}(P_{DG}) \\ \text{s.t. } (\boldsymbol{A}_{PDN})_a P_{DG} + (\boldsymbol{D}_{PDN})_a P^{EVC} = (\boldsymbol{c}_{PDN})_a \end{gathered} \tag{50}$$

The Lagrange function is expressed as:

$$L(P_{DG}, \mu) = \frac{1}{2} P_{DG} A_{\mathrm{Re}}^T P_{DG} + B_{\mathrm{Re}}^T P_{DG} + \lambda_{MAIN} P_1 + \eta^T \left((\boldsymbol{A}_{PDN})_a P_{DG} + (\boldsymbol{D}_{PDN})_a P^{EVC} - (\boldsymbol{c}_{PDN})_a \right) \tag{51}$$

By applying complementary slackness conditions of KKT theory,

$$\frac{\partial L(P_{DG}, \mu)}{\partial P_{DG}} = A_{\mathrm{Re}}^T P_{DG} + B_{\mathrm{Re}}^T + (\boldsymbol{A}_{PDN})_a^T \mu = 0 \tag{52}$$

Simultaneously ensuring the satisfaction of the active constraints of Equation (49) gives:

$$\begin{pmatrix} A_{\mathrm{Re}}^T & (\boldsymbol{A}_{PDN})_a^T \\ (\boldsymbol{A}_{PDN})_a & 0 \end{pmatrix} \begin{pmatrix} P_{DG} \\ \eta \end{pmatrix} = \begin{pmatrix} -B_{\mathrm{Re}}^T \\ (\boldsymbol{c}_{PDN})_a - (\boldsymbol{D}_{PDN})_a P^{EVC} \end{pmatrix} \tag{53}$$

By solving Equation (53), the values P_{DG}, η are

$$\begin{cases} P_{DG} = (\boldsymbol{A}_{PDN})_a^{-1} \left((\boldsymbol{c}_{PDN})_a - (\boldsymbol{D}_{PDN})_a P_k^{EVC} \right) \\ \eta = -A_{\mathrm{Re}}^T \left((\boldsymbol{A}_{PDN})_a (\boldsymbol{A}_{PDN})_a^T \right)^{-1} \left((\boldsymbol{c}_{PDN})_a - (\boldsymbol{D}_{PDN})_a P^{EVC} \right) \\ \quad - \left((\boldsymbol{A}_{PDN})_a^T \right)^{-1} B_{\mathrm{Re}}^T \end{cases} \tag{54}$$

There exists a linear relationship between the active power from the main grid and the charging power (i.e., P^{EVC}). It is shown that f_{PDN} (P^{EVC}) is a quadratic function.

Nonetheless, the above derivation process should be performed in the neighbour interval of the given charging power value (i.e., $\hat{P}^{EVC}$), and the active and inactive constraints in Equations (49) and (50) remain unchanged. Hence, it is essential to analyse the conditions affecting the neighbour interval.

Condition 1: Lagrange multipliers η remain non-negative, ensuring that the inequality constraints of the problem Equation (49) are satisfied, i.e.,

$$\eta \geq 0 \tag{55}$$

Condition 2: To guarantee the feasibility of Equation (47), the solution of Equation (51) must satisfy the inactive constraints of Equation (49).

To obtain the optimal cost function of the PDN under the feasible region, a piecewise approach is proposed. Specifically, based on the constraints set in Equation (46), the upper and lower bounds of the charging power (i.e., P^{EVC}) are expressed as

$$\begin{cases} \underline{P}^{EVC} = \min P^{EVC} \\ \overline{P}^{EVC} = \max P^{EVC} \end{cases}$$
$$\text{s.t.} \begin{cases} h_{PDN}^{eq}(\boldsymbol{y}_{PDN}, u, P^{EVC}) = 0 \\ h_{PDN}^{ineq}(\boldsymbol{y}_{PDN}, u, P^{EVC}) \leq 0 \\ \boldsymbol{A}_{PDN}\boldsymbol{y}_{PDN} + \boldsymbol{D}_{PDN}P^{EVC} = \boldsymbol{c}_{PDN} \end{cases} \tag{56}$$

where $\underline{P}^{EVC}, \overline{P}^{EVC}$ denote the upper and lower bounds of the charging power, respectively.

After conducting the analysis mentioned above, it is clear that the deduction can only be valid within a neighbour interval. To achieve this, the feasible region interval $\left[\underline{P}^{EVC}, \overline{P}^{EVC}\right]$ is divided into multiple sub-intervals, with each sub-interval determined by conditions 1 and 2. One significant advantage of determining the interval width based on active and inactive constraints is the ability to enumerate all sub-intervals, which satisfies the following constraint, i.e.,

$$d_{PDN}^{m-1} \leq d_{PDN}^{m} \tag{57}$$

where the m-th upper bound denotes d_{PDN}^{m}, and the m-th sub-interval is expressed as $\left[d_{PDN}^{m-1}, d_{PDN}^{m}\right]$.

As the exploration of sub-intervals continues, the right endpoint d_{PDN}^{m} of the sub-interval is equal to the upper bound $\overline{P}^{EVC}$ of the feasible region, which is used as a stopping criterion to express the end of the exploration process. The specific form of the optimal cost in the PDN is expressed as

$$f_{PDN}(P^{EVC}) = \sum_{m \in M} f_{PDN}^{m}(d_{PDN}^{m}) \tag{58}$$

where M is the overall count of sub-intervals. $f_{PDN}^{m}(d_{PDN}^{m})$ denotes the optimal cost of the m-th sub-interval in the PDN.

According to the definition of LMPs, the charging price is

$$LMP = \frac{\partial f_{PDN}(P^{EVC})}{\partial P^{EVC}} \tag{59}$$

5. Case Study

5.1. Basic Settings

This part develops a connected electric–transportation system for the purpose of conducting simulation analysis. Figure 3 displays the topology of the traffic network. There are a total of six EVCSs and four distinct types of highways in both the outer and inner loops. The trip demand, free travel time, and traffic capacity for each origin–destination pair (O-D pair) are previously known, and specific parameter configurations are supplied in Tables 1 and 2. The EVs have an average charging power of 50 kW, and it is estimated that the average charging time is 30 min. The journey time is valued at USD 10 per hour.

Table 1. Link parameters of roads in the 12-node transportation network.

Road	Type 1	Type 2	Type 3	Type 4	EVCS
$c_a(p.u)$	100	100	80	60	15
$t_a^0(min)$	5	8	5	7	20

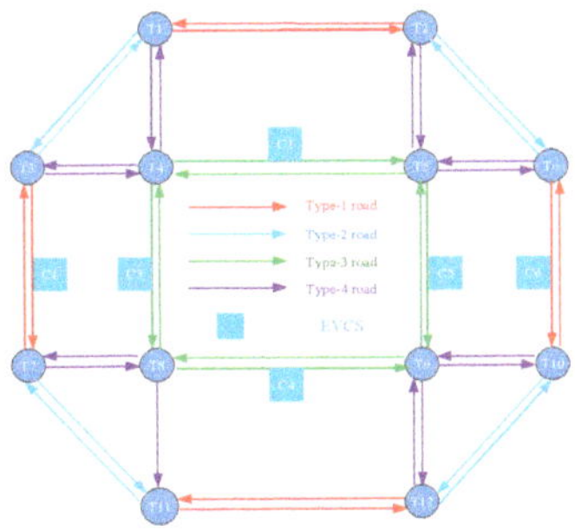

Figure 3. Topology of the UTN.

Table 2. O-D pairs and trip rates of t = 1 (in p.u.).

O-D Pair	q_g^{rs}	q_e^{rs}
T1–T6	9	1
T1–T10	36	4
T1–T11	18	2
T1–T12	27	3
T3–T6	9	1
T3–T10	27	3
T3–T11	18	2
T3–T12	27	3

Figure 4 displays the configuration of the updated IEEE 33-bus system. Node 1 was linked to the primary power network for the purpose of procuring electricity. Nodes 8, 15, and 31 were linked to electric vehicle (EV) charging stations, which have a maximum charging capacity of 400 kW. Controllable generations with voltage regulation facilities were connected to nodes 18 and 33. The scheduling period was fixed at 24 h. The simulations in this part were performed on the lenovo Y9000P laptop equipped with an AMD Ryzen 7 5800H processor with Radeon Graphics, running at a clock speed of 3.20 GHz, and 16 GB of RAM. The models were solved using MATLAB R2021b and YALMIP toolboxes, with the solvers IBM CPLEX 12.8 and IPOPT.

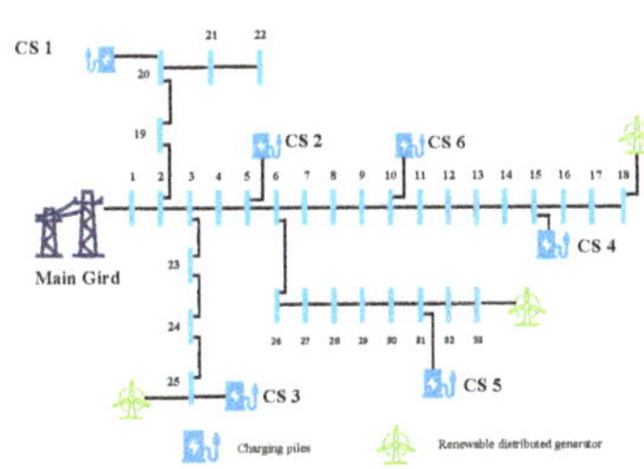

Figure 4. Topology of the PDN.

In order to assess the efficacy of the suggested method utilising equivalent projection, three specific scenarios were constructed for thorough investigation.

M1: The independent scheduling operation of the power–traffic network without coupling.

M2: The coordinated scheduling operation of the power–traffic network with the centralised method.

M3: The coordinated scheduling operation of the power–traffic network with equivalent projection.

5.2. Analysis Discussion

The cost of the power–traffic linked system is clearly lower than operating the power grid and traffic network separately, as shown in Figure 5. Furthermore, as the adoption of EVs continues to grow, the disparity between the overall cost of the integrated power–traffic system and the individual operation managed by M1 in the power–traffic networks becomes more pronounced. The effectiveness of coupled system scheduling in significantly reducing operational expenses is proved when compared to standalone scheduling. Furthermore, the overall expense of the power–traffic coupled network with M2 is equal to the one computed by M3 (refer to Figure 5). This suggests that by employing the cost function of the PDNs, the cost coupling in PTNs may be correctly aligned.

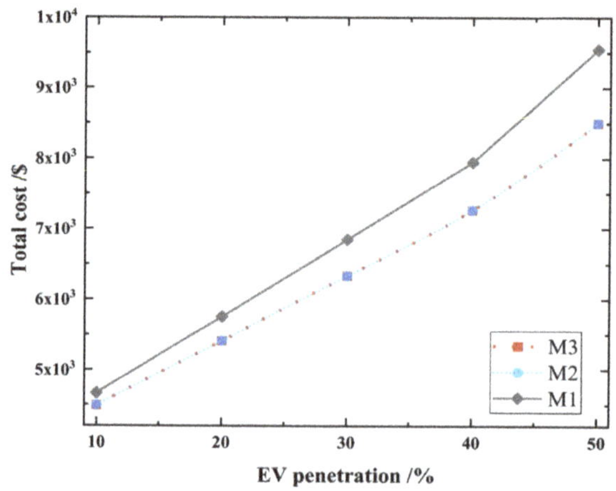

Figure 5. Total costs of the PTNs under different EV penetration levels.

Figure 6 examines the charging frequency of EVs at charging stations within the transportation network during the peak time (t = 10). In a scenario where there is a 10% penetration rate, the scheduling using M1 does not require a direct connection between the UTN and the PDN. Due to their higher capacity and lower waiting times, the majority of EVs prefer to charge at EVCS 2 and 6. The number of charging vehicles at Station 2 and Station 6 is 4.95 and 4.68, respectively, which together make up nearly 75% of the total charging vehicle count. The independent scheduling of EV charging results in uneven distribution of EV traffic flow, leading to high saturation levels at EVCS 2 and 6. Additionally, the paucity of charging stations at Stations 1, 4, and 5 exacerbates this issue. Unlike the uncoordinated scheduling, the coordinated scheduling of the power–traffic network efficiently reduces the concentrated distribution of EV traffic flow, leading to a more balanced traffic flow distribution and an improved utilisation rate for charging stations. Furthermore, it is important to mention that, regardless of the different levels of EV adoption, the number of times EVs need to be charged at charging stations, as calculated using M2 and M3, is consistent with the results obtained from the centralised coordinated scheduling of the power–traffic system.

Significantly, if the penetration rate falls below 40%, the aggregate demand for electric vehicles remains relatively modest, which does not pose any risk to the secure functioning limits of the power distribution networks. Nevertheless, the scenario will undergo a transformation once the EV adoption rate surpasses 40%. Figure 7 illustrates the voltage magnitudes of the PDNs when the EV penetration rate is 50%. Through a thorough examination of Figures 7 and 8, it becomes evident that EVs have a tendency to gather or cluster near EVCSs 1 and 4. Consequently, the voltage magnitudes at nodes 15 and 16 in the distribution network decrease to 0.9186 and 0.9190, respectively, indicating voltage violations in the power grid. The increased traffic in the road segments where charging stations are situated results in higher impact loads on the distribution grid nodes that are connected to the charging stations.

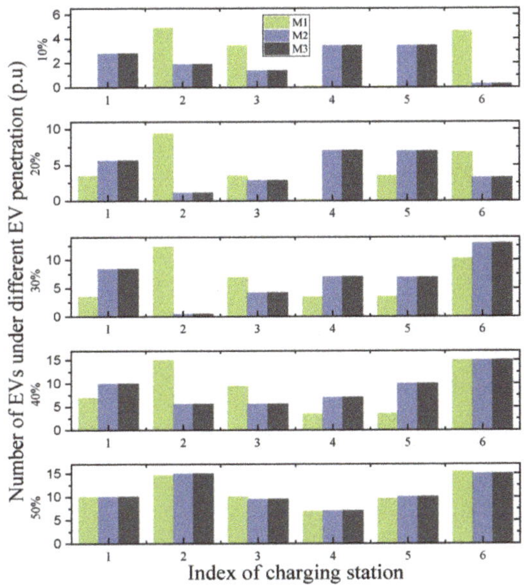

Figure 6. Distribution of EV charging numbers among EVCSs under different EV penetration levels.

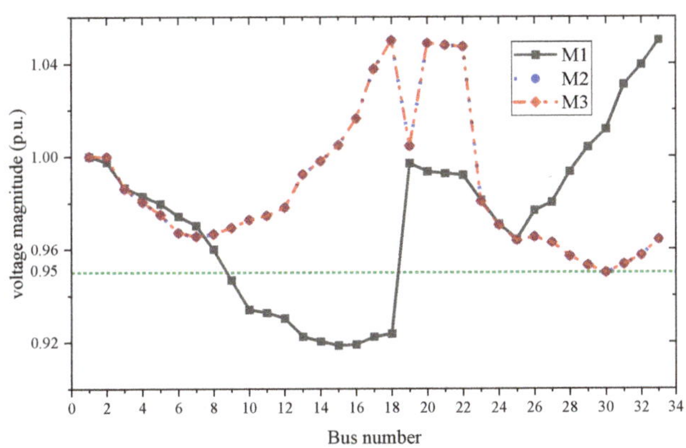

Figure 7. The voltage magnitude in the PDN when EVs penetrate 50%.

It is evident that there is an enhancement in the voltage magnitude, particularly at node 18, for M2 and M3. Additionally, the voltage distribution in the PDNs is effectively maintained. The coordinated strategy uses the charging price for the PDNs to incentivise EVs to charge at specific stations, such as EVCS 1 and 4. Therefore, it can efficiently mitigate traffic congestion and the low voltage problem at distribution network nodes generated by the independent scheme. This guarantees the secure and effective functioning of the power–traffic network.

After conducting a thorough examination of operational costs, EV charging numbers, and voltage magnitude, we evaluated the calculation time of three approaches in two common scenarios, peak and off-peak, to further examine their usefulness. Table 3 clearly demonstrates that M1 has the shortest computation time. Nevertheless, it fails to accomplish the most efficient functioning of the PTNs, which could potentially jeopardise the secure operating of the PDNs. As M2 approaches its optimal state, it heavily depends on the transmission of large amounts of data between the power and traffic systems. This raises concerns over the privacy of information and the practicality of its implementation. When comparing M2 with M3, it is evident that M3 achieves a better balance between maintaining system privacy and providing precise solutions. Additionally, M3 is able to provide accuracy even when working with little data.

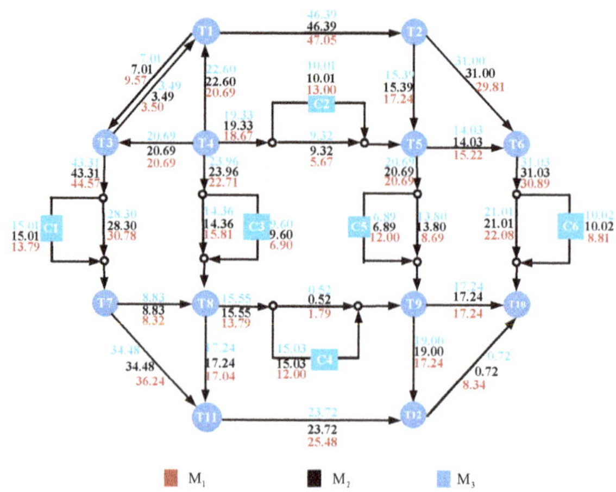

Figure 8. The traffic flow of the UTN when EVs penetrate 50%.

Table 3. Computation times of different methods.

Different EV Penetrations	Time of M_1 (s)	Time of M_2 (s)	Time of M_3 (s)
10%	14.42	46.14	169.34
20%	15.13	47.39	170.05
30%	15.98	49.36	171.96
40%	16.56	50.47	172.56
50%	17.26	52.30	173.31

Figure 8 displays the traffic patterns during peak hours (t = 10) in the UTN for in-depth examination. The figure reveals that the majority of vehicles opt to travel on the inner and outer loops due to their higher capacities and fewer traffic signals (t0) compared to other connections. The traffic flow of the charging link (T4–T5) in the coordinated system is 10.01 p.u., which indicates a 20% reduction compared to the autonomous scheme. The coordinated scheduling strategy considers the secure operation of the PDNs. By modifying the charging fees, voltage losses in the vicinity can be reduced, thereby diverting charging vehicles to other EVCSs. Furthermore, the traffic patterns achieved by M3 are identical to the centralised optimisation of the interconnected PTNs, which is solved by M2. This suggests that the use of boundary information mapping and cost functions can effectively replace the original power grid model with M3. By employing the suggested approach, the correctness of optimisation outcomes for the interconnected PTNs is guaranteed, while also safeguarding the confidentiality of sensitive system information.

5.3. Comparative Analysis of the Case Studies

This section examines the influence of various charging station connection sites on the power–traffic coupling system. Figure 9 illustrates the topology for scenario 2, which is distinct from the topologies shown in Figures 3 and 4. The parameter configurations align with those outlined in Section 5.2.

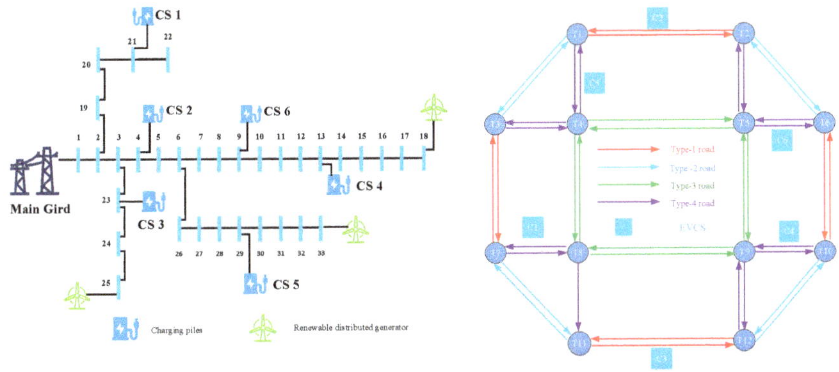

(a) (b)

Figure 9. (**a**) Comparative case topology of the PDN with different charging station connections. (**b**) Comparative case topology of the UTN with different charging station connections.

Figure 10 displays the overall expenses of the PTNs at various locations. In contrast to instance 1, the bulk of the charging stations are situated within or in close proximity to the outer circle. Figure 10 demonstrates that the overall operational expenses in case 2 are often more than those in case 1. The greater operational costs are caused by the positioning of charging stations along the outer loops and capacity, which forces EVs to detour from the optimal path and travel extra distances to access the charging stations. From these data, it can be inferred that placing charging stations along the outer ring has an adverse effect on the overall operational expenses of the EVs. In order to maximise cost-effectiveness, it is advisable to strategically position charging stations along the inner loop, so minimising any extra trip time.

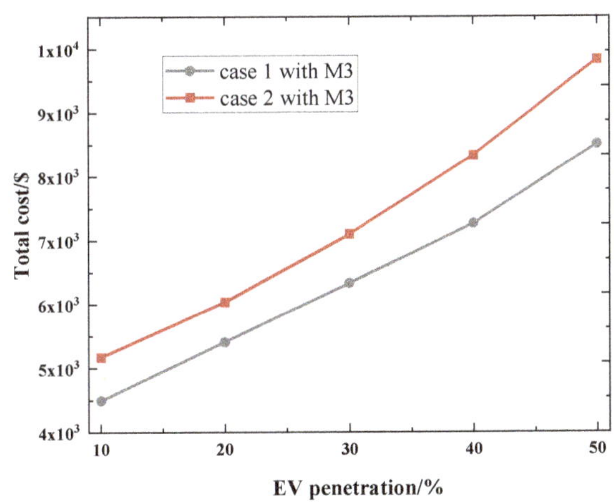

Figure 10. Comparison of total costs under different charging station locations.

Figure 11 illustrates the voltage magnitudes for various charging station sites when the EV penetration rate is 50%. Case 2 exhibits fewer voltage variations compared to case 1. In scenario 1, a substantial quantity of electric vehicles congregate at EVCS 5, which is situated within the inner circle. This efficiently decreases the amount of time spent travelling while still satisfying the need for travel. The voltage near node 30 approaches its limit as a result. In case 2, the charging stations are strategically placed in the outer ring or near the outer loop. This arrangement helps to decentralise the charging of EVs and minimise the strain on the PDN, resulting in a reduction in voltage fluctuations.

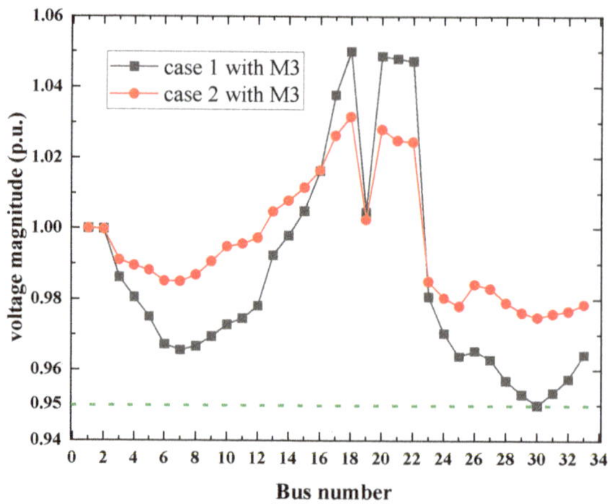

Figure 11. Comparison of voltage magnitudes under different charging station locations.

6. Conclusions

This research introduces a new coordinated scheduling technique for PTNs that addresses the limitations of previous coordinated optimisation methods. The method utilises boundary information mapping and implements a non-iterative framework for the PTNs. Using this proposed approach, the efficient functioning of the PDNs is translated into a mathematical equation and a safety operation set with minimum data, enabling the PDNs to attain external equivalence. The suggested model replaces the PDNs by utilising partial boundary information, allowing for coordinated optimisation of the PTNs. This approach eliminates the need for iterative solutions between the linked systems. By maintaining uniformity in the highest level of effectiveness, it simultaneously ensures confidentiality and protection across many systems. A case analysis is performed on a system that consists of a 12-node traffic network and an IEEE 33-bus system in order to verify the efficacy of the suggested method. The main discoveries are as follows: (1) The proposed method for mapping the feasible domain can effectively align the operational limitations of the power grid, enabling coordinated scheduling of the power–traffic network while safeguarding the confidentiality of power grid information. (2) In comparison to independently scheduling the two networks, the coordinated scheduling of the power–traffic networks can optimise the distribution of electric vehicle traffic flow and charging load in both the power grid and the traffic network, thereby enhancing the safety and efficiency of the power–traffic networks.

Author Contributions: Conceptualization, W.D.; Methodology, W.D.; Software, Z.Z. (Zhihong Zeng); Validation, Z.Z. (Zhihong Zeng); Investigation, Z.Z. (Zhihong Zeng) and J.X.; Resources, C.W.; Data curation, C.W. and Y.G.; Writing—original draft, Z.Z. (Zhihong Zeng); Writing—review & editing, W.D.; Supervision, Z.Z. (Zhijie Zhang); Funding acquisition, W.D. All authors have read and agreed to the published version of the manuscript.

Funding: This research was funded by the National Natural Science Foundation of China under grant No.52107082 and the Specific Research Project of Guangxi for Research Bases and Talents under grant No.AD22080052. The funder is Wei Dai.

Data Availability Statement: The raw data supporting the conclusions of this article will be made available by the authors on request.

Acknowledgments: The authors would like to thank Huihwang Goh for his help in our research. We appreciate the discussions and comments from all of the editors and anonymous reviewers.

Conflicts of Interest: Author Cheng Wang was employed by the Gui'an Power Supply Bureau of Guizhou Power Grid Co., Ltd. The remaining authors declare that the research was conducted in the absence of any commercial or financial relationships that could be construed as a potential conflict of interest.

References

1. Lopes, J.A.P.; Soares, F.J.; Almeida, P.M.R. Integration of Electric Vehicles in the Electric Power System. *Proc. IEEE* **2011**, *99*, 168–183. [CrossRef]
2. Global EV Outlook 2022 [EB/OL]. [Online]. Available online: https://www.iea.org/reports/global-ev-outlook-2022 (accessed on 25 December 2023).
3. Kezunovic, M.; Waller, S.T.; Damnjanovic, I. Framework for studying emerging policy issues associated with phevs in managing coupled power and transportation systems. In Proceedings of the 2010 IEEE Green Technologies Conference, Grapevine, TX, USA, 15–16 April 2010; pp. 1–8.
4. Yang, W.; Liu, W.; Chung, C.Y.; Wen, F. Joint planning of EV fast charging stations and power distribution systems with balanced traffic flow assignment. *IEEE Trans. Ind. Inform.* **2020**, *17*, 1795–1809.
5. Zhang, H.; Hu, Z.; Song, Y. Power and transport nexus: Routing electric vehicles to promote renewable power integration. *IEEE Trans. Smart Grid* **2020**, *11*, 3291–3301. [CrossRef]
6. Wei, W.; Wu, L.; Wang, J.; Mei, S. Network equilibrium of coupled transportation and power distribution systems. *IEEE Trans. Smart Grid* **2017**, *9*, 6764–6779. [CrossRef]
7. Shi, B.; Dai, W.; Luo, C.; Goh, H.H.; Li, J. Modeling and Impact Analysis for Solar Road Integration in Distribution Networks. *IEEE Trans. Sustain. Energy* **2022**, *14*, 935–947. [CrossRef]
8. Qian, T.; Shao, C.; Li, X.; Wang, X.; Shahidehpour, M. Enhanced coordinated operations of electric power and transportation networks via EV charging services. *IEEE Trans. Smart Grid* **2020**, *11*, 3019–3030. [CrossRef]
9. Sun, Y.; Chen, Z.; Li, Z.; Tian, W.; Shahidehpour, M. EV charging schedule in coupled constrained networks of transportation and power system. *IEEE Trans. Smart Grid* **2018**, *10*, 4706–4716. [CrossRef]
10. Baghali, S.; Guo, Z.; Wei, W.; Shahidehpour, M. Electric Vehicles for Distribution System Load Pickup Under Stressed Conditions: A Network Equilibrium Approach. *IEEE Trans. Power Syst.* **2023**, *38*, 2304–2317. [CrossRef]
11. Guo, Z.; Afifah, F.; Qi, J.; Baghali, S. A stochastic multiagent optimization framework for interdependent transportation and power system analyses. *IEEE Trans. Transp. Electrif.* **2021**, *7*, 1088–1098. [CrossRef]
12. Shao, C.; Li, K.; Qian, T.; Shahidehpour, M.; Wang, X. Generalized user equilibrium for coordination of coupled power-transportation network. *IEEE Trans. Smart Grid* **2022**, *14*, 2140–2151. [CrossRef]
13. Sadhu, K.; Haghshenas, K.; Rouhani, M.; Aiello, M. Optimal joint operation of coupled transportation and power distribution urban networks. *Energy Inform.* **2022**, *5*, 35. [CrossRef]
14. Sun, G.; Li, G.; Xia, S.; Shahidehpour, M.; Lu, X.; Chan, K.W. ALADIN-based coordinated operation of power distribution and traffic networks with electric vehicles. *IEEE Trans. Ind. Appl.* **2020**, *56*, 5944–5954. [CrossRef]
15. Liu, J.; Lin, G.; Huang, S.; Zhou, Y.; Rehtanz, C.; Li, Y. Collaborative EV routing and charging scheduling with power distribution and traffic networks interaction. *IEEE Trans. Power Syst.* **2022**, *37*, 3923–3936. [CrossRef]
16. Aghajan-Eshkevari, S.; Ameli, M.T.; Azad, S. Optimal routing and power management of electric vehicles in coupled power distribution and transportation systems. *Appl. Energy* **2023**, *341*, 121126. [CrossRef]
17. Lu, Z.; Xu, X.; Yan, Z.; Shahidehpour, M. Multistage robust optimization of routing and scheduling of mobile energy storage in coupled transportation and power distribution networks. *IEEE Trans. Transp. Electrif.* **2021**, *8*, 2583–2594. [CrossRef]
18. Lv, S.; Chen, S.; Wei, Z. Coordinating Urban Power-Traffic Networks: A Subsidy-Based Nash–Stackelberg–Nash Game Model. *IEEE Trans. Ind. Inform.* **2022**, *19*, 1778–1790. [CrossRef]
19. Xie, S.; Xu, Y.; Zheng, X. On dynamic network equilibrium of a coupled power and transportation network. *IEEE Trans. Smart Grid* **2021**, *13*, 1398–1411. [CrossRef]
20. Xie, S.; Wu, Q.; Hatziargyriou, N.D.; Zhang, M.; Zhang, Y.; Xu, Y. Collaborative pricing in a power-transportation coupled network: A variational inequality approach. *IEEE Trans. Power Syst.* **2022**, *38*, 783–795. [CrossRef]
21. Lai, X.; Xie, L.; Xia, Q.; Zhong, H.; Kang, C. Decentralized multi-area economic dispatch via dynamic multiplier-based Lagrangian relaxation. *IEEE Trans. Power Syst.* **2014**, *30*, 3225–3233. [CrossRef]
22. Jiang, H.; Zhang, Y.; Chen, Y.; Zhao, C.; Tan, J. Power-traffic coordinated operation for bi-peak shaving and bi-ramp smoothing–A hierarchical data-driven approach. *Appl. Energy* **2018**, *229*, 756–766. [CrossRef]
23. Xie, S.; Chen, Z.; Zhang, Y.; Cao, S.; Chen, K. Decentralized optimization of multi-area power-transportation coupled systems based on variational inequalities. *CSEE J. Power Energy Syst.* **2023**. *early access*. [CrossRef]
24. Huang, S.; Wu, Q. Dynamic tariff-subsidy method for PV and V2G congestion management in distribution networks. *IEEE Trans. Smart Grid* **2019**, *10*, 5851–5860. [CrossRef]
25. Alizadeh, M.; Wai, H.T.; Chowdhury, M.; Goldsmith, A.; Scaglione, A.; Javidi, T. Optimal pricing to manage electric vehicles in coupled power and transportation networks. *IEEE Trans. Control Netw. Syst.* **2016**, *4*, 863–875. [CrossRef]
26. Zhou, Z.; Liu, Z.; Su, H.; Zhang, L. Integrated pricing strategy for coordinating load levels in coupled power and transportation networks. *Appl. Energy* **2022**, *307*, 118100. [CrossRef]

27. Mouna, K.-B. Charging station location problem: A comprehensive review on models and solution approaches. *Transp. Res. Part C Emerg. Technol.* **2021**, *132*, 103376.
28. Loaiza, Q.C.; Arbelaez, A.; Climent, L. Robust ebuses charging location problem. *IEEE Open J. Intell. Transp. Syst.* **2022**, *3*, 856–871. [CrossRef]
29. Efthymiou, D.; Chrysostomou, K.; Morfoulaki, M.; Aifantopoulou, G. Electric vehicles charging infrastructure location: A genetic algorithm approach. *Eur. Transp. Res. Rev.* **2017**, *9*, 27.
30. Majhi, R.C.; Ranjitkar, P.; Sheng, M.; Covic, G.A.; Wilson, D.J. A systematic review of charging infrastructure location problem for electric vehicles. *Transp. Rev.* **2021**, *41*, 432–455. [CrossRef]
31. Burak, K.Ö.; Gzara, F.; Alumur, S.A. Full cover charging station location problem with routing. *Transp. Res. Part B Methodol.* **2021**, *144*, 1–22.
32. Kun, A. Battery electric bus infrastructure planning under demand uncertainty. *Transp. Res. Part C Emerg. Technol.* **2020**, *111*, 572–587.
33. An, Y.; Gao, Y.; Wu, N.; Zhu, J.; Li, H.; Yang, J. Optimal scheduling of electric vehicle charging operations considering real-time traffic condition and travel distance. *Expert Syst. Appl.* **2023**, *213*, 118941. [CrossRef]
34. Padraigh, J.; Climent, L.; Arbelaez, A. *Smart and Sustainable Scheduling of Charging Events for Electric Buses. EU Cohesion Policy Implementation—Evaluation Challenges Opportunities*; Springer Nature: Berlin/Heidelberg, Germany, 2023; pp. 121–129.
35. Manshadi, S.D.; Khodayar, M.E.; Abdelghany, K.; Üster, H. Wireless charging of electric vehicles in electricity and transportation networks. *IEEE Trans. Smart Grid* **2017**, *9*, 4503–4512. [CrossRef]
36. Alizadeh, M.; Wai, H.T.; Goldsmith, A.; Scaglione, A. Retail and wholesale electricity pricing considering electric vehicle mobility. *IEEE Trans. Control Netw. Syst.* **2018**, *6*, 249–260. [CrossRef]
37. Qiao, W.; Han, Y.; Zhao, Q.; Si, F.; Wang, J. A distributed coordination method for coupled traffic-power network equilibrium incorporating behavioral theory. *IEEE Trans. Veh. Technol.* **2022**, *71*, 12588–12601. [CrossRef]
38. Tao, D.; Bie, Z. Parallel augmented Lagrangian relaxation for dynamic economic dispatch using diagonal quadratic approximation method. *IEEE Trans. Power Syst.* **2016**, *32*, 1115–1126.
39. Benders, J.F. Partitioning procedures for solving mixed-variables programming problems. *Comput. Manag. Sci.* **2005**, *2*, 3–19. [CrossRef]
40. Li, Z.; Wu, W.; Zhang, B.; Wang, B. Decentralized multi-area dynamic economic dispatch using modified generalized benders decomposition. *IEEE Trans. Power Syst.* **2015**, *31*, 526–538. [CrossRef]
41. Lin, C.; Wu, W. A nested decomposition method and its application for coordinated operation of hierarchical electrical power grids. *arXiv* **2020**, arXiv:2007.02214.
42. Tan, Z.; Yan, Z.; Zhong, H.; Xia, Q. Non-Iterative Solution for Coordinated Optimal Dispatch via Equivalent Projection—Part II: Method and Applications. *IEEE Trans. Power Syst.* **2023**, *39*, 899–908. [CrossRef]
43. Tan, Z.; Yan, Z.; Zhong, H.; Xia, Q. Non-Iterative Solution for Coordinated Optimal Dispatch via Equivalent Projection—Part I: Theory. *IEEE Trans. Power Syst.* **2024**, *39*, 890–898. [CrossRef]
44. Dai, W.; Wang, C.; Goh, H.; Zhao, J.; Jian, J. Hosting Capacity Evaluation Method for Power Distribution Networks Integrated with Electric Vehicles. *J. Mod. Power Syst. Clean Energy* **2023**, *11*, 1564–1575. [CrossRef]
45. Bolognani, S.; Zampieri, S. On the existence and linear approximation of the power flow solution in power distribution networks. *IEEE Trans. Power Syst.* **2015**, *31*, 163–172. [CrossRef]
46. Yang, J.; Zhang, N.; Kang, C.; Xia, Q. A state-independent linear power flow model with accurate estimation of voltage magnitude. *IEEE Trans. Power Syst.* **2016**, *32*, 3607–3617. [CrossRef]

MDPI
St. Alban-Anlage 66
4052 Basel
Switzerland
www.mdpi.com

Energies Editorial Office
E-mail: energies@mdpi.com
www.mdpi.com/journal/energies

www.ingramcontent.com/pod-product-compliance
Lightning Source LLC
Chambersburg PA
CBHW040750250926
51850CB00007BB/1460
* 9 7 8 3 7 2 5 8 1 2 2 7 1 *